Multiplicative Analytic Geometry

This book is devoted to multiplicative analytic geometry. The book reflects recent investigations into the topic. The reader can use the main formulae for investigations of multiplicative differential equations, multiplicative integral equations and multiplicative geometry.

The authors summarize the most recent contributions in this area. The goal of the authors is to bring the most recent research on the topic to capable senior undergraduate students, beginning graduate students of engineering and science and researchers in a form to advance further study. The book contains eight chapters. The chapters in the book are pedagogically organized. Each chapter concludes with a section with practical problems.

Two operations, differentiation and integration, are basic in calculus and analysis. In fact, they are the infinitesimal versions of the subtraction and addition operations on numbers, respectively. In the period from 1967 till 1970, Michael Grossman and Robert Katz gave definitions of a new kind of derivative and integral, moving from the roles of subtraction and addition to division and multiplication, and thus established a new calculus, called multiplicative calculus. Multiplicative calculus can especially be useful as a mathematical tool for economics and finance.

Multiplicative Analytic Geometry builds upon multiplicative calculus and advances the theory to the topics of analytic and differential geometry.

Svetlin G. Georgiev (born 05 April 1974, Rouse, Bulgaria) is a mathematician who has worked in various areas of mathematics. He currently focuses on harmonic analysis, functional analysis, partial differential equations, ordinary differential equations, Clifford and quaternion analysis, integral equations and dynamic calculus on time scales.

Khaled Zennir was born in Skikda, Algeria, in 1982. He received his PhD in Mathematics in 2013 from Sidi Bel Abbès University, Algeria (Assist. Professor). He obtained his highest diploma in Algeria (Habilitation, Mathematics) from Constantine University, Algeria, in May 2015 (Assoc. Professor). He is now Associate Professor at Qassim University, KSA. His research interests lie in nonlinear hyperbolic partial differential equations: global existence, blow-up and long time behavior.

Aissa Boukarou received his PhD in Mathematics in 2021 from Ghardaia University, Algeria (Assist. Professor). His research interests lie in partial differential equations, harmonic analysis, stochastic PDE and numerical analysis.

Multiplicative Analytic Geometry

Svetlin G. Georgiev
Khaled Zennir
Aissa Boukarou

CRC Press
Taylor & Francis Group
Boca Raton London New York

CRC Press is an imprint of the
Taylor & Francis Group, an **informa** business
A CHAPMAN & HALL BOOK

First edition published 2023
by CRC Press
6000 Broken Sound Parkway NW, Suite 300, Boca Raton, FL 33487-2742

and by CRC Press
4 Park Square, Milton Park, Abingdon, Oxon, OX14 4RN

CRC Press is an imprint of Taylor & Francis Group, LLC

ISBN: 9781032350981 (hbk)
ISBN: 9781032350998 (pbk)
ISBN: 9781003325284 (ebk)

DOI: 10.1201/9781003325284

Typeset in Nimbus
by KnowledgeWorks Global Ltd.

Contents

Preface

Differential and integral calculus, the most applicable mathematical theory, was created independently by Isaac Newton and Gottfried Wilhelm Leibnitz in the second half of the seventeenth century. Two operations, differentiation and integration, are basic in calculus and analysis. In fact, they are the infinitesimal versions of the subtraction and addition operations on numbers, respectively. In the period from 1967 till 1970, Michael Grossman and Robert Katz gave definitions of a new kind of derivative and integral, moving from the roles of subtraction and addition to division and multiplication, and thus established a new calculus, called multiplicative calculus. Sometimes, it is called an alternative or non-Newtonian calculus as well. Multiplicative calculus can especially be useful as a mathematical tool for economics and finance.

This book is devoted to the multiplicative Euclidean and non-Euclidean geometry. It summarizes the most recent contributions in this area. The book is intended for senior undergraduate students and beginning graduate students of engineering and science courses. The book contains eight chapters. The chapters in the book are pedagogically organized. Each chapter concludes with a section with practical problems.

Chapter 1 introduces the field $\mathbb{R}_\star$ and defines the basic multiplicative arithmetic operations: multiplicative addition, multiplicative subtraction, multiplicative multiplication and multiplicative division and some of their properties are deduced. The basic elementary multiplicative functions are defined and studied. Chapter 2 investigates the multiplicative vector space $\mathbb{R}^2_\star$, the multiplicative inner product space and the multiplicative Euclidean space $E^2_\star$. Multiplicative lines are defined and their equations are deduced. Perpendicular, parallel and intersecting multiplicative lines are introduced. Multiplicative isometries, multiplicative translations, multiplicative rotations and multiplicative glide reflections are investigated. Chapter 3 deals with multiplicative affine transformations, multiplicative affine reflections, multiplicative affine symmetries, multiplicative shears, multiplicative dilatations and multiplicative similarities, and some of their properties are investigated. Multiplicative segments, multiplicative angles and multiplicative rectilinear figures are defined. Some criteria for existence of multiplicative affine transformations that leave multiplicative lines and multiplicative points fixed are given. A multiplicative barycentric coordinate system is introduced and some of its applications are given. Some criteria for congruence of multiplicative angles and triangles are deduced. Chapter 4 is devoted to cyclic and dihedral subgroups of $O_\star(e^2)$ and some of their properties are investigated. Conjugate subgroups, orbits and stabilizers are introduced. In the chapter, regular multiplicative polygons are defined. Chapter 5 introduces multiplicative spheres and multiplicative lines on multiplicative spheres, multiplicative reflections on multiplicative spheres and multiplicative rectilinear figures on multiplicative spheres, and some of their properties are deduced. Chapter 6 investigates the projective multiplicative plane $P^2_\star$ and multiplicative perpendicular points, multiplicative lines, multiplicative perpendicular multiplicative lines, multiplicative poles, multiplicative polarities, multiplicative conics, multiplicative tangents, multiplicative secants and a multiplicative cross product are defined and some of their properties are investigated. Multiplicative analogues of the Desargues and Pappus theorems are proved, as well, and the fundamental theorem of the projective multiplicative geometry is given. Chapter 7 is devoted to the multiplicative distance geometry on $P^2_\star$. Multiplicative orthogonal transformations, multiplicative reflections, multiplicative rotations and multiplicative translations are introduced. Chapter 8 deals with the multiplicative hyperbolic plane $H^2_\star$. Multiplicative lines, multiplicative segments, multiplicative triangles, multiplicative quadrilateral figures, multiplicative circles, multiplicative horocycles and multiplicative equidistant curves are defined and investigated. Multiplicative isometries, multiplicative reflections, multiplicative rotations and multiplicative translations in the multiplicative hyperbolic plane are studied.

This book is addressed to a wide audience of specialists such as mathematicians, physicists, engineers and biologists. It can be used as a textbook at the graduate level and as a reference book for several disciplines.

Paris, March 2022
Svetlin G. Georgiev, Khaled Zennir and Aissa Boukarou

1

The Field $\mathbb{R}_\star$

In this chapter, the field $\mathbb{R}_\star$ is introduced, and the basic multiplicative arithmetic operations: multiplicative addition, multiplicative subtraction, multiplicative multiplication and multiplicative division are defined and some of their properties are deduced. The basic elementary multiplicative functions are introduced and studied.

1.1 Definition

Let $\mathbb{R}_\star = (0, \infty)$.

Definition 1.1. *In the set $\mathbb{R}_\star$, we define the multiplicative addition or $\star$ addition $+_\star$ in the following manner*

$$a +_\star b = ab, \quad a, b \in \mathbb{R}_\star.$$

Example 1.1. *Let $a = 1$, $b = 3$. Then*

$$\begin{aligned} a +_\star b &= 1 +_\star 3 \\ &= 1 \cdot 3 \\ &= 3. \end{aligned}$$

Definition 1.2. *In the set $\mathbb{R}_\star$, we define the multiplicative multiplication or $\star$ multiplication $\cdot_\star$ as follows:*

$$a \cdot_\star b = e^{\log a \log b}.$$

Example 1.2. *Let $a = 1$ and $b = e$. Then*

$$\begin{aligned} a \cdot_\star b &= e^{\log 1 \log e} \\ &= 1. \end{aligned}$$

Example 1.3. *Let $a = 2$, $b = \frac{1}{3}$, $c = 4$. We find*

$$A = (a +_\star b) \cdot_\star c.$$

We have

$$\begin{aligned} a +_\star b &= 2 \cdot \frac{1}{3} \\ &= \frac{2}{3}. \end{aligned}$$

DOI: 10.1201/9781003325284-1

Then

$$\begin{aligned} A &= e^{\log(a+_\star b)\log c} \\ &= e^{\log\frac{2}{3}\log 4} \\ &= e^{2\log 2\log\frac{2}{3}}. \end{aligned}$$

Exercise 1.1. *Let* $a = e^3$, $b = e^4$, $c = e^{10}$. *Find*

$$A = (a +_\star b) \cdot_\star c.$$

Answer 1.1. $A = e^{70}$.

Definition 1.3. *In the set* $\mathbb{R}_\star$ *we define the multiplicative zero(* $\star$ *zero) and multiplicative unit(*$\star$ *unit) as follows:*

$$0_\star = 1 \quad \textit{and} \quad 1_\star = e.$$

Below, we list some of the properties of the multiplicative addition and multiplicative multiplication.

1. **Commutativity of $\star$ Addition.** Let $x, y \in \mathbb{R}_\star$ be arbitrarily chosen. Then

$$\begin{aligned} x +_\star y &= xy \\ &= yx \\ &= y +_\star x. \end{aligned}$$

2. **Associativity of $\star$ Addition.** Let $x, y, z \in \mathbb{R}_\star$ be arbitrarily chosen. Then

$$\begin{aligned} x +_\star (y +_\star z) &= x +_\star (yz) \\ &= xyz \\ &= (xy)z \\ &= (x +_\star y)z \\ &= (x +_\star y) +_\star z. \end{aligned}$$

3. **$\star$ Identity Element of $\star$ Addition.** Let $x \in \mathbb{R}_\star$ be arbitrarily chosen. Then

$$\begin{aligned} x +_\star 0_\star &= x +_\star 1 \\ &= x \cdot 1 \\ &= x. \end{aligned}$$

4. **$\star$ Inverse Elements of $\star$ Addition.** Let $x \in \mathbb{R}_\star$ be arbitrarily chosen. Define

$$-_\star x = \frac{1}{x}.$$

Then

$$\begin{aligned} x +_\star (-_\star x) &= x +_\star \left(\frac{1}{x}\right) \\ &= x \cdot \frac{1}{x} \\ &= 1 \\ &= 0_\star. \end{aligned}$$

5. $\star$ **Identity Element of $\star$ Multiplication.** Let $x \in \mathbb{R}_\star$ be arbitrarily chosen. Then

$$\begin{aligned} x \cdot_\star 1_\star &= x \cdot_\star e \\ &= e^{\log x \log e} \\ &= e^{\log x} \\ &= x. \end{aligned}$$

6. $\star$ **Inverse Elements of $\star$ Multiplication.** Let $x \in \mathbb{R}_\star$ be arbitrarily chosen. Take

$$x^{-1_\star} = e^{\frac{1}{\log x}}.$$

Then

$$\begin{aligned} x \cdot_\star x^{-1_\star} &= x \cdot_\star \left(e^{\frac{1}{\log x}}\right) \\ &= e^{\log x \log e^{\frac{1}{\log x}}} \\ &= e^{\frac{\log x}{\log x}} \\ &= e \\ &= 1_\star. \end{aligned}$$

7. **Distributivity.** Let $x, y, z \in \mathbb{R}_\star$ be arbitrarily chosen. Then

$$\begin{aligned} (x +_\star y) \cdot_\star z &= (xy) \cdot_\star z \\ &= e^{\log(xy) \log z} \\ &= e^{(\log x + \log y) \log z} \\ &= e^{\log x \log z + \log y \log z} \\ &= e^{\log x \log z} e^{\log y \log z} \\ &= (x \cdot_\star y) \cdot (y \cdot_\star z) \\ &= (x \cdot_\star z) +_\star (y \cdot_\star z). \end{aligned}$$

Definition 1.4. *For any $x \in \mathbb{R}_\star$, the number*

$$-_\star x = \frac{1}{x},$$

will be called the multiplicative opposite number or $\star$ opposite number of the number x.

We have

$$\begin{aligned} -_\star(-_\star x) &= -_\star\left(\frac{1}{x}\right) \\ &= \frac{1}{\frac{1}{x}} \\ &= x, \end{aligned}$$

for any $x \in \mathbb{R}_\star$.

Definition 1.5. *For $x, y \in \mathbb{R}_\star$, define multiplicative subtraction or $\star$ subtraction $-_\star$ as follows:*

$$\begin{aligned} x -_\star y &= x +_\star (-_\star y) \\ &= x(-_\star y) \\ &= x \cdot \frac{1}{y} \\ &= \frac{x}{y}. \end{aligned}$$

Definition 1.6. *For $x \in \mathbb{R}_\star$, $x \neq 0_\star$, the number*

$$x^{-1_\star} = e^{\frac{1}{\log x}},$$

will be called the multiplicative reciprocal or $\star$ reciprocal of the number x.

We have

$$\begin{aligned} \left(x^{-1_\star}\right)^{-1_\star} &= \left(e^{\frac{1}{\log x}}\right)^{-1_\star} \\ &= e^{\frac{1}{\log e^{\frac{1}{\log x}}}} \\ &= e^{\log x} \\ &= x, \end{aligned}$$

for any $x \in \mathbb{R}_\star$, $x \neq 0_\star$.

Definition 1.7. *For $x, y \in \mathbb{R}_\star$, define multiplicative division or $\star$ division $/_\star$ as follows:*

$$\begin{aligned} x /_\star y &= x \cdot_\star \left(y^{-1_\star}\right) \\ &= x \cdot_\star \left(e^{\frac{1}{\log y}}\right) \\ &= e^{\log x \log e^{\frac{1}{\log y}}} \\ &= e^{\frac{\log x}{\log y}}. \end{aligned}$$

Example 1.4. *We find*

$$A = (2 +_\star 3) \cdot_\star 4 -_\star (3 +_\star 1) /_\star 5.$$

We have

$$\begin{aligned} A &= (2 \cdot 3) \cdot_\star 4 -_\star (3 \cdot 1) /_\star 5 \\ &= 6 \cdot_\star 4 -_\star 3 /_\star 5 \\ &= e^{\log 6 \log 4} -_\star e^{\frac{\log 3}{\log 5}} \\ &= e^{2 \log 6 \log 2 - \frac{\log 3}{\log 5}}. \end{aligned}$$

Exercise 1.2. *Find*

1. $A = (3 -_\star 5) /_\star 2 +_\star (4 +_\star 2) \cdot_\star e.$
2. $A = 3 +_\star 2 -_\star 3 \cdot_\star (2 +_\star 4).$
3. $A = 1 -_\star 3 +_\star 4 \cdot_\star (1 +_\star 5).$

Answer 1.2.

1. $e^{\frac{\log\frac{3}{5}}{\log 2}+3\log 2}$.
2. $\frac{6}{e^{3\log 3\log 2}}$.
3. $\frac{1}{3}e^{2\log 2\log 5}$.

Theorem 1.3. *For any $a, b \in \mathbb{R}_\star$, the equation*

$$a +_\star x = b, \tag{1.1}$$

has at least one solution.

Proof. Let

$$x = b -_\star a.$$

Then

$$\begin{aligned} a +_\star (b -_\star a) &= a +_\star \left(\frac{b}{a}\right) \\ &= a \cdot \frac{b}{a} \\ &= b. \end{aligned}$$

This completes the proof. □

Corollary 1.1. *Any solution $x \in \mathbb{R}_\star$ of the equation*

$$a +_\star x = a, \quad a \in \mathbb{R}_\star, \tag{1.2}$$

is a solution of the equation

$$b +_\star x = b, \quad b \in \mathbb{R}_\star. \tag{1.3}$$

Proof. By Theorem 1.3, it follows that the equation (1.2) and the equation

$$a +_\star y = b$$

have at least one solution x and y, respectively. Then

$$\begin{aligned} b +_\star x &= (a +_\star y) +_\star x \\ &= a +_\star (y +_\star x) \\ &= a +_\star (x +_\star y) \\ &= (a +_\star x) +_\star y \\ &= a +_\star y \\ &= b, \end{aligned}$$

i.e., x is a solution of the equation (1.3). This completes the proof. □

Corollary 1.2. *The equation* (1.2) *has a unique solution.*

Proof. By Theorem 1.3, it follows that the equation (1.2) has at least one solution. Assume that the equation (1.2) has two solutions x and y. Then

$$a +_\star x = a$$

and

$$a +_\star y = a.$$

By Corollary 1.1, it follows

$$y +_\star x = y$$

and

$$x +_\star y = x.$$

Hence,

$$\begin{aligned} y &= y +_\star x \\ &= x +_\star y \\ &= x. \end{aligned}$$

This completes the proof. □

Remark 1.1. *By Corollary 1.2, it follows that the multiplicative zero $0_\star$ is unique.*

Corollary 1.3. *For any $a, b \in \mathbb{R}_\star$, the equation* (1.1) *has a unique solution.*

Proof. By Theorem 1.3, it follows that the equation (1.1) has at least one solution. Assume that the equation (1.1) has two solutions x and y. Then

$$a +_\star x = b$$

and

$$a +_\star y = b.$$

By Theorem 1.3, it follows that the equation

$$a +_\star z = 0_\star,$$

has at least one solution. Hence,

$$\begin{aligned} y &= y +_\star 0_\star \\ &= y +_\star (a +_\star z) \\ &= (y +_\star a) +_\star z \\ &= (a +_\star y) +_\star z \\ &= b +_\star z \\ &= (a +_\star x) +_\star z \\ &= a +_\star (x +_\star z) \\ &= a +_\star (z +_\star x) \\ &= (a +_\star z) +_\star x \\ &= 0_\star +_\star x \\ &= x. \end{aligned}$$

This completes the proof. □

Theorem 1.4. *For any $a \in \mathbb{R}_\star$, $a \neq 0_\star$, the equation*

$$a \cdot_\star x = b, \tag{1.4}$$

has a solution.

Proof. Let

$$x = b/_\star a.$$

Then

$$\begin{aligned}
a \cdot_\star x &= a \cdot_\star (b/_\star a) \\
&= a \cdot_\star e^{\frac{\log b}{\log a}} \\
&= e^{\log a \log e^{\frac{\log b}{\log a}}} \\
&= e^{\log a \cdot \frac{\log b}{\log a}} \\
&= e^{\log b} \\
&= b.
\end{aligned}$$

This completes the proof. □

Corollary 1.4. *If $a \in \mathbb{R}_\star$, $a \neq 0_\star$, then any solution of the equation*

$$a \cdot_\star x = a, \tag{1.5}$$

is a solution to the equation

$$b \cdot_\star x = b, \quad b \in \mathbb{R}_\star. \tag{1.6}$$

Proof. By Theorem 1.4, it follows that the equation (1.5) and the equation

$$a \cdot_\star y = b,$$

have at least one solution x and y, respectively. Then

$$\begin{aligned}
b \cdot_\star x &= (a \cdot_\star y) \cdot_\star x \\
&= a \cdot_\star (y \cdot_\star x) \\
&= a \cdot_\star (x \cdot_\star y) \\
&= (a \cdot_\star x) \cdot_\star y \\
&= a \cdot_\star y \\
&= b,
\end{aligned}$$

i.e., x is a solution to the equation (1.6). This completes the proof. □

Corollary 1.5. *Let $a \in \mathbb{R}_\star$, $a \neq 0_\star$. Then the equation* (1.5) *has a unique solution.*

Proof. By Theorem 1.4, it follows that the equation (1.5) has at least one solution. Let x and y be two solutions to the equation (1.5). By Corollary 1.4, it follows that

$$y \cdot_\star x = y$$

and

$$x \cdot_\star y = x.$$

Hence,

$$\begin{aligned}
y &= y \cdot_\star x \\
&= x \cdot_\star y \\
&= x.
\end{aligned}$$

This completes the proof. □

Remark 1.2. *By Corollary 1.5, it follows that the multiplicative unit is unique.*

Corollary 1.6. *Let $a \in \mathbb{R}_\star$, $a \neq 0_\star$. Then the equation* (1.4) *has a unique solution.*

Proof. By Theorem 1.4, it follows that the equation (1.4) has at least one solution. Assume that the equation (1.4) has two solutions x and y. By Theorem 1.4, we have that the equation

$$a \cdot_\star z = 1_\star,$$

has at least one solution. Then

$$\begin{aligned} y &= y \cdot_\star 1_\star \\ &= y \cdot_\star (a \cdot_\star z) \\ &= (y \cdot_\star a) \cdot_\star z \\ &= (a \cdot_\star y) \cdot_\star z \\ &= b \cdot_\star z \\ &= (a \cdot_\star x) \cdot_\star z \\ &= (x \cdot_\star a) \cdot_\star z \\ &= x \cdot_\star (a \cdot_\star z) \\ &= x \cdot_\star 1_\star \\ &= x. \end{aligned}$$

This completes the proof. □

Corollary 1.7. *Let $a, b \in \mathbb{R}_\star$. Then*

1. $0_\star = -_\star 0_\star$.
2. $a -_\star 0_\star = a$.
3. $a \cdot_\star 0_\star = 0_\star$.
4. $(-_\star 1_\star) \cdot_\star a = -_\star a$.
5. $(-_\star 1_\star) \cdot_\star (-_\star 1_\star) = 1_\star$.
6. $-_\star (a -_\star b) = b -_\star a$.

Proof.

1. We have that $0_\star$ is a solution to the equation

$$0_\star +_\star x = 0_\star. \tag{1.7}$$

Since

$$0_\star +_\star (-_\star 0_\star) = 0_\star,$$

we obtain that $-_\star 0_\star$ is also a solution of the equation (1.7). By Corollary 1.2, it follows that the equation (1.7) has a unique solution. Therefore

$$0_\star = -_\star 0_\star.$$

2. We have

$$\begin{aligned} a -_\star 0_\star &= a +_\star (-_\star 0_\star) \\ &= a +_\star 0_\star \\ &= a +_\star 1 \\ &= a \cdot 1 \\ &= a. \end{aligned}$$

3. We have

$$\begin{aligned} a \cdot_\star 0_\star &= a \cdot_\star 1 \\ &= e^{\log a \log 1} \\ &= 1 \\ &= 0_\star. \end{aligned}$$

4. We have

$$\begin{aligned} (-_\star 1_\star) \cdot_\star a &= (-_\star e) \cdot_\star a \\ &= \frac{1}{e} \cdot_\star a \\ &= e^{\log \frac{1}{e} \log a} \\ &= e^{-\log a} \\ &= \frac{1}{a} \\ &= -_\star a. \end{aligned}$$

5. We have

$$\begin{aligned} (-_\star 1_\star) \cdot_\star (-_\star 1_\star) &= (-_\star e) \cdot_\star (-_\star e) \\ &= \frac{1}{e} \cdot_\star \frac{1}{e} \\ &= e^{\log \frac{1}{e} \log \frac{1}{e}} \\ &= e \\ &= 1_\star. \end{aligned}$$

6. We have

$$\begin{aligned} -_\star (a -_\star b) &= -_\star (a +_\star (-_\star b)) \\ &= -_\star \left(a +_\star \frac{1}{b} \right) \\ &= -_\star \frac{a}{b} \\ &= \frac{1}{\frac{a}{b}} \\ &= \frac{b}{a} \\ &= b -_\star a. \end{aligned}$$

This completes the proof.

□

1.2 An Order in $\mathbb{R}_\star$

Definition 1.8. *We say that a number $a \in \mathbb{R}_\star$ is multiplicative positive or $\star$ positive if $a > 1$. We write $a >_\star 0_\star$.*

Example 1.5. *5 is $\star$ positive.*

Definition 1.9. *A number $a \in \mathbb{R}_\star$ is said to be multiplicative negative or $\star$ negative if it is not equal to $0_\star$ and it is not $\star$ positive. We write $a <_\star 0_\star$.*

Example 1.6. *$\frac{1}{2}$ is a $\star$ negative number.*

Definition 1.10. *Let $a, b \in \mathbb{R}_\star$. We say that a is multiplicative greater than b or $\star$ greater than b and we write $a >_\star b$ if $a -_\star b >_\star 0_\star$. We denote $a \geq_\star b$ if $a >_\star b$ or $a = b$.*

Remark 1.3. *Let $a, b \in \mathbb{R}_\star$. Then*

$$a >_\star b \iff a -_\star b >_\star 0_\star \iff \frac{a}{b} > 1 \iff a > b.$$

Definition 1.11. *Let $a, b \in \mathbb{R}_\star$. We say that a is multiplicative less than b or $\star$ less than b and we write $a <_\star b$ if $a -_\star b <_\star 0_\star$. We denote $a \leq_\star b$ if $a <_\star b$ or $a = b$.*

Remark 1.4. *Let $a, b \in \mathbb{R}_\star$. Then*

$$a <_\star b \iff a -_\star b <_\star 0_\star \iff \frac{a}{b} < 1 \iff a < b.$$

Theorem 1.5. *Let $a, b, c, d \in \mathbb{R}_\star$. If*

$$a >_\star b \quad \text{and} \quad c >_\star d, \tag{1.8}$$

then

$$a +_\star c >_\star b +_\star d.$$

Proof. By (1.8), it follows that

$$a > b > 0 \quad \text{and} \quad c > d > 0.$$

Then

$$\begin{aligned} a +_\star c &= ac \\ &> bd \\ &= b +_\star d. \end{aligned}$$

This completes the proof. □

Theorem 1.6. *Let $a, b, c, d \in \mathbb{R}_\star$. If*

$$a >_\star b, \quad c >_\star d, \quad c >_\star 0_\star, \quad b >_\star 0_\star, \tag{1.9}$$

then

$$a \cdot_\star c >_\star b \cdot_\star d.$$

Proof. By (1.9), it follows that

$$a > b, \quad c > d, \quad c > 1, \quad b > 1.$$

Then

$$\begin{aligned} \log a &> \log b, \\ \log c &> \log d, \\ \log c &> 0, \\ \log b &> 0, \end{aligned}$$

whereupon

$$\begin{aligned}\log a \log c &> \log b \log c\\ &> \log b \log d\end{aligned}$$

and

$$e^{\log a \log c} > e^{\log b \log d},$$

i.e.,

$$a \cdot_\star c >_\star b \cdot_\star d.$$

This completes the proof. □

Theorem 1.7 (Chebyschev Multiplicative Inequality). *Let* $a_1, a_2, b_1, b_2 \in \mathbb{R}_\star$. *Then the inequality*

$$a_1 \cdot_\star b_1 +_\star a_2 \cdot_\star b_2 \leq_\star e^{\frac{1}{2}} \cdot_\star (a_1 +_\star a_2) \cdot_\star (b_1 +_\star b_2)$$

holds if and only if the inequality

$$\log \frac{a_2}{a_1} \log \frac{b_2}{b_1} \leq 0,$$

holds.

Proof. Note that

$$\log \frac{a_2}{a_1} \log \frac{b_2}{b_1} \leq 0,$$

if and only if

$$\log a_1 \log b_1 + \log a_2 \log b_2 \leq \frac{1}{2} \log(a_1 a_2) \log(b_1 b_2),$$

if and only if

$$\begin{aligned} e^{\log a_1 \log b_1 + \log a_2 \log b_2} &\leq e^{\frac{1}{2} \log(a_1 a_2) \log(b_1 b_2)}\\ &= e^{\log e^{\frac{1}{2}} \log(a_1 a_2) \log(b_1 b_2)}, \end{aligned}$$

if and only if

$$a_1 \cdot_\star b_1 +_\star a_2 \cdot_\star b_2 \leq_\star e^{\frac{1}{2}} \cdot_\star (a_1 +_\star a_2) \cdot_\star (b_1 +_\star b_2).$$

This completes the proof. □

Theorem 1.8. *Let* $a, b, c, \lambda \in \mathbb{R}_\star$.

1. *If* $a >_\star b$, *then* $a +_\star c >_\star b +_\star c$.
2. *If* $a >_\star b$ *and* $b >_\star c$, *then* $a >_\star c$.
3. *If* $\lambda >_\star 0_\star$ *and* $a >_\star b$, *then* $\lambda \cdot_\star a >_\star \lambda \cdot_\star b$.
4. *If* $\lambda <_\star 0_\star$ *and* $a >_\star b$, *then* $\lambda \cdot_\star a <_\star \lambda \cdot_\star b$.

Proof.

1. Let $a >_\star b$. Then $a > b$ and hence,

$$ac > bc,$$

or

$$a +_\star c >_\star b +_\star c.$$

2. Let $a >_\star b$ and $b >_\star c$. Then $a > b$ and $b > c$. Hence, $a > c$ and $a >_\star c$.
3. Let $\lambda >_\star 0_\star$ and $a >_\star b$. Then $\lambda > 1$ and $\log\lambda > 0$, and

$$\log a > \log b.$$

Hence,

$$\log\lambda \log a > \log\lambda \log b$$

and

$$e^{\log\lambda \log a} > e^{\log\lambda \log b},$$

or

$$\lambda \cdot_\star a >_\star \lambda \cdot_\star b.$$

4. Let $\lambda <_\star 0_\star$ and $a >_\star b$. Then $\lambda \in (0,1)$ and

$$\log a > \log b, \quad \log\lambda < 0.$$

Hence,

$$\log\lambda \log a < \log\lambda \log b$$

and

$$e^{\log\lambda \log a} < e^{\log\lambda \log b},$$

or

$$\lambda \cdot_\star a <_\star \lambda \cdot_\star b.$$

This completes the proof.

□

1.3 Multiplicative Absolute Value

In this section, we define the multiplicative absolute value and we deduce some of its properties.

Definition 1.12. *Let $x \in \mathbb{R}_\star$. The multiplicative absolute value is defined as follows:*

$$|x|_\star = \begin{cases} x & \text{if } x \geq_\star 0_\star, \\ \frac{1}{x} & \text{if } x \leq_\star 0_\star, \end{cases}$$

or

$$|x|_\star = \begin{cases} x & \text{if } x \geq 1, \\ \frac{1}{x} & \text{if } x \leq 1. \end{cases}$$

Example 1.7. *Let $x = 5$. Then $|x|_\star = 5$.*

Example 1.8. *Let $x = \frac{1}{6}$. Then $|x|_\star = 6$.*

Below, we deduce some of the properties of the multiplicative absolute value. Let $a, b \in \mathbb{R}_\star$.

1. $|a|_\star \geq_\star 0_\star$.

Proof. If $a \geq_\star 0_\star$, then $a \geq 1$ and

$$\begin{aligned} |a|_\star &= a \\ &\geq 1 \\ &\geq_\star 0_\star. \end{aligned}$$

Let $a \leq_\star 0_\star$. Then $a \leq 1$ and

$$\begin{aligned} |a|_\star &= \frac{1}{a} \\ &\geq 1 \\ &\geq_\star 0_\star. \end{aligned}$$

This completes the proof. □

2. $|a|_\star = |-_\star a|_\star$.

Proof. We have

$$-_\star a = \frac{1}{a}.$$

Then

$$|a|_\star = \begin{cases} a & \text{if} \quad a \geq 1, \\ \frac{1}{a} & \text{if} \quad a \leq 1 \end{cases}$$

and

$$|-_\star a|_\star = \begin{cases} \frac{1}{a} & \text{if} \quad a \leq 1, \\ a & \text{if} \quad a \geq 1. \end{cases}$$

Therefore,

$$|a|_\star = |-_\star a|_\star.$$

This completes the proof. □

3. $|a +_\star b|_\star \leq |a|_\star +_\star |b|_\star$.

Proof. We have

$$a +_\star b = ab.$$

We consider the following cases.

(a) Let $ab \geq 1$.

i. Let $a \geq 1$, $b \leq 1$. Then

$$\begin{aligned} |a|_\star &= a, \\ |b|_\star &= \frac{1}{b} \end{aligned}$$

and

$$\begin{aligned} |a +_\star b|_\star &= |ab|_\star \\ &= ab \\ &\leq a\frac{1}{b} \\ &= a +_\star \frac{1}{b} \\ &= |a|_\star +_\star |b|_\star. \end{aligned}$$

ii. The case $a \leq 1$ and $b \geq 1$, we leave to the reader as an exercise.
iii. Let $a \geq 1$ and $b \geq 1$. Then

$$\begin{aligned} |a|_\star &= a, \\ |b|_\star &= b. \end{aligned}$$

Hence,

$$\begin{aligned} |a +_\star b|_\star &= |ab|_\star \\ &= ab \\ &= a +_\star b \\ &= |a|_\star +_\star |b|_\star. \end{aligned}$$

(b) Let $ab \leq 1$.
i. Let $a \leq 1$ and $b \leq 1$. Then

$$\begin{aligned} |a|_\star &= \frac{1}{a}, \\ |b|_\star &= \frac{1}{b}, \\ |ab|_\star &= \frac{1}{ab}. \end{aligned}$$

Hence,

$$\begin{aligned} |a +_\star b|_\star &= |ab|_\star \\ &= \frac{1}{ab} \\ &= |a|_\star |b|_\star \\ &= |a|_\star +_\star |b|_\star. \end{aligned}$$

ii. Let $a \geq 1$, $b \leq 1$. Then

$$\begin{aligned} |a|_\star &= a, \\ |b|_\star &= \frac{1}{b}, \\ |ab|_\star &= \frac{1}{ab}, \\ \frac{1}{a} &\leq 1. \end{aligned}$$

Hence,

$$\begin{aligned} |a +_\star b|_\star &= |ab|_\star \\ &= \frac{1}{ab} \\ &\leq \frac{a}{b} \\ &= |a|_\star |b|_\star \\ &= |a|_\star +_\star |b|_\star. \end{aligned}$$

iii. The case $a \leq 1$ and $b \geq 1$ we leave to the reader as an exercise. This completes the proof.

□

4. $|a -_\star b|_\star \geq_\star |a|_\star -_\star |b|_\star$.

   Proof. We have

$$\begin{aligned} |a|_\star -_\star |b|_\star &= |a -_\star b +_\star b|_\star -_\star |b|_\star \\ &\leq |a -_\star b|_\star +_\star |b|_\star -_\star |b|_\star \\ &= |a -_\star b|_\star. \end{aligned}$$

   This completes the proof. □

Exercise 1.3. *Let $a, b \in \mathbb{R}_*$. Prove that*

1. $||a|_\star -_\star |b|_\star|_\star \leq_\star |a -_\star b|_\star$.
2. $|a \cdot_\star b|_\star = |a|_\star \cdot_\star |b|_\star$.
3. $|a /_\star b|_\star = |a|_\star /_\star |b|_\star$.

1.4 The Power Function

In this section, we define the power function and we deduce some of its properties.

Definition 1.13. *Let $x \in \mathbb{R}_\star$ and $k \in \mathbb{N}$. Define*

$$x^{k_\star} = \underbrace{x \cdot_\star \cdot_\star \cdots \cdot_\star x}_{k}.$$

By Definition 1.13, it follows

$$\begin{aligned} x^{2_\star} &= x \cdot_\star x \\ &= e^{\log x \log x} \\ &= e^{(\log x)^2}. \end{aligned}$$

Assume that

$$x^{k_\star} = e^{(\log x)^k},$$

for some $k \in \mathbb{N}$. We prove that

$$x^{k+1_\star} = e^{(\log x)^{k+1}}.$$

Really,

$$\begin{aligned} x^{k+1_\star} &= \underbrace{x \cdot_\star \cdots \cdot_\star x}_{k} \cdot_\star x \\ &= e^{(\log x)^k} \cdot_\star x \\ &= e^{\log e^{(\log x)^k} \log x} \\ &= e^{(\log x)^k \log x} \\ &= e^{(\log x)^{k+1}}. \end{aligned}$$

Example 1.9. *We have*

$$2^{3_\star} = e^{(\log 2)^3}.$$

Example 1.10. *We compute*

$$A = (3 +_\star 4) \cdot_\star 3^{2_\star}.$$

We have

$$\begin{aligned} 3 +_\star 4 &= 3 \cdot 4 \\ &= 12, \\ 3^{2_\star} &= e^{(\log 3)^2}. \end{aligned}$$

Then

$$\begin{aligned} A &= 12 \cdot_\star e^{(\log 3)^2} \\ &= e^{(\log(12))(\log 3)^2} \\ &= e^{(2\log 2 + \log 3)(\log 3)^2}. \end{aligned}$$

Exercise 1.4. *Compute*

1. $A = ((2 -_\star 1) /_\star 3)^{2_\star}$.
2. $A = ((4 +_\star 2) \cdot_\star 2)^{2_\star} -_\star 2$.
3. $A = ((5 -_\star 3) /_\star 4)^{2_\star} +_\star (3 \cdot_\star 2)$.

Answer 1.9.

1. $e^{\left(\frac{\log 2}{\log 3}\right)^2}$.
2. $\frac{1}{2} e^{9(\log 2)^4}$.
3. $e^{\log 3 \log 2 + \frac{\left(\log \frac{5}{3}\right)^2}{4(\log 2)^2}}$.

Definition 1.14. *Let $x \in \mathbb{R}_\star$ and $k \in \mathbb{N}$. Define*

$$x^{-k_\star} = e^{\left(\frac{1}{\log x}\right)^k}.$$

Note that

$$\begin{aligned} x^{-k_\star} \cdot_\star x^{k_\star} &= e^{\left(\frac{1}{\log x}\right)^k} \cdot_\star e^{(\log x)^k} \\ &= e^{\log e^{\left(\frac{1}{\log x}\right)^k} \log e^{(\log x)^k}} \\ &= e^{\left(\frac{1}{\log x}\right)^k (\log x)^k} \\ &= e \\ &= 1_\star. \end{aligned}$$

Example 1.11. *We have*

$$5^{-4_\star} = e^{\left(\frac{1}{\log 5}\right)^4}.$$

Definition 1.15. *Let $x \in \mathbb{R}_\star$, $p, q \in \mathbb{N}$. Define*

$$x^{\frac{p}{q}_\star} = e^{(\log x)^{\frac{p}{q}}}.$$

Note that

$$\begin{aligned}\left(x^{\frac{p}{q}\star}\right)^{q_\star} &= \left(e^{(\log x)^{\frac{p}{q}}}\right)^{q_\star}\\ &= e^{\left(\log e^{(\log x)^{\frac{p}{q}}}\right)^q}\\ &= e^{\left((\log x)^{\frac{p}{q}}\right)^q}\\ &= e^{(\log x)^p}\\ &= x^{p_\star}.\end{aligned}$$

Example 1.12. *We have*

$$2^{\frac{3}{4}\star} = e^{(\log 2)^{\frac{3}{4}}}.$$

Definition 1.16. *Let $x \in \mathbb{R}_\star$ and $k \in \mathbb{R}$. Define*

$$x^{k_\star} = e^{(\log x)^k}.$$

Note that for any $k \in \mathbb{R}$, we have

$$\begin{aligned}1_\star^{k_\star} &= e^{(\log e)^k}\\ &= e\\ &= 1_\star.\end{aligned}$$

Now, we deduce some useful formulas.

1. $(x +_\star y)^{2_\star} = x^{2_\star} +_\star e^2 \cdot_\star x \cdot_\star y +_\star y^{2_\star}$, $x, y \in \mathbb{R}_\star$.

   Proof. We have

$$\begin{aligned}(x +_\star y)^{2_\star} &= (xy)^{2_\star}\\ &= e^{(\log(xy))^2}\\ &= e^{(\log x + \log y)^2}\\ &= e^{(\log x)^2 + 2\log x \log y + (\log y)^2}\\ &= e^{(\log x)^2} e^{2\log x \log y} e^{(\log y)^2}\\ &= e^{(\log x)^2} +_\star e^{2\log x \log y} +_\star e^{(\log y)^2}\\ &= x^{2_\star} +_\star e^2 \cdot_\star e^{\log x \log y} +_\star e^{(\log y)^2}\\ &= x^{2_\star} +_\star e^2 \cdot_\star x \cdot_\star y +_\star y^{2_\star}.\end{aligned}$$

   This completes the proof. □

2. $(x -_\star y)^{2_\star} = x^{2_\star} -_\star e^2 \cdot_\star x \cdot_\star y +_\star y^{2_\star}$, $x, y \in \mathbb{R}_\star$.

Proof. We have

$$\begin{aligned}
(x -_\star y)^{2_\star} &= \left(\frac{x}{y}\right)^{2_\star} \\
&= e^{(\log \frac{x}{y})^2} \\
&= e^{(\log x - \log y)^2} \\
&= e^{(\log x)^2 - 2\log x \log y + (\log y)^2} \\
&= e^{(\log x)^2} e^{-2\log x \log y} e^{(\log y)^2} \\
&= e^{(\log x)^2} -_\star e^{2\log x \log y} +_\star e^{(\log y)^2} \\
&= x^{2_\star} -_\star e^2 \cdot_\star e^{\log x \log y} +_\star e^{(\log y)^2} \\
&= x^{2_\star} -_\star e^2 \cdot_\star x \cdot_\star y +_\star y^{2_\star}.
\end{aligned}$$

This completes the proof. □

3. $(x +_\star y)^{3_\star} = x^{3_\star} +_\star e^3 \cdot_\star x^{2_\star} \cdot_\star y +_\star e^3 \cdot_\star x \cdot_\star y^{2_\star} +_\star y^{3_\star}$, $x, y \in \mathbb{R}_\star$.

Proof. We have

$$\begin{aligned}
(x +_\star y)^{3_\star} &= (xy)^{3_\star} \\
&= e^{(\log(xy))^3} \\
&= e^{(\log x + \log y)^3} \\
&= e^{(\log x)^3 + 3(\log x)^2 \log y + 3\log x (\log y)^2 + (\log y)^3} \\
&= e^{(\log x)^3} +_\star e^{3(\log x)^2 \log y} +_\star e^{3\log x (\log y)^2} +_\star e^{(\log y)^3} \\
&= x^{3_\star} +_\star e^3 \cdot_\star e^{(\log x)^2} \cdot_\star e^{\log y} +_\star e^3 \cdot_\star e^{\log x} \cdot_\star e^{(\log y)^2} \\
&\quad +_\star e^{(\log y)^3} \\
&= x^{3_\star} +_\star e^3 \cdot_\star x^{2_\star} \cdot_\star y +_\star e^3 \cdot_\star x \cdot_\star y^{2_\star} +_\star y^{3_\star}.
\end{aligned}$$

This completes the proof. □

4. $(x -_\star y)^{3_\star} = x^{3_\star} -_\star e^3 \cdot_\star x^{2_\star} \cdot_\star y +_\star e^3 \cdot_\star x \cdot_\star y^{2_\star} -_\star y^{3_\star}$, $x, y \in \mathbb{R}_\star$.

Proof. We have

$$\begin{aligned}
(x -_\star y)^{3_\star} &= \left(\frac{x}{y}\right)^{3_\star} \\
&= e^{(\log \frac{x}{y})^3} \\
&= e^{(\log x - \log y)^3} \\
&= e^{(\log x)^3 - 3(\log x)^2 \log y + 3\log x (\log y)^2 - (\log y)^3} \\
&= e^{(\log x)^3} -_\star e^{3(\log x)^2 \log y} +_\star e^{3\log x (\log y)^2} -_\star e^{(\log y)^3} \\
&= x^{3_\star} -_\star e^3 \cdot_\star e^{(\log x)^2} \cdot_\star e^{\log y} +_\star e^3 \cdot_\star e^{\log x} \cdot_\star e^{(\log y)^2} \\
&\quad -_\star e^{(\log y)^3} \\
&= x^{3_\star} -_\star e^3 \cdot_\star x^{2_\star} \cdot_\star y +_\star e^3 \cdot_\star x \cdot_\star y^{2_\star} -_\star y^{3_\star}.
\end{aligned}$$

This completes the proof. □

Exercise 1.5. *Prove*

1. $(x +_\star y) \cdot_\star (x -_\star y) = x^{2_\star} -_\star y^{2_\star}$, $x, y \in \mathbb{R}_\star$.
2. $(x +_\star y) \cdot_\star (x^{2_\star} -_\star x \cdot_\star y +_\star y^{2_\star}) = x^{3_\star} +_\star y^{3_\star}$, $x, y \in \mathbb{R}_\star$.
3. $(x -_\star y) \cdot_\star (x^{2_\star} -_\star x \cdot_\star y +_\star y^{2_\star}) = x^{3_\star} -_\star y^{3_\star}$, $x, y \in \mathbb{R}_\star$.

1.5 Multiplicative Trigonometric Functions

Definition 1.17. *For $x \in \mathbb{R}_\star$, define multiplicative sine, multiplicative cosine, multiplicative tangent and multiplicative cotangent as follows:*

1. $\sin_\star x = e^{\sin \log x}$.
2. $\cos_\star x = e^{\cos \log x}$.
3. $\tan_\star x = e^{\tan \log x}$.
4. $\cot_\star x = e^{\cot \log x}$.

Example 1.13. *We have*

$$\begin{aligned} \sin_\star e^{\frac{\pi}{2}} &= e^{\sin \log e^{\frac{\pi}{2}}} \\ &= e^{\sin \frac{\pi}{2}} \\ &= e^1 \\ &= 1_\star \end{aligned}$$

and

$$\begin{aligned} \cos_\star e^{\frac{\pi}{2}} &= e^{\cos \log e^{\frac{\pi}{2}}} \\ &= e^{\cos \frac{\pi}{2}} \\ &= 1 \\ &= 0_\star. \end{aligned}$$

Example 1.14. *We compute*

$$A = 2 \cdot_\star \sin_\star e^{\frac{\pi}{4}} +_\star \cos_\star e^{\frac{\pi}{4}}.$$

We have

$$\begin{aligned} \sin_\star e^{\frac{\pi}{4}} &= e^{\sin \log e^{\frac{\pi}{4}}} \\ &= e^{\sin \frac{\pi}{4}} \\ &= e^{\frac{\sqrt{2}}{2}}, \\ 2 \cdot_\star \sin_\star e^{\frac{\pi}{4}} &= e^{\log 2 \log e^{\frac{\sqrt{2}}{2}}} \\ &= e^{\frac{\sqrt{2}\log 2}{2}}, \\ \cos_\star e^{\frac{\pi}{4}} &= e^{\cos\left(\log e^{\frac{\pi}{4}}\right)} \\ &= e^{\cos \frac{\pi}{4}} \\ &= e^{\frac{\sqrt{2}}{2}}. \end{aligned}$$

Therefore,

$$\begin{aligned} A &= e^{\frac{\sqrt{2}\log 2}{2}} +_\star e^{\frac{\sqrt{2}}{2}} \\ &= e^{\frac{\sqrt{2}}{2}\log 2} e^{\frac{\sqrt{2}}{2}} \\ &= e^{\frac{\sqrt{2}}{2}(1+\log 2)}. \end{aligned}$$

Exercise 1.6. *Find*

$$A = 3 \cdot_\star \cos_\star e^{\frac{\pi}{4}} -_\star \sin_\star e^{\frac{\pi}{2}} +_\star 2 \cdot_\star \cos_\star e^{\frac{\pi}{2}} +_\star \cot_\star e^{\frac{\pi}{4}}.$$

Answer 1.10. $e^{\frac{\sqrt{2}}{2}\log 3}$.

We deduce some of the basic properties of the multiplicative trigonometric functions.

1. $\sin_\star x +_\star \sin_\star y = e^2 \cdot_\star \sin_\star \left((x +_\star y)/_\star e^2\right) \cdot_\star \cos_\star \left((x -_\star y)/_\star e^2\right)$.

   Proof. We have

$$\begin{aligned} \sin_\star x +_\star \sin_\star y &= e^{\sin\log x} +_\star e^{\sin\log y} \\ &= e^{\sin\log x + \sin\log y} \\ &= e^{2\sin\frac{\log x + \log y}{2}\cos\frac{\log x - \log y}{2}} \\ &= e^{2\sin\left(\log\sqrt{xy}\right)\cos\left(\log\sqrt{\frac{x}{y}}\right)} \\ &= e^2 \cdot_\star e^{\sin\left(\log\sqrt{xy}\right)\cos\left(\log\sqrt{\frac{x}{y}}\right)} \\ &= e^2 \cdot_\star e^{\sin\left(\log\sqrt{xy}\right)} \cdot_\star e^{\cos\left(\log\sqrt{\frac{x}{y}}\right)} \\ &= e^2 \cdot_\star e^{\sin\left(\log\left((x+_\star y)/_\star e^2\right)\right)} \cdot_\star e^{\cos\left(\log\left((x-_\star y)/_\star e^2\right)\right)} \\ &= e^2 \cdot_\star \sin_\star \left((x +_\star y)/_\star e^2\right) \cdot_\star \cos_\star \left((x -_\star y)/_\star e^2\right). \end{aligned}$$

   This completes the proof. □

2. $\sin_\star x -_\star \sin_\star y = e^2 \cdot_\star \sin_\star \left((x -_\star y)/_\star e^2\right) \cdot_\star \cos_\star \left((x +_\star y)/_\star e^2\right)$.

   Proof. We have

$$\begin{aligned} \sin_\star x -_\star \sin_\star y &= e^{\sin\log x} -_\star e^{\sin\log y} \\ &= e^{\sin\log x - \sin\log y} \\ &= e^{2\sin\frac{\log x - \log y}{2}\cos\frac{\log x + \log y}{2}} \\ &= e^{2\cos\left(\log\sqrt{xy}\right)\sin\left(\log\sqrt{\frac{x}{y}}\right)} \\ &= e^2 \cdot_\star e^{\cos\left(\log\sqrt{xy}\right)\sin\left(\log\sqrt{\frac{x}{y}}\right)} \\ &= e^2 \cdot_\star e^{\cos\left(\log\sqrt{xy}\right)} \cdot_\star e^{\sin\left(\log\sqrt{\frac{x}{y}}\right)} \\ &= e^2 \cdot_\star e^{\cos\left(\log\left((x+_\star y)/_\star e^2\right)\right)} \cdot_\star e^{\sin\left(\log\left((x-_\star y)/_\star e^2\right)\right)} \\ &= e^2 \cdot_\star \sin_\star \left((x -_\star y)/_\star e^2\right) \cdot_\star \cos_\star \left((x +_\star y)/_\star e^2\right). \end{aligned}$$

   This completes the proof. □

3. $(\sin_\star x)^{2_\star} +_\star (\cos_\star x)^{2_\star} = 1_\star$.

Proof. We have

$$\begin{aligned}
(\sin_\star x)^{2_\star} +_\star (\cos_\star x)^{2_\star} &= \left(e^{\sin\log x}\right)^{2_\star} +_\star \left(e^{\cos\log x}\right)^{2_\star} \\
&= e^{\left(\log e^{\sin\log x}\right)^2} +_\star e^{\left(\log e^{\cos\log x}\right)^2} \\
&= e^{(\sin\log x)^2} +_\star e^{(\cos\log x)^2} \\
&= e^{(\sin\log x)^2+(\cos\log x)^2} \\
&= e \\
&= 1_\star .
\end{aligned}$$

This completes the proof. □

4. $(\sin_\star x)^{2_\star} = \left(1_\star -_\star \cos_\star\left(e^2 \cdot_\star x\right)\right) /_\star e^2.$

   Proof. We have

$$\begin{aligned}
\left(1_\star -_\star \cos_\star\left(e^2 \cdot_\star x\right)\right) /_\star e^2 &= \left(e -_\star e^{\cos\left(\log\left(e^2\cdot_\star x\right)\right)}\right) /_\star e^2 \\
&= \left(e -_\star e^{\cos\left(\log e^{2\log x}\right)}\right) /_\star e^2 \\
&= \left(e -_\star e^{\cos(2\log x)}\right) /_\star e^2 \\
&= \left(\frac{e}{e^{\cos(2\log x)}}\right) /_\star e^2 \\
&= \left(e^{1-\cos(2\log x)}\right) /_\star e^2 \\
&= \left(e^{2(\sin\log x)^2}\right) /_\star e^2 \\
&= e^{\frac{\log e^{2(\sin\log x)^2}}{\log e^2}} \\
&= e^{(\sin\log x)^2} \\
&= e^{\left(\log e^{\sin\log x}\right)^2} \\
&= \left(e^{\sin\log x}\right)^{2_\star} \\
&= (\sin_\star x)^{2_\star} .
\end{aligned}$$

   This completes the proof. □

5. $(\cos_\star x)^{2_\star} = \left(1_\star +_\star \cos_\star\left(e^2 \cdot_\star x\right)\right) /_\star e^2.$

   Proof. We have

$$\begin{aligned}
\left(1_\star +_\star \cos_\star\left(e^2 \cdot_\star x\right)\right) /_\star e^2 &= \left(e +_\star e^{\cos\left(\log\left(e^2\cdot_\star x\right)\right)}\right) /_\star e^2 \\
&= \left(e +_\star e^{\cos\left(\log e^{2\log x}\right)}\right) /_\star e^2 \\
&= \left(e +_\star e^{\cos(2\log x)}\right) /_\star e^2 \\
&= \left(e^{1+\cos(2\log x)}\right) /_\star e^2 \\
&= \left(e^{2(\cos\log x)^2}\right) /_\star e^2
\end{aligned}$$

$$
\begin{aligned}
&= e^{\frac{\log e^{2(\cos\log x)^2}}{\log e^2}}\\
&= e^{(\cos\log x)^2}\\
&= e^{\left(\log e^{\cos\log x}\right)^2}\\
&= \left(e^{\cos\log x}\right)^{2_\star}\\
&= (\cos_\star x)^{2_\star}.
\end{aligned}
$$

This completes the proof. □

6. $\cos_\star x +_\star \cos_\star y = e^2 \cdot_\star \cos_\star\left((x+_\star y)/_\star e^2\right) \cdot_\star \cos_\star\left((x-_\star y)/_\star e^2\right)$.

Proof. We have

$$
\begin{aligned}
\cos_\star x +_\star \cos_\star y &= e^{\cos\log x} +_\star e^{\cos\log y}\\
&= e^{\cos\log x+\cos\log y}\\
&= e^{2\cos\frac{\log x+\log y}{2}\cos\frac{\log x-\log y}{2}}\\
&= e^{2\cos\left(\log\sqrt{xy}\right)\cos\left(\log\sqrt{\frac{x}{y}}\right)}\\
&= e^2 \cdot_\star e^{\cos\left(\log\sqrt{xy}\right)\cos\left(\log\sqrt{\frac{x}{y}}\right)}\\
&= e^2 \cdot_\star e^{\cos\left(\log\sqrt{xy}\right)} \cdot_\star e^{\cos\left(\log\sqrt{\frac{x}{y}}\right)}\\
&= e^2 \cdot_\star e^{\cos\left(\log\left((x+_\star y)/_\star e^2\right)\right)} \cdot_\star e^{\cos\left(\log\left((x-_\star y)/_\star e^2\right)\right)}\\
&= e^2 \cdot_\star \cos_\star\left((x+_\star y)/_\star e^2\right) \cdot_\star \cos_\star\left((x-_\star y)/_\star e^2\right).
\end{aligned}
$$

This completes the proof. □

7. $\cos_\star x -_\star \cos_\star y = -_\star e^2 \cdot_\star \sin_\star\left((x-_\star y)/_\star e^2\right) \cdot_\star \sin_\star\left((x+_\star y)/_\star e^2\right)$.

Proof. We have

$$
\begin{aligned}
\cos_\star x -_\star \cos_\star y &= e^{\cos\log x} -_\star e^{\cos\log y}\\
&= e^{\cos\log x-\cos\log y}\\
&= e^{-2\sin\frac{\log x-\log y}{2}\sin\frac{\log x+\log y}{2}}\\
&= e^{-2\sin\left(\log\sqrt{xy}\right)\sin\left(\log\sqrt{\frac{x}{y}}\right)}\\
&= -_\star e^2 \cdot_\star e^{\sin\left(\log\sqrt{xy}\right)\sin\left(\log\sqrt{\frac{x}{y}}\right)}\\
&= -_\star e^2 \cdot_\star e^{\sin\left(\log\sqrt{xy}\right)} \cdot_\star e^{\sin\left(\log\sqrt{\frac{x}{y}}\right)}\\
&= -_\star e^2 \cdot_\star e^{\sin\left(\log\left((x+_\star y)/_\star e^2\right)\right)} \cdot_\star e^{\sin\left(\log\left((x-_\star y)/_\star e^2\right)\right)}\\
&= -_\star e^2 \cdot_\star \sin_\star\left((x-_\star y)/_\star e^2\right) \cdot_\star \sin_\star\left((x+_\star y)/_\star e^2\right).
\end{aligned}
$$

This completes the proof. □

8. $\tan_\star x = \sin_\star x /_\star \cos_\star x$.

   Proof. We have

$$\begin{aligned}
\tan_\star x &= e^{\tan\log x} \\
&= e^{\frac{\sin\log x}{\cos\log x}} \\
&= e^{\frac{\log e^{\sin\log x}}{\log e^{\cos\log x}}} \\
&= \left(e^{\sin\log x}\right) /_\star \left(e^{\cos\log x}\right) \\
&= \sin_\star x /_\star \cos_\star x.
\end{aligned}$$

   This completes the proof. □

9. $\cot_\star x = \cos_\star x /_\star \sin_\star x$.

   Proof. We have

$$\begin{aligned}
\cot_\star x &= e^{\cot\log x} \\
&= e^{\frac{\cos\log x}{\sin\log x}} \\
&= e^{\frac{\log e^{\cos\log x}}{\log e^{\sin\log x}}} \\
&= \left(e^{\cos\log x}\right) /_\star \left(e^{\sin\log x}\right) \\
&= \cos_\star x /_\star \sin_\star x.
\end{aligned}$$

   This completes the proof. □

1.6 Multiplicative Inverse Trigonometric Functions

In this section, we define the multiplicative inverse trigonometric functions.

Definition 1.18. *For $x \in \mathbb{R}_*$, we define the inverse trigonometric functions as follows:*

1. $\arcsin_\star x = e^{\arcsin(\log x)}$.
2. $\arccos_\star x = e^{\arccos(\log x)}$.
3. $\arctan_\star x = e^{\arctan(\log x)}$.
4. $arccot_\star x = e^{arccot(\log x)}$.

Example 1.15. *We have*

$$\begin{aligned}
\arcsin_\star e^{\frac{\sqrt{2}}{2}} &= e^{\arcsin\left(\log e^{\frac{\sqrt{2}}{2}}\right)} \\
&= e^{\arcsin\left(\frac{\sqrt{2}}{2}\right)} \\
&= e^{\frac{\pi}{4}},
\end{aligned}$$

and

$$\begin{aligned}\arctan_\star e &= e^{\arctan \log e}\\ &= e^{\arctan 1}\\ &= e^{\frac{\pi}{4}}.\end{aligned}$$

Now, we list some of the properties of the inverse trigonometric functions.

1. For $x \in \mathbb{R}_\star$, we have

$$\begin{aligned}\sin_\star(\arcsin_\star x) &= e^{\sin(\log(\arcsin_\star x))}\\ &= e^{\sin\left(\log\left(e^{\arcsin(\log x)}\right)\right)}\\ &= e^{\sin(\arcsin(\log x))}\\ &= e^{\log x}\\ &= x.\end{aligned}$$

2. For $x \in \mathbb{R}_\star$, we have

$$\begin{aligned}\arcsin_\star(\sin_\star x) &= e^{\arcsin(\log(\sin_\star x))}\\ &= e^{\arcsin\left(\log\left(e^{\sin(\log x)}\right)\right)}\\ &= e^{\arcsin(\sin(\log x))}\\ &= e^{\log x}\\ &= x.\end{aligned}$$

3. For $x \in \mathbb{R}_\star$, we have

$$\begin{aligned}\cos_\star(\arccos_\star x) &= e^{\cos(\log(\arccos_\star x))}\\ &= e^{\cos\left(\log e^{\arccos(\log x)}\right)}\\ &= e^{\cos(\arccos(\log x))}\\ &= e^{\log x}\\ &= x.\end{aligned}$$

Exercise 1.7. *Prove that*

$$\arccos_\star(\cos_\star x) = x, \quad x \in \mathbb{R}_\star.$$

Example 1.16. *We find*

$$\arcsin_\star(\cos_\star x), \quad x \in \mathbb{R}_\star.$$

We have

$$\begin{aligned}\arcsin_\star(\cos_\star x) &= e^{\arcsin(\log(\cos_\star x))}\\ &= e^{\arcsin\left(\log\left(e^{\cos(\log x)}\right)\right)}\\ &= e^{\arcsin(\cos(\log x))}\\ &= e^{\arcsin\left(\sin\left(\frac{\pi}{2}-\log x\right)\right)}\\ &= e^{\frac{\pi}{2}-\log x}\\ &= \frac{e^{\frac{\pi}{2}}}{x}\\ &= e^{\frac{\pi}{2}} -_\star x, \quad x \in \mathbb{R}_\star.\end{aligned}$$

4. For $x \in \mathbb{R}_\star$, we have

$$\begin{aligned}\tan_\star(\arctan_\star x) &= e^{\tan(\log(\arctan_\star x))}\\ &= e^{\tan\left(\log e^{\arctan(\log x)}\right)}\\ &= e^{\tan(\arctan(\log x))}\\ &= e^{\log x}\\ &= x.\end{aligned}$$

Exercise 1.8. *Prove that*

$$\arctan_\star(\tan_\star x) = x, \quad x \in \mathbb{R}_\star.$$

5. For $x \in \mathbb{R}_\star$, we have

$$\begin{aligned}\cos_\star(\arccos_\star x) &= e^{\cos(\log(\arccos_\star x))}\\ &= e^{\cos\left(\log e^{\arccos(\log x)}\right)}\\ &= e^{\cos(\arccos(\log x))}\\ &= e^{\log x}\\ &= x.\end{aligned}$$

6. For $x \in \mathbb{R}_\star$, we have

$$\begin{aligned}\cot_\star(\operatorname{arccot}_\star x) &= e^{\cot(\log(\operatorname{arccot}_\star x))}\\ &= e^{\cot\left(\log e^{\operatorname{arccot}(\log x)}\right)}\\ &= e^{\cot(\operatorname{arccot}(\log x))}\\ &= e^{\log x}\\ &= x.\end{aligned}$$

Exercise 1.9. *Prove that*

$$arccot_\star(\cot_\star x) = x, \quad x \in \mathbb{R}_\star.$$

Example 1.17. *We find*

$$\sin_\star(\arctan_\star x), \quad x \in \mathbb{R}_\star.$$

We have

$$\begin{aligned}\sin_\star(\arctan_\star x) &= e^{\sin(\log(\arctan_\star x))}\\ &= e^{\sin\left(\log e^{\arctan(\log x)}\right)}\\ &= e^{\sin(\arctan(\log x))}, \quad x \in \mathbb{R}_\star.\end{aligned}$$

Let

$$y = \arctan(\log x), \quad x \in \mathbb{R}_\star.$$

Then $y \in \left(-\frac{\pi}{2}, \frac{\pi}{2}\right)$ *and*

$$\tan y = \log x, \quad x \in \mathbb{R}_\star,$$

and

$$\sin y = (\log x)\cos y, \quad x \in \mathbb{R}_\star.$$

Hence,

$$\begin{aligned} 1 &= \sin^2 y + \cos^2 y \\ &= (\log x)^2 \cos^2 y + \cos^2 y \\ &= \left(1 + (\log x)^2\right) \cos^2 y, \quad x \in \mathbb{R}_\star, \end{aligned}$$

and

$$\cos^2 y = \frac{1}{1 + (\log x)^2}, \quad x \in \mathbb{R}_\star,$$

or

$$\begin{aligned} \cos y &= \frac{1}{\sqrt{1 + (\log x)^2}}, \\ \sin y &= \frac{\log x}{\sqrt{1 + (\log x)^2}}, \quad x \in \mathbb{R}_\star. \end{aligned}$$

Consequently,

$$\begin{aligned} \sin_\star(\arctan_\star x) &= e^{\sin y} \\ &= e^{\frac{\log x}{\sqrt{1+(\log x)^2}}}, \quad x \in \mathbb{R}_\star. \end{aligned}$$

Exercise 1.10. *Find*

1. $\cos_\star(\arcsin_\star x)$, $x \in \mathbb{R}_\star$.
2. $\tan_\star(\arcsin_\star x)$, $x \in \mathbb{R}_\star$.
3. $\cot_\star(\arccos_\star x)$, $x \in \mathbb{R}_\star$.

Answer 1.11.

1. $e^{\sqrt{1-(\log x)^2}}$, $x \in \left[e^{-1}, e\right]$.
2. $e^{\frac{\log x}{\sqrt{1-(\log x)^2}}}$, $x \in \left(e^{-1}, e\right)$.
3. $e^{\frac{\log x}{\sqrt{1-(\log x)^2}}}$, $x \in \left(e^{-1}, e\right)$.

1.7 Multiplicative Hyperbolic Functions

In this section, we define the multiplicative hyperbolic functions and we deduce some of their properties.

Definition 1.19. *For $x \in \mathbb{R}_\star$, define*

1. $\sinh_\star x = e^{\sinh(\log x)}$.
2. $\cosh_\star x = e^{\cosh(\log x)}$.
3. $\tanh_\star x = e^{\tanh(\log x)}$.
4. $\coth_\star x = e^{\coth(\log x)}$.

Now, we list some of the properties of the multiplicative hyperbolic functions.

1. For $x \in \mathbb{R}_\star$, we have

$$\tanh_\star x = \sinh_\star x /_\star \cosh_\star x.$$

Proof. We have

$$\begin{aligned} \sinh_\star x /_\star \cosh_\star x &= e^{\frac{\log(\sinh_\star x)}{\log(\cosh_\star x)}} \\ &= e^{\frac{\log\left(e^{\sinh(\log x)}\right)}{\log\left(e^{\cosh(\log x)}\right)}} \\ &= e^{\frac{\sinh(\log x)}{\cosh(\log x)}} \\ &= e^{\tanh(\log x)} \\ &= \tanh_\star x. \end{aligned}$$

This completes the proof. □

2. For $x \in \mathbb{R}_\star$, we have

$$\coth_\star x = \cosh_\star x /_\star \sinh_\star x.$$

Proof. We have

$$\begin{aligned} \cosh_\star x /_\star \sinh_\star x &= e^{\frac{\log(\cosh_\star x)}{\log(\sinh_\star x)}} \\ &= e^{\frac{\log\left(e^{\cosh(\log x)}\right)}{\log\left(e^{\sinh(\log x)}\right)}} \\ &= e^{\frac{\cosh(\log x)}{\sinh(\log x)}} \\ &= e^{\coth(\log x)} \\ &= \coth_\star x. \end{aligned}$$

This completes the proof. □

3. For $x, y \in \mathbb{R}_\star$, we have

$$\sinh_\star(x +_\star y) = \sinh_\star x \cdot_\star \cosh_\star y +_\star \sinh_\star y \cdot_\star \cosh_\star x.$$

Proof. We have

$$\begin{aligned} \sinh_\star(x +_\star y) &= \sinh_\star(xy) \\ &= e^{\sinh(\log(xy))} \\ &= e^{\sinh(\log x + \log y)} \\ &= e^{\sinh(\log x)\cosh(\log y) + \sinh(\log y)\cosh(\log x)} \\ &= e^{\sinh(\log x)\cosh(\log y)} e^{\sinh(\log y)\cosh(\log x)}, \end{aligned}$$

and

$$\begin{aligned} &\sinh_\star x \cdot_\star \cosh_\star y +_\star \sinh_\star y \cdot_\star \cosh_\star x \\ &= e^{\log(\sinh_\star x)\log(\cosh_\star y)} +_\star e^{\log(\sinh_\star y)\log(\cosh_\star x)} \\ &= e^{\log\left(e^{\sinh(\log x)}\right)\log\left(e^{\cosh(\log y)}\right)} \\ &\quad +_\star e^{\log\left(e^{\sinh(\log y)}\right)\log\left(e^{\cosh(\log x)}\right)} \\ &= e^{\sinh(\log x)\cosh(\log y)} +_\star e^{\sinh(\log y)\cosh(\log x)} \\ &= e^{\sinh(\log x)\cosh(\log y)} e^{\sinh(\log y)\cosh(\log x)} \\ &= e^{\sinh(\log x)\cosh(\log y) + \sinh(\log y)\cosh(\log x)}. \end{aligned}$$

Consequently,

$$\sinh_\star(x +_\star y) = \sinh_\star x \cdot_\star \cosh_\star y +_\star \sinh_\star y \cdot_\star \cosh_\star x.$$

This completes the proof. □

4. For $x, y \in \mathbb{R}_\star$, we have

$$\cosh_\star(x +_\star y) = \cosh_\star x \cdot_\star \cosh_\star y +_\star \sinh_\star y \cdot_\star \sinh_\star x.$$

Proof. We have

$$\begin{aligned}
\cosh_\star(x +_\star y) &= \cosh_\star(xy) \\
&= e^{\cosh(\log(xy))} \\
&= e^{\cosh(\log x + \log y)} \\
&= e^{\cosh(\log x)\cosh(\log y) + \sinh(\log y)\sinh(\log x)} \\
&= e^{\cosh(\log x)\cosh(\log y)} e^{\sinh(\log y)\sinh(\log x)},
\end{aligned}$$

and

$$\begin{aligned}
&\cosh_\star x \cdot_\star \cosh_\star y +_\star \sinh_\star y \cdot_\star \sinh_\star x \\
&= e^{\log(\cosh_\star x)\log(\cosh_\star y)} +_\star e^{\log(\sinh_\star y)\log(\sinh_\star x)} \\
&= e^{\log\left(e^{\cosh(\log x)}\right)\log\left(e^{\cosh(\log y)}\right)} \\
&\quad +_\star e^{\log\left(e^{\sinh(\log y)}\right)\log\left(e^{\sinh(\log x)}\right)} \\
&= e^{\cosh(\log x)\cosh(\log y)} +_\star e^{\sinh(\log y)\sinh(\log x)} \\
&= e^{\cosh(\log x)\cosh(\log y)} e^{\sinh(\log y)\sinh(\log x)} \\
&= e^{\cosh(\log x)\cosh(\log y) + \sinh(\log y)\sinh(\log x)}.
\end{aligned}$$

Consequently,

$$\cosh_\star(x +_\star y) = \cosh_\star x \cdot_\star \cosh_\star y +_\star \sinh_\star y \cdot_\star \sinh_\star x.$$

This completes the proof. □

1.8 Multiplicative Inverse Hyperbolic Functions

In this section, we define the multiplicative inverse hyperbolic functions and we deduce some of their properties.

Definition 1.20. *For $x \in \mathbb{R}_\star$, define*

1. $\sinh_\star^{-1_\star} x = e^{\sinh^{-1}(\log x)}$.
2. $\cosh_\star^{-1_\star} x = e^{\cosh^{-1}(\log x)}$.
3. $\tanh_\star^{-1_\star} x = e^{\tanh^{-1}(\log x)}$.
4. $\coth_\star^{-1_\star} x = e^{\coth^{-1}(\log x)}$.

Now, we list some of the properties of the multiplicative inverse hyperbolic functions.

1. $\sinh_\star\left(\sinh_\star^{-1_\star} x\right) = x,\, x \in \mathbb{R}_\star$.

   Proof. We have

$$\begin{aligned}\sinh_\star\left(\sinh_\star^{-1_\star} x\right) &= e^{\sinh\left(\log\left(\sinh_\star^{-1_\star} x\right)\right)}\\ &= e^{\sinh\left(\log e^{\sinh^{-1}(\log x)}\right)}\\ &= e^{\sinh\left(\sinh^{-1}(\log x)\right)}\\ &= e^{\log x}\\ &= x, \quad x \in \mathbb{R}_\star.\end{aligned}$$

   This completes the proof. □

2. $\cosh_\star\left(\cosh_\star^{-1_\star} x\right) = x,\, x \in \mathbb{R}_\star$.

   Proof. We have

$$\begin{aligned}\cosh_\star\left(\cosh_\star^{-1_\star} x\right) &= e^{\cosh\left(\log\left(\cosh_\star^{-1_\star} x\right)\right)}\\ &= e^{\cosh\left(\log e^{\cosh^{-1}(\log x)}\right)}\\ &= e^{\cosh\left(\cosh^{-1}(\log x)\right)}\\ &= e^{\log x}\\ &= x, \quad x \in \mathbb{R}_\star.\end{aligned}$$

   This completes the proof. □

3. $\tanh_\star\left(\tanh_\star^{-1_\star} x\right) = x,\, x \in \mathbb{R}_\star$.

   Proof. We have

$$\begin{aligned}\tanh_\star\left(\tanh_\star^{-1_\star} x\right) &= e^{\tanh\left(\log\left(\tanh_\star^{-1_\star} x\right)\right)}\\ &= e^{\tanh\left(\log e^{\tanh^{-1}(\log x)}\right)}\\ &= e^{\tanh\left(\tanh^{-1}(\log x)\right)}\\ &= e^{\log x}\\ &= x, \quad x \in \mathbb{R}_\star.\end{aligned}$$

   This completes the proof. □

4. $\coth_\star\left(\coth_\star^{-1_\star} x\right) = x,\, x \in \mathbb{R}_\star$.

   Proof. We have

$$\begin{aligned}\coth_\star\left(\coth_\star^{-1_\star} x\right) &= e^{\coth\left(\log\left(\coth_\star^{-1_\star} x\right)\right)}\\ &= e^{\coth\left(\log e^{\coth^{-1}(\log x)}\right)}\\ &= e^{\coth\left(\coth^{-1}(\log x)\right)}\\ &= e^{\log x}\\ &= x, \quad x \in \mathbb{R}_\star.\end{aligned}$$

   This completes the proof. □

1.9 Multiplicative Matrices

Definition 1.21. *A matrix whose entries are elements of* $\mathbb{R}_\star$ *will be called a multiplicative matrix. The set of all multiplicative matrices* $m \times n$ *will be denoted by* $\mathscr{M}_{\star m\times n}$.

Definition 1.22. *Let* $A = (a_{ij}) \in \mathscr{M}_{\star m\times n}$ *and* $B = (b_{ij}) \in \mathscr{M}_{\star m\times n}$ *be multiplicative matrices and* $\lambda \in \mathbb{R}_\star$. *Then, we define*

$$A \pm_\star B = (a_{ij} \pm_\star b_{ij}) \quad \text{and} \quad \lambda \cdot_\star A = (\lambda \cdot_\star a_{ij}).$$

Example 1.18. *Let*

$$A = \begin{pmatrix} 2 & \frac{1}{3} \\ 4 & \frac{1}{8} \end{pmatrix} \quad \text{and} \quad \lambda = 3.$$

Then

$$\begin{aligned} 3 \cdot_\star 2 &= e^{\log 3 \log 2}, \\ 3 \cdot_\star \frac{1}{3} &= e^{\log 3 \log \frac{1}{3}} \\ &= e^{-(\log 3)^2}, \\ 3 \cdot_\star 4 &= e^{\log 3 \log 4} \\ &= e^{2 \log 3 \log 2}, \\ 3 \cdot_\star \frac{1}{8} &= e^{\log 3 \log \frac{1}{8}} \\ &= e^{-3 \log 3 \log 2}. \end{aligned}$$

Hence,

$$\begin{aligned} \lambda \cdot_\star A &= \begin{pmatrix} 3 \cdot_\star 2 & 3 \cdot_\star \frac{1}{3} \\ 3 \cdot_\star 4 & 3 \cdot_\star \frac{1}{8} \end{pmatrix} \\ &= \begin{pmatrix} e^{\log 3 \log 2} & e^{-(\log 3)^2} \\ e^{2 \log 3 \log 2} & e^{-3 \log 3 \log 2} \end{pmatrix}. \end{aligned}$$

Example 1.19. *Let*

$$A = \begin{pmatrix} 1 & e \\ 2 & 3 \end{pmatrix} \quad \text{and} \quad B = \begin{pmatrix} 2 & 3 \\ \frac{1}{2} & \frac{1}{3} \end{pmatrix}.$$

Then

$$\begin{aligned} 1 +_\star 2 &= 1 \cdot 2 \\ &= 2, \\ e +_\star 3 &= e \cdot 3 \\ &= 3e, \\ 2 +_\star \frac{1}{2} &= 2 \cdot \frac{1}{2} \\ &= 1, \\ 3 +_\star \frac{1}{3} &= 3 \cdot \frac{1}{3} \\ &= 1. \end{aligned}$$

Hence,

$$\begin{aligned} A +_\star B &= \begin{pmatrix} 1 +_\star 2 & e +_\star 3 \\ 2 +_\star \frac{1}{2} & 3 +_\star \frac{1}{3} \end{pmatrix} \\ &= \begin{pmatrix} 2 & 3e \\ 1 & 1 \end{pmatrix}. \end{aligned}$$

Exercise 1.11. *Let*

$$A = \begin{pmatrix} \frac{1}{3} & \frac{1}{2} & 4 \\ 2 & 1 & e \\ 3 & 4 & 5 \end{pmatrix} \quad and \quad B = \begin{pmatrix} \frac{1}{3} & \frac{1}{5} & 6 \\ 2 & 3 & 1 \\ e & 2 & 4 \end{pmatrix}.$$

Find

1. $A +_\star B$.
2. $A -_\star B$.
3. $(3 \cdot_\star A) +_\star \left(\frac{1}{2} \cdot_\star B\right)$.

Answer 1.12.

1. $\begin{pmatrix} \frac{1}{9} & \frac{1}{10} & 24 \\ 4 & 3 & e \\ 3e & 8 & 20 \end{pmatrix}$.
2. $\begin{pmatrix} 1 & \frac{5}{2} & \frac{2}{3} \\ 1 & \frac{1}{3} & e \\ \frac{3}{e} & 2 & \frac{5}{4} \end{pmatrix}$.
3. $\begin{pmatrix} e^{\log 2 \log \frac{2}{3}} & e^{\log 2 \log \frac{5}{3}} & e^{\log 2 \log \frac{3}{2}} \\ e^{\log 2 \log \frac{3}{2}} & e^{-\log 2 \log 3} & 3 \\ e^{(\log 3)^2 - \log 2} & e^{\log 2 \log \frac{9}{2}} & e^{\log 3 \log 5 - 2(\log 2)^2} \end{pmatrix}$.

Definition 1.23. *The matrix $I_\star \in \mathscr{M}_{n\times n}$, given by*

$$I_\star = \begin{pmatrix} 1_\star & 0_\star & \cdots & 0_\star \\ 0_\star & 1_\star & \cdots & 0_\star \\ \vdots & \vdots & \vdots & \vdots \\ 0_\star & 0_\star & \cdots & 1_\star \end{pmatrix},$$

will be called the multiplicative unit matrix.

We have

$$I_\star = \begin{pmatrix} e & 1 & \cdots & 1 \\ 1 & e & \cdots & 1 \\ \vdots & \vdots & \vdots & \vdots \\ 1 & 1 & \cdots & e \end{pmatrix}.$$

Definition 1.24. *The matrix $O_\star \in \mathscr{M}_{\star n\times n}$, defined by*

$$O_\star = \begin{pmatrix} 0_\star & 0_\star & \cdots & 0_\star \\ 0_\star & 0_\star & \cdots & 0_\star \\ \vdots & \vdots & \vdots & \vdots \\ 0_\star & 0_\star & \cdots & 0_\star \end{pmatrix},$$

will be called the multiplicative zero matrix.

We have

$$O_\star = \begin{pmatrix} 1 & 1 & \cdots & 1 \\ 1 & 1 & \cdots & 1 \\ \vdots & \vdots & \vdots & \vdots \\ 1 & 1 & \cdots & 1 \end{pmatrix}.$$

If $A = (a_{ij}) \in \mathscr{M}_{\star m\times n}$, then

$$\begin{aligned} A -_\star A &= (a_{ij}) -_\star (a_{ij}) \\ &= \left(\frac{a_{ij}}{a_{ij}}\right) \\ &= (1) \\ &= (0_\star) \\ &= O_\star. \end{aligned}$$

Definition 1.25. *Let $A \in \mathscr{M}_{\star r\times n}$, $A = (a_{ij})$, $B \in \mathscr{M}_{\star n\times m}$, $B = (b_{ij})$. Then, we define $A \cdot_\star B \in \mathscr{M}_{\star r\times m}$ as follows:*

$$A \cdot_\star B = (a_{k1} \cdot_\star b_{1l} +_\star a_{k2} \cdot_\star b_{2l} +_\star \cdots +_\star a_{kn} \cdot_\star b_{nl}).$$

Example 1.20. *Let*

$$A = \begin{pmatrix} 1 & \frac{1}{2} & 3 \\ 2 & e & 1 \end{pmatrix}, \quad B = \begin{pmatrix} 1 & 2 \\ e & 3 \\ \frac{1}{4} & 5 \end{pmatrix}.$$

We find $A \cdot_\star B$. We have

$$\begin{aligned} & 1 \cdot_\star 1 +_\star \frac{1}{2} \cdot_\star e +_\star 3 \cdot_\star \frac{1}{4} \\ &= e^{\log 1 \log 1} +_\star e^{\log \frac{1}{2} \log e} +_\star e^{\log 3 \log \frac{1}{4}} \\ &= 1 +_\star e^{-\log 2} +_\star e^{-2\log 3 \log 2} \\ &= 1 \cdot e^{-\log 2} \cdot e^{-2\log 3 \log 2} \\ &= \frac{1}{2} e^{-2\log 3 \log 2} \end{aligned}$$

and

$$\begin{aligned} & 1 \cdot_\star 2 +_\star \frac{1}{2} \cdot_\star 3 +_\star 3 \cdot_\star 5 \\ &= e^{\log 1 \log 2} +_\star e^{\log \frac{1}{2} \log 3} +_\star e^{\log 3 \log 5} \\ &= 1 +_\star e^{-\log 2 \log 3} +_\star e^{\log 3 \log 5} \\ &= 1 \cdot e^{-\log 2 \log 3} \cdot e^{\log 3 \log 5} \\ &= e^{\log 3 (\log 5 - \log 2)} \\ &= e^{\log 3 \log \frac{5}{2}}, \end{aligned}$$

and

$$\begin{aligned} & 2 \cdot_\star 1 +_\star e \cdot_\star e +_\star 1 \cdot_\star \frac{1}{4} \\ &= e^{\log 2 \log 1} +_\star e^{\log e \log e} +_\star e^{\log 1 \log \frac{1}{4}} \\ &= 1 +_\star e^{\log 2} +_\star 1 \\ &= e^{\log 2} \\ &= 2, \end{aligned}$$

and

$$\begin{aligned} & 2 \cdot_\star 2 +_\star e \cdot_\star 3 +_\star 1 \cdot_\star 5 \\ &= e^{(\log 2)^2} +_\star e^{\log e \log 3} +_\star e^{\log 1 \log 5} \\ &= e^{(\log 2)^2} +_\star e^{\log 3} \\ &= 3e^{(\log 2)^2}. \end{aligned}$$

Hence,

$$A \cdot_\star B = \begin{pmatrix} \frac{1}{2}e^{-2\log 3 \log 2} & e^{\log 3 \log \frac{5}{2}} \\ 2 & 3e^{(\log 2)^2} \end{pmatrix}.$$

Exercise 1.12. *Find $A \cdot_\star B$, where*

$$A = \begin{pmatrix} 2 & 3 \\ 4 & 5 \\ 6 & 7 \end{pmatrix} \quad \text{and} \quad B = \begin{pmatrix} 2 & 1 & 3 & 6 & 7 \\ \frac{1}{2} & \frac{1}{4} & \frac{1}{3} & \frac{1}{8} & 5 \end{pmatrix}.$$

Definition 1.26. *Let*

$$A = \begin{pmatrix} a_{11} & a_{12} \\ a_{21} & a_{22} \end{pmatrix} \in \mathscr{M}_{\star 2\times 2}.$$

We define the multiplicative determinant of A as follows:

$$\det_\star A = a_{11} \cdot_\star a_{22} -_\star a_{21} \cdot_\star a_{12}.$$

We have

$$\begin{aligned} \det_\star A &= a_{11} \cdot_\star a_{22} -_\star a_{21} \cdot_\star a_{12} \\ &= e^{\log a_{11} \log a_{22}} -_\star e^{\log a_{12} \log a_{21}} \\ &= \frac{e^{\log a_{11} \log a_{22}}}{e^{\log a_{21} \log a_{12}}} \\ &= e^{\log a_{11} \log a_{22} - \log a_{12} \log a_{21}}. \end{aligned}$$

Example 1.21. *Let*

$$A = \begin{pmatrix} 2 & e \\ 1 & 3 \end{pmatrix}.$$

We find $\det_\star A$. *We have*

$$\begin{aligned} \det_\star A &= e^{\log 2 \log 3 - \log 1 \log 2} \\ &= e^{\log 2 \log 3}. \end{aligned}$$

Exercise 1.13. *Find* $\det_\star A$, *where*

$$A = \begin{pmatrix} \frac{1}{2} & 3 \\ 4 & \frac{1}{5} \end{pmatrix}.$$

Answer 1.13. $e^{\log 2 \log \frac{5}{9}}$.

Definition 1.27. *Let*

$$A = \begin{pmatrix} a_{11} & a_{12} \\ a_{21} & a_{22} \end{pmatrix} \in \mathscr{M}_{\star 2\times 2}$$

and

$$\begin{aligned} \det_\star A &= e^{\log a_{11}\log a_{22}-\log a_{12}\log a_{21}} \\ &\neq 0_\star. \end{aligned}$$

Then, we define the multiplicative inverse matrix $A^{-1_\star}$ of the matrix A as follows:

$$A^{-1_\star} = (1_\star /_\star \det_\star A)\cdot_\star \begin{pmatrix} a_{22} & \frac{1}{a_{12}} \\ \frac{1}{a_{21}} & a_{11} \end{pmatrix}.$$

We have

$$\begin{aligned} A^{-1_\star} &= (e/_\star \det_\star A)\cdot_\star \begin{pmatrix} a_{22} & \frac{1}{a_{12}} \\ \frac{1}{a_{21}} & a_{11} \end{pmatrix} \\ &= e^{\frac{1}{\log a_{11}\log a_{22}-\log a_{12}\log a_{21}}}\cdot_\star \begin{pmatrix} a_{22} & \frac{1}{a_{12}} \\ \frac{1}{a_{21}} & a_{11} \end{pmatrix} \\ &= \begin{pmatrix} e^{\frac{\log a_{22}}{\log a_{11}\log a_{22}-\log a_{12}\log a_{21}}} & e^{-\frac{\log a_{12}}{\log a_{11}\log a_{22}-\log a_{12}\log a_{21}}} \\ e^{-\frac{\log a_{21}}{\log a_{11}\log a_{22}-\log a_{12}\log a_{21}}} & e^{\frac{\log a_{11}}{\log a_{11}\log a_{22}-\log a_{12}\log a_{21}}} \end{pmatrix}. \end{aligned}$$

Hence,

$$\begin{aligned} A^{-1_\star}\cdot_\star A &= A\cdot_\star A^{-1_\star} \\ &= \begin{pmatrix} e^{\frac{\log a_{22}}{\log a_{11}\log a_{22}-\log a_{12}\log a_{21}}} & e^{-\frac{\log a_{12}}{\log a_{11}\log a_{22}-\log a_{12}\log a_{21}}} \\ e^{-\frac{\log a_{21}}{\log a_{11}\log a_{22}-\log a_{12}\log a_{21}}} & e^{\frac{\log a_{11}}{\log a_{11}\log a_{22}-\log a_{12}\log a_{21}}} \end{pmatrix} \\ &\quad \cdot_\star \begin{pmatrix} a_{11} & a_{12} \\ a_{21} & a_{22} \end{pmatrix} \\ &= \begin{pmatrix} e^{\frac{\log a_{11}\log a_{22}-\log a_{12}\log a_{21}}{\log a_{11}\log a_{22}-\log a_{12}\log a_{21}}} & e^{-\frac{\log a_{21}\log a_{12}-\log a_{21}\log a_{12}}{\log a_{11}\log a_{22}-\log a_{12}\log a_{21}}} \\ e^{-\frac{\log a_{12}\log a_{21}-\log a_{12}\log a_{21}}{\log a_{11}\log a_{22}-\log a_{12}\log a_{21}}} & e^{\frac{\log a_{11}\log a_{22}-\log a_{12}\log a_{21}}{\log a_{11}\log a_{22}-\log a_{12}\log a_{21}}} \end{pmatrix} \\ &= \begin{pmatrix} e & 1 \\ 1 & e \end{pmatrix} \\ &= \begin{pmatrix} 1_\star & 0_\star \\ 0_\star & 1_\star \end{pmatrix} \\ &= I_\star. \end{aligned}$$

Example 1.22. *Let*

$$A = \begin{pmatrix} 2 & 3 \\ 3 & 4 \end{pmatrix}.$$

We find $A^{-1_\star}$. We have

$$\begin{aligned} A^{-1_\star} &= \begin{pmatrix} e^{\frac{\log 4}{\log 2\log 4-(\log 3)^2}} & e^{-\frac{\log 3}{\log 2\log 4-(\log 3)^2}} \\ e^{-\frac{\log 3}{\log 2\log 4-(\log 3)^2}} & e^{\frac{\log 2}{\log 2\log 4-(\log 3)^2}} \end{pmatrix} \\ &= \begin{pmatrix} e^{\frac{2\log 2}{2(\log 2)^2-(\log 3)^2}} & e^{-\frac{\log 3}{2(\log 2)^2-(\log 3)^2}} \\ e^{-\frac{\log 3}{2(\log 2)^2-(\log 3)^2}} & e^{\frac{\log 2}{2(\log 2)^2-(\log 3)^2}} \end{pmatrix}. \end{aligned}$$

Exercise 1.14. *Find* $A^{-1_\star}$, *where*

$$A = \begin{pmatrix} \frac{1}{2} & 4 \\ \frac{1}{3} & 9 \end{pmatrix}.$$

Answer 1.14. $A^{-1_\star}$ *does not exist because* $\det_\star A = 0_\star$.

Definition 1.28. *Let*

$$A = \begin{pmatrix} a_{11} & a_{12} & a_{13} \\ a_{21} & a_{22} & a_{23} \\ a_{31} & a_{32} & a_{33} \end{pmatrix} \in \mathscr{M}_{\star 3\times 3}.$$

Define the multiplicative determinant of A as follows:

$$\begin{aligned} \det_\star A &= a_{11} \cdot_\star a_{22} \cdot_\star a_{33} +_\star a_{31} \cdot_\star a_{12} \cdot_\star a_{23} \\ &\quad +_\star a_{21} \cdot_\star a_{32} \cdot_\star a_{13} -_\star a_{31} \cdot_\star a_{22} \cdot_\star a_{13} \\ &\quad -_\star a_{11} \cdot_\star a_{32} \cdot_\star a_{23} -_\star a_{21} \cdot_\star a_{12} \cdot_\star a_{33}. \end{aligned}$$

We have

$$\begin{aligned} \det_\star A &= e^{\log a_{11} \log a_{22} \log a_{33}} +_\star e^{\log a_{31} \log a_{12} \log a_{23}} \\ &\quad +_\star e^{\log a_{21} \log a_{32} \log a_{13}} -_\star e^{\log a_{31} \log a_{22} \log a_{13}} \\ &\quad -_\star e^{\log a_{11} \log a_{32} \log a_{23}} -_\star e^{\log a_{21} \log a_{12} \log a_{33}} \\ &= \exp\Big(\log a_{11} \log a_{22} \log a_{33} + \log a_{31} \log a_{12} \log a_{23} \\ &\quad + \log a_{21} \log a_{32} \log a_{13} - \log a_{31} \log a_{22} \log a_{13} \\ &\quad - \log a_{11} \log a_{32} \log a_{23} - \log a_{21} \log a_{12} \log a_{33} \Big). \end{aligned}$$

Example 1.23. *Let*

$$A = \begin{pmatrix} e & e^2 & e^4 \\ e^3 & e^7 & e^{11} \\ e^{15} & e^8 & e^9 \end{pmatrix}.$$

Then

$$\begin{aligned} \det_\star A &= e^{1\cdot 7\cdot 9 + 2\cdot 11\cdot 15 + 3\cdot 8\cdot 4 - 4\cdot 7\cdot 15 - 1\cdot 11\cdot 8 - 2\cdot 3\cdot 9} \\ &= e^{63+330+96-420-88-54} \\ &= e^{159-90-142} \\ &= e^{69-142} \\ &= e^{-73}. \end{aligned}$$

Exercise 1.15. *Let*

$$A = \begin{pmatrix} 1 & e^3 & e^7 \\ 4 & 5 & e^8 \\ 6 & 7 & e^{-1} \end{pmatrix}.$$

Find $\det_\star A$.

Answer 1.15. $\det_\star A = e^{24\log 6 + 14\log 2\log 7 - 7\log 5\log 6 + 6\log 2}$.

Definition 1.29. *Let*

$$A = \begin{pmatrix} a_{11} & a_{12} & a_{13} \\ a_{21} & a_{22} & a_{23} \\ a_{31} & a_{32} & a_{33} \end{pmatrix} \in \mathscr{M}_{\star 3\times 3}.$$

Suppose that $\det_\star A \neq 0_\star$. *Set*

$$\begin{aligned} b_{11} &= e^{\frac{\log a_{22}\log a_{33}}{\log a_{32}\log a_{23}}}, \\ b_{12} &= e^{\frac{\log a_{32}\log a_{13}}{\log a_{12}\log a_{33}}}, \\ b_{13} &= e^{\frac{\log a_{12}\log a_{23}}{\log a_{13}\log a_{22}}}, \\ b_{21} &= e^{\frac{\log a_{31}\log a_{23}}{\log a_{21}\log a_{33}}}, \\ b_{22} &= e^{\frac{\log a_{11}\log a_{33}}{\log a_{13}\log a_{31}}}, \\ b_{23} &= e^{\frac{\log a_{21}\log a_{13}}{\log a_{11}\log a_{23}}}, \\ b_{31} &= e^{\frac{\log a_{21}\log a_{32}}{\log a_{31}\log a_{22}}}, \\ b_{32} &= e^{\frac{\log a_{31}\log a_{12}}{\log a_{11}\log a_{32}}}, \\ b_{33} &= e^{\frac{\log a_{11}\log a_{22}}{\log a_{12}\log a_{21}}}. \end{aligned}$$

The matrix

$$A^{-1_\star} = (1_\star /_\star \det_\star A) \cdot_\star \begin{pmatrix} b_{11} & b_{12} & b_{13} \\ b_{21} & b_{22} & b_{23} \\ b_{31} & b_{32} & b_{33} \end{pmatrix}$$

is said to be the multiplicative inverse matrix of the matrix A.

Exercise 1.16. *Let* $A \in \mathscr{M}_{\star 3\times 3}$ *and* $\det_\star A \neq 0_\star$. *Prove that*

$$A \cdot_\star A^{-1_\star} = A^{-1_\star} \cdot_\star A = I_\star.$$

Exercise 1.17. *Let*

$$A = \begin{pmatrix} 2 & e^3 & e^7 \\ 3 & 11 & 5 \\ e^2 & e & e^4 \end{pmatrix}.$$

Find $A^{-1_\star}$.

Answer 1.16.

$$\begin{aligned} A^{-1_\star} &= e^{\frac{4\log 2\log 11 + 6\log 5 + 7\log 3}{14\log 11 + \log 2\log 5 + 12\log 3}} \\ &\quad \begin{pmatrix} e^{\frac{4\log 11}{\log 5}} & e^{\frac{7}{12}} & e^{\frac{3\log 5}{7\log 11}} \\ e^{\frac{2\log 5}{4\log 3}} & e^{\frac{2\log 2}{7}} & e^{\frac{7\log 3}{\log 2\log 5}} \\ e^{\frac{\log 3}{2\log 11}} & e^{\frac{6}{\log 2}} & e^{\frac{\log 2\log 11}{3\log 3}} \end{pmatrix}. \end{aligned}$$

Definition 1.30. *A matrix* $A \in \mathscr{M}_{\star n\times n}$ *is said to be multiplicative orthogonal matrix if*

$$A \cdot_\star A^T = I_\star.$$

1.10 Advanced Practical Problems

Problem 1.1. *Find*

1. $A = (3 \cdot_\star 4) -_\star (1 +_\star 5) /_\star 2$.
2. $A = (2 -_\star 4) \cdot_\star 7 +_\star (3 +_\star 1) /_\star 4$.
3. $A = (1 +_\star 2) /_\star 3 +_\star (4 +_\star 2) /_\star 7$.

Answer 1.17.

1. $e^{2\log 2 \log 3 - \frac{\log 5}{\log 2}}$.
2. $e^{-\log 2 \log 7 + \frac{\log 3}{2\log 2}}$.
3. $e^{\frac{\log 2}{\log 3} + \frac{3\log 2}{\log 7}}$.

Problem 1.2. *Solve the equations*

1. $3 -_\star x = 1$.
2. $4 +_\star x = 7$.
3. $7 +_\star x = 18$.

Answer 1.18.

1. $x = 3$.
2. $x = \frac{7}{4}$.
3. $x = \frac{18}{7}$.

Problem 1.3. *Solve the equations*

1. $2 \cdot_\star x = 7$.
2. $4 \cdot_\star x = 8$.
3. $3 /_\star x = 5$.

Answer 1.19.

1. $x = e^{\frac{\log 7}{\log 2}}$.
2. $x = e^{\frac{3}{2}}$.
3. $x = e^{\frac{\log 3}{\log 5}}$.

Problem 1.4. *Solve the equations*

1. $3 +_\star (4 \cdot_\star x) = 5$.
2. $(2 -_\star x) \cdot_\star 7 = 1$.
3. $(4 -_\star x) \cdot_\star 2 = 1$.

Answer 1.20.

1. $x = e^{\frac{\log \frac{5}{3}}{2\log 2}}$.
2. $x = 2$.
3. $x = 4$.

Problem 1.5. *Compute*

1. $A = |3 -_\star 4|_\star \cdot_\star 4$.
2. $A = |2 +_\star 1|_\star \cdot_\star |2/_\star 3|_\star$.
3. $A = |2 \cdot_\star 4|_\star \cdot_\star |3 -_\star 1|_\star$.

Answer 1.21.

1. $e^{4\log 2 - \log 3}$.
2. 3.
3. $e^{3\log 2 + \log 3}$.

Problem 1.6. *Compute*

1. $A = ((7 +_\star 2) \cdot_\star 3)^{2_\star}$.
2. $A = ((6 -_\star 3) /_\star 4)^{2_\star} +_\star 2$.
3. $A = ((5 +_\star 1) \cdot_\star 7)^{3_\star} -_\star (2 \cdot_\star 4)^{4_\star}$.

Answer 1.22.

1. $e^{(\log 14 \log 3)^2}$.
2. $2e^{\frac{1}{4}}$.
3. $e^{(\log 5 \log 7)^3 - 16(\log 2)^8}$.

Problem 1.7. *Find*

$$A = e^2 \cdot_\star \tan e^{\frac{\pi}{4}} -_\star \sin_\star e^{\frac{\pi}{2}} +_\star 2 \cdot_\star \cos e^{\frac{3\pi}{4}}.$$

Answer 1.23. $e^{1 - \frac{\sqrt{2}}{2}\log 2}$.

Problem 1.8. *Find*

1. $\tan_\star(\arccos_\star x)$.
2. $\tan_\star(3 \arccos_\star x)$, $x \in \mathbb{R}_\star$.

Answer 1.24.

1. $e^{\frac{\sqrt{1-(\log x)^2}}{\log x}}$, $x \in [e^{-1}, e]$.
2. $e^{\frac{\log x}{\sqrt{1-\log x}}}$, $x \in (e^{-1}, e)$.

Problem 1.9. *Let*

$$A = \begin{pmatrix} 4 & 2 & 3 \\ 3 & 1 & e \\ 2 & 5 & \frac{1}{2} \end{pmatrix} \quad \text{and} \quad \begin{pmatrix} 2 & 3 & 4 \\ 2 & e & e \\ 3 & 1 & \frac{1}{4} \end{pmatrix}.$$

Find

1. $A +_\star B$.
2. $A -_\star B$.
3. $A \cdot_\star B$.

Answer 1.25.

1. $$\begin{pmatrix} 8 & 6 & 12 \\ 6 & e & e^2 \\ 6 & 5 & \frac{1}{8} \end{pmatrix}.$$

2. $$\begin{pmatrix} 2 & \frac{2}{3} & \frac{3}{4} \\ \frac{3}{2} & \frac{1}{e} & 1 \\ \frac{2}{3} & 5 & 2 \end{pmatrix}.$$

3. $$\begin{pmatrix} e^{3(\log 2)^2+(\log 3)^2} & e^{\log 2(1+2\log 3)} & e^{4(\log 2)^2+\log 2-2\log 2\log 3} \\ e^{\log 3(1+\log 2)} & e^{(\log 3)^2} & e^{2\log 2(\log 3-1)} \\ e^{(\log 2)^2+\log 2\log\frac{5}{3}} & e^{\log 2\log 3+\log 5} & e^{4(\log 2)^2+\log 5} \end{pmatrix}.$$

Problem 1.10. *Find* $\det_\star A$, *where*

$$A = \begin{pmatrix} 4 & \frac{1}{8} \\ \frac{1}{9} & \frac{1}{11} \end{pmatrix}.$$

Answer 1.26. $e^{-2\log 2\log 297}$.

Problem 1.11. *Let*

$$A = \begin{pmatrix} 4 & e^{-2} & 7 \\ 1 & 3 & 8 \\ e^7 & e^3 & e \end{pmatrix}.$$

Find $\det_\star A$.

Answer 1.27. $e^{2\log 2\log 3-42\log 2-7\log 3\log 7-18(\log 2)^2}$.

Problem 1.12. *Find* $A^{-1_\star}$, *where*

$$A = \begin{pmatrix} 3 & \frac{1}{5} \\ 4 & e \end{pmatrix}.$$

Answer 1.28. $$A^{-1_\star} = \begin{pmatrix} e^{\frac{1}{\log 3+2\log 2\log 5}} & e^{\frac{\log 5}{\log 3+2\log 2\log 5}} \\ e^{-\frac{2\log 2}{\log 3+2\log 2\log 5}} & e^{\frac{\log 3}{\log 3+2\log 2\log 5}} \end{pmatrix}$$

2

Multiplicative Plane Euclidean Geometry

In this chapter, the multiplicative vector space $\mathbb{R}^2_\star$, the multiplicative inner product space $\mathbb{R}^2_\star$ and the multiplicative Euclidean space $E^2_\star$ are introduced and investigated. Multiplicative lines are defined and their equations are deduced. Perpendicular, parallel and intersecting multiplicative lines are defined and studied. In the chapter, multiplicative isometries, multiplicative translations, multiplicative rotations and multiplicative glide reflections are developed.

2.1 The Multiplicative Vector Space $\mathbb{R}^2_\star$

Definition 2.1. *Each ordered pair (p_1, p_2) of elements of $\mathbb{R}_\star$ determines exactly one point of the multiplicative plane. The point $(0_\star, 0_\star) = (1,1)$ is called the multiplicative origin. The ordered pair (p_1, p_2) is also referred to as the multiplicative coordinate vector (shortly vector) of P.*

Definition 2.2. *Let $c \in \mathbb{R}_\star$, $x = (x_1, x_2) \in \mathbb{R}^2_\star$, $y = (y_1, y_2) \in \mathbb{R}^2_\star$. Define*

$$\begin{aligned} x +_\star y &= (x_1 +_\star y_1, x_2 +_\star y_2) \\ &= (x_1 y_1, x_2 y_2), \\ c \cdot_\star x &= (c \cdot_\star x_1, c \cdot_\star x_2) \\ &= \left(e^{\log c \log x_1}, e^{\log c \log x_1}\right), \\ x -_\star y &= (x_1 -_\star y_1, x_2 -_\star y_2) \\ &= \left(\frac{x_1}{y_1}, \frac{x_2}{y_2}\right), \\ -_\star x &= (-_\star x_1, -_\star x_2) \\ &= \left(\frac{1}{x_1}, \frac{1}{x_2}\right). \end{aligned}$$

Example 2.1. *Let $x = (1,2)$, $y = (1,4) \in \mathbb{R}^2_\star$, $c = 2$. Then*

$$\begin{aligned} x +_\star y &= (1 +_\star 1, 2 +_\star 4) \\ &= (1 \cdot 1, 2 \cdot 4) \\ &= (1, 8), \\ x -_\star y &= (1 -_\star 1, 2 -_\star 4) \\ &= \left(\frac{1}{1}, \frac{2}{4}\right) \\ &= \left(1, \frac{1}{2}\right), \end{aligned}$$

DOI: 10.1201/9781003325284-2

and

$$\begin{aligned} 2\cdot_\star x &= (2\cdot_\star 1, 2\cdot_\star 2) \\ &= \left(e^{\log 2\log 1}, e^{(\log 2)^2}\right) \\ &= \left(1, e^{(\log 2)^2}\right). \end{aligned}$$

Example 2.2. *Let* $x = (2,4) \in \mathbb{R}^2_\star$. *Then* $-_\star x = \left(\frac{1}{2}, \frac{1}{4}\right)$.

Exercise 2.1. *Let* $x = (3,5)$, $y = (1,4) \in \mathbb{R}^2_\star$. *Find*

1. $x +_\star y$.
2. $3\cdot_\star (x -_\star y)$.

Answer 2.1.

1. $(3,20)$.
2. $\left(e^{(\log 3)^2}, e^{\log 3\log \frac{5}{4}}\right)$.

Below, we deduce some of the properties of the defined operations. Let

$$x = (x_1, x_2), \quad y = (y_1, y_2), \quad z = (z_1, z_2) \in \mathbb{R}^2_\star, \quad c, d \in \mathbb{R}_\star.$$

1. $(x +_\star y) +_\star z = x +_\star (y +_\star z)$.

   Proof. We have

$$\begin{aligned} (x +_\star y) +_\star z &= (x_1 +_\star y_1, x_2 +_\star y_2) +_\star (z_1, z_2) \\ &= (x_1 y_1, x_2 y_2) +_\star (z_1, z_2) \\ &= ((x_1 y_1) +_\star z_1, (x_2 y_2) +_\star z_2) \\ &= ((x_1 y_1) z_1, (x_2 y_2) z_2) \\ &= (x_1 (y_1 z_1), x_2 (y_2 z_2)) \\ &= (x_1 +_\star (y_1 z_1), x_2 +_\star (y_2 z_2)) \\ &= (x_1, x_2) +_\star (y_1 z_1, y_2 z_2) \\ &= x +_\star (y_1 +_\star z_1, y_2 +_\star z_2) \\ &= x +_\star ((y_1, y_2) +_\star (z_1, z_2)) \\ &= x +_\star (y +_\star z). \end{aligned}$$

   This completes the proof. □

2. $x +_\star y = y +_\star x$.

   Proof. We have

$$\begin{aligned} x +_\star y &= (x_1, x_2) +_\star (y_1, y_2) \\ &= (x_1 +_\star y_1, x_2 +_\star y_2) \\ &= (x_1 y_1, x_2 y_2) \\ &= (y_1 x_1, y_2 x_2) \\ &= (y_1 +_\star x_1, y_2 +_\star x_2) \\ &= (y_1, y_2) +_\star (x_1, x_2) \\ &= y +_\star x. \end{aligned}$$

   This completes the proof. □

3. $x +_\star 0_\star = x$.

Proof. We have

$$\begin{aligned} x +_\star 0_\star &= (x_1, x_2) +_\star (1, 1) \\ &= (x_1 +_\star 1, x_2 +_\star 1) \\ &= (x_1, x_2) \\ &= x. \end{aligned}$$

This completes the proof. □

4. $x +_\star (-_\star x) = 0_\star$.

Proof. We have

$$\begin{aligned} x +_\star (-_\star x) &= (x_1, x_2) +_\star \left(\frac{1}{x_1}, \frac{1}{x_2}\right) \\ &= \left(x_1 +_\star \frac{1}{x_1}, x_2 +_\star \frac{1}{x_2}\right) \\ &= (1, 1) \\ &= (0_\star, 0_\star) \\ &= 0_\star. \end{aligned}$$

This completes the proof. □

5. $e \cdot_\star x = x$.

Proof. We have

$$\begin{aligned} e \cdot_\star x &= e \cdot_\star (x_1, x_2) \\ &= (e \cdot_\star x_1, e \cdot_\star x_2) \\ &= \left(e^{\log e \log x_1}, e^{\log e \log x_2}\right) \\ &= (x_1, x_2) \\ &= x. \end{aligned}$$

This completes the proof. □

6. $c \cdot_\star (x +_\star y) = c \cdot_\star x +_\star c \cdot_\star y$.

Proof. We have

$$\begin{aligned} c \cdot_\star (x +_\star y) &= c \cdot_\star ((x_1, x_2) +_\star (y_1, y_2)) \\ &= c \cdot_\star (x_1 +_\star y_1, x_2 +_\star y_2) \\ &= c \cdot_\star (x_1 y_1, x_2 y_2) \\ &= (c \cdot_\star (x_1 y_1), c \cdot_\star (x_2 y_2)) \\ &= \left(e^{\log c \log(x_1 y_1)}, e^{\log c \log(x_2 y_2)}\right) \\ &= \left(e^{\log c (\log x_1 + \log y_1)}, e^{\log c (\log x_2 + \log y_2)}\right) \end{aligned}$$

$$
\begin{aligned}
&= \left(e^{\log c \log x_1} e^{\log c \log y_1}, e^{\log c \log x_2 + \log c \log y_2}\right) \\
&= ((c \cdot_\star x_1) +_\star (c \cdot_\star y_1), (c \cdot_\star x_2) +_\star (c \cdot_\star y_2)) \\
&= (c \cdot_\star x_1, c \cdot_\star x_2) +_\star (c \cdot_\star y_1, c \cdot_\star y_2) \\
&= c \cdot_\star (x_1, x_2) +_\star c \cdot_\star (y_1, y_2) \\
&= c \cdot_\star x + c \cdot_\star y.
\end{aligned}
$$

This completes the proof. □

7. $(c +_\star d) \cdot_\star x = c \cdot_\star x +_\star d \cdot_\star x.$

Proof. We have

$$
\begin{aligned}
(c +_\star d) \cdot_\star x &= (cd) \cdot_\star x \\
&= (cd) \cdot_\star (x_1, x_2) \\
&= ((cd) \cdot_\star x_1, (cd) \cdot_\star x_2) \\
&= \left(e^{\log(cd) \log x_1}, e^{\log(cd) \log x_2}\right) \\
&= \left(e^{(\log c + \log d) \log x_1}, e^{(\log c + \log d) \log x_2}\right) \\
&= \left(e^{\log c \log x_1} e^{\log d \log x_1}, e^{\log c \log x_2} e^{\log d \log x_2}\right) \\
&= (c \cdot_\star x_1 +_\star d \cdot_\star x_1, c \cdot_\star x_2 +_\star d \cdot_\star x_2) \\
&= (c \cdot_\star x_1, c \cdot_\star x_2) +_\star (d \cdot_\star x_1, d \cdot_\star x_2) \\
&= c \cdot_\star (x_1, x_2) +_\star d \cdot_\star (x_1, x_2) \\
&= c \cdot_\star x +_\star d \cdot_\star x.
\end{aligned}
$$

This completes the proof. □

8. $c \cdot_\star (d \cdot_\star x) = (c \cdot_\star d) \cdot_\star x.$

Proof. We have

$$
\begin{aligned}
c \cdot_\star (d \cdot_\star x) &= c \cdot_\star (d \cdot_\star x_1, d \cdot_\star x_2) \\
&= (c \cdot_\star (d \cdot_\star x_1), c \cdot_\star (d \cdot_\star x_2)) \\
&= ((c \cdot_\star d) \cdot_\star x_1, (c \cdot_\star d) \cdot_\star x_2) \\
&= (c \cdot_\star d) \cdot_\star x.
\end{aligned}
$$

This completes the proof. □

2.2 The Multiplicative Inner Product Space $\mathbb{R}^2_\star$

Definition 2.3. *Let $x = (x_1, x_2)$, $y = (y_1, y_2) \in \mathbb{R}^2_\star$. Define the multiplicative inner product of x and y as follows:*

$$\langle x, y \rangle_\star = e^{\log x_1 \log y_1 + \log x_2 \log y_2}.$$

Example 2.3. *Let*

$$x = (2,3), \quad y = (3,4) \in \mathbb{R}^2_\star.$$

Then

$$\begin{aligned} \langle x,y\rangle_\star &= e^{\log 2\log 3+\log 3\log 4} \\ &= e^{\log 2\log 3+2\log 2\log 3} \\ &= e^{3\log 2\log 3}. \end{aligned}$$

Exercise 2.2. *Let*

$$x = (3,8), \quad y = (2,4).$$

Find

1. $x +_\star y$.
2. $x -_\star y$.
3. $\langle x,y\rangle_\star$.

Answer 2.2.

1. $(6,32)$.
2. $\left(\frac{3}{2},2\right)$.
3. $e^{\log 3\log 2+6(\log 2)^2}$.

Below, we deduce some of the properties of the multiplicative inner product.

Theorem 2.3. *Let*

$$x = (x_1,x_2), \quad y = (y_1,y_2), \quad z = (z_1,z_2) \in \mathbb{R}^2_\star.$$

Then

$$\langle x, y +_\star z\rangle_\star = \langle x,y\rangle_\star +_\star \langle x,z\rangle_\star.$$

Proof. We have

$$\begin{aligned} y +_\star z &= (y_1,y_2) +_\star (z_1,z_2) \\ &= (y_1 +_\star z_1, y_2 +_\star z_2) \\ &= (y_1z_1, y_2z_2), \\ \langle x,y\rangle_\star &= e^{\log x_1\log y_1+\log x_2\log y_2}, \\ \langle x,z\rangle_\star &= e^{\log x_1\log z_1+\log x_2\log z_2}. \end{aligned}$$

Hence,

$$\begin{aligned} \langle x,y\rangle_\star +_\star \langle x,z\rangle_\star &= e^{\log x_1\log y_1+\log x_2\log y_2}e^{\log x_1\log z_1+\log x_2\log z_2} \\ &= e^{\log x_1\log y_1+\log x_2\log y_2+\log x_1\log z_1+\log x_2\log z_2} \\ &= e^{\log x_1(\log y_1+\log z_1)+\log x_2(\log y_2+\log z_2)} \\ &= e^{\log x_1\log(y_1z_1)+\log x_2\log(y_2z_2)} \\ &= \langle x, y +_\star z\rangle_\star. \end{aligned}$$

This completes the proof. □

Theorem 2.4. *Let*

$$x = (x_1,x_2), \quad y = (y_1,y_2) \in \mathbb{R}^2_\star, \quad c \in \mathbb{R}_\star.$$

Then

$$c \cdot_\star \langle x,y\rangle_\star = \langle c \cdot_\star x, y\rangle_\star = \langle x, c \cdot_\star y\rangle_\star. \tag{2.1}$$

Proof. We have

$$\begin{aligned} c\cdot_\star x &= c\cdot_\star(x_1,x_2)\\ &= (c\cdot_\star x_1, c\cdot_\star x_2)\\ &= \left(e^{\log c\log x_1}, e^{\log c\log x_2}\right), \end{aligned}$$

and as above,

$$c\cdot_\star y = \left(e^{\log c\log y_1}, e^{\log c\log y_2}\right).$$

Next,

$$\langle x,y\rangle = e^{\log x_1\log y_1+\log x_2\log y_2},$$

and

$$\begin{aligned} c\cdot_\star\langle x,y\rangle_\star &= c\cdot_\star e^{\log x_1\log y_1+\log x_2\log y_2}\\ &= e^{\log c e^{\log x_1\log y_1+\log x_2\log y_2}}\\ &= e^{\log c(\log x_1\log y_1+\log x_2\log y_2)}, \end{aligned} \tag{2.2}$$

and

$$\begin{aligned} \langle c\cdot_\star x,y\rangle_\star &= e^{\left(\log e^{\log c\log x_1}\right)\log y_1+\left(\log e^{\log c\log x_2}\right)\log y_2}\\ &= e^{\log c\log x_1\log y_1+\log c\log x_2\log y_2}\\ &= e^{\log c(\log x_1\log y_1+\log x_2\log y_2)}, \end{aligned} \tag{2.3}$$

and

$$\begin{aligned} \langle x,c\cdot_\star y\rangle_\star &= e^{\log x_1\log e^{\log c\log y_1}+\log x_2\log e^{\log c\log y_2}}\\ &= e^{\log x_1\log c\log y_1+\log x_2\log c\log y_2}\\ &= e^{\log c(\log x_1\log y_1+\log x_2\log y_2)}. \end{aligned} \tag{2.4}$$

By (2.2), (2.3) and (2.4), we get (2.1). This completes the proof. □

Corollary 2.1. *Suppose that all conditions of Theorem 2.4 hold. Then*

$$\langle x,y\rangle_\star = \langle y,x\rangle_\star.$$

Proof. We apply the definition for multiplicative inner product and we get the desired result. This completes the proof. □

Theorem 2.5. *If* $\langle x,y\rangle_\star = 0_\star$ *for any* $x\in\mathbb{R}^2_\star$*, then* $y = 0_\star$.

Proof. We have

$$\begin{aligned} 0_\star &= 1\\ &= \langle x,y\rangle_\star\\ &= e^{\log x_1\log y_1+\log x_2\log y_2}\\ &= e^0, \end{aligned}$$

for any $x_1,x_2\in\mathbb{R}_\star$. For $x_1=x_2=e$, we get

$$\begin{aligned} e^0 &= e^{\log y_1+\log y_2}\\ &= e^{\log(y_1y_2)}, \end{aligned}$$

whereupon

$$y_1y_2 = 1. \tag{2.5}$$

For $x_1=e$, $x_2=e^{-1}$, we find

$$e^{\log y_1-\log y_2} = e^0,$$

or

$$e^{\log \frac{y_1}{y_2}} = e^0.$$

Therefore,

$$\frac{y_1}{y_2} = 1.$$

By the last equation and (2.5), we arrive at the system

$$\begin{aligned} \frac{y_1}{y_2} &= 1 \\ y_1 y_2 &= 1. \end{aligned}$$

Hence,

$$y_1 = y_2 = 1,$$

and

$$\begin{aligned} y &= (y_1, y_2) \\ &= (1,1) \\ &= 0_\star. \end{aligned}$$

This completes the proof. □

Definition 2.4. *For a vector $x \in \mathbb{R}^2_\star$, define the multiplicative length of x as follows:*

$$|x|_\star = e^{\left((\log x_1)^2 + (\log x_2)^2\right)^{\frac{1}{2}}}.$$

Note that

$$\langle x, x\rangle_\star = e^{(\log x_1)^2 + (\log x_2)^2},$$

and

$$\begin{aligned} \left(\langle x, x\rangle_\star\right)^{\frac{1}{2}\star} &= e^{\left(\log e^{(\log x_1)^2 + (\log x_2)^2}\right)^{\frac{1}{2}}} \\ &= e^{\left((\log x_1)^2 + (\log x_2)^2\right)^{\frac{1}{2}}} \\ &= |x|_\star. \end{aligned}$$

Example 2.4. *Let $x = (2,4)$. Then*

$$\begin{aligned} |x|_\star &= e^{\left((\log 2)^2 + (\log 4)^2\right)^{\frac{1}{2}}} \\ &= e^{\left((\log 2)^2 + 4(\log 2)^2\right)^{\frac{1}{2}}} \\ &= e^{5^{\frac{1}{2}} \log 2}. \end{aligned}$$

Theorem 2.6. *The multiplicative length has the following properties.*

1. $|x|_\star \geq_\star 0_\star$.
2. *If* $|x|_\star = 0_\star$, *then* $x = 0_\star$.
3. $|c \cdot_\star x|_\star = |c|_\star \cdot_\star |x|_\star$.

Proof.

1. We have

$$\begin{aligned} |x|_\star &= e^{\left((\log x_1)^2+(\log x_2)^2\right)^{\frac{1}{2}}} \\ &\geq e^0 \\ &= 1 \\ &= 0_\star. \end{aligned}$$

2. Let $|x|_\star = 0_\star$. Then

$$\begin{aligned} e^{\left((\log x_1)^2+(\log x_2)^2\right)^{\frac{1}{2}}} &= 0_\star \\ &= 1 \\ &= e^0. \end{aligned}$$

Hence,

$$(\log x_1)^2 + (\log x_2)^2 = 0$$

and

$$\begin{aligned} \log x_1 &= 0 \\ \log x_2 &= 0. \end{aligned}$$

Therefore,

$$x_1 = x_2 = 1,$$

and $x = 0_\star$.

3. Let $c \geq 1$. Then $|c|_\star - c$ and

$$c \cdot_\star x = \left(e^{\log c \log x_1}, e^{\log c \log x_2}\right),$$

and

$$\begin{aligned} |c \cdot_\star x|_\star &= e^{\left(\left(\log e^{\log c \log x_1}\right)^2+\left(\log e^{\log c_1 \log x_2}\right)^2\right)^{\frac{1}{2}}} \\ &= e^{\left((\log c \log x_1)^2+(\log c \log x_2)^2\right)^{\frac{1}{2}}} \\ &= e^{\log c\left((\log x_1)^2+(\log x_2)^2\right)^{\frac{1}{2}}}, \end{aligned}$$

and

$$\begin{aligned} c \cdot_\star |x|_\star &= c \cdot_\star e^{\left((\log x_1)^2+(\log x_2)^2\right)^{\frac{1}{2}}} \\ &= e^{\log c \log e^{\left((\log x_1)^2+(\log x_2)^2\right)^{\frac{1}{2}}}} \\ &= e^{\log c\left((\log x_1)^2+(\log x_2)^2\right)^{\frac{1}{2}}}. \end{aligned}$$

Therefore,

$$|c \cdot_\star x|_\star = c \cdot_\star |x|_\star.$$

The case $c \in (0, 1]$ we leave to the reader as an exercise. This completes the proof.

□

Theorem 2.7 (The Multiplicative Cauchy–Schwartz Inequality). *For any* $x, y \in \mathbb{R}^2_\star$, *we have*

$$|\langle x,y\rangle_\star|_\star \le |x|_\star \cdot_\star |y|_\star.$$

The equality holds if and only if x and y are multiplicative proportional.

Proof. Let $x = (x_1, x_2)$ and $y = (y_1, y_2)$. Then

$$\langle x,y\rangle_\star = e^{\log x_1 \log y_1 + \log x_2 \log y_2},$$

and

$$\begin{aligned} |\langle x,y\rangle_\star|_\star &= \begin{cases} e^{\log x_1 \log y_1 + \log x_2 \log y_2} \\ \text{if} \quad \log x_1 \log y_1 + \log x_2 \log y_2 \ge 0 \\ e^{-(\log x_1 \log y_1 + \log x_2 \log y_2)} \\ \text{if} \quad \log x_1 \log y_1 + \log x_2 \log y_2 < 0. \end{cases} \\ &\le e^{\left((\log x_1)^2 + (\log x_2)^2\right)^{\frac{1}{2}} \left((\log y_1)^2 + (\log y_2)^2\right)^{\frac{1}{2}}} \\ &= |x|_\star \cdot_\star |y|_\star. \end{aligned}$$

Next,

$$|\langle x,y\rangle|_\star = |x|_\star \cdot_\star |y|_\star,$$

if and only if

$$e^{|\log x_1 \log y_1 + \log x_2 \log y_2|} = e^{\left((\log x_1)^2 + (\log x_2)^2\right)^{\frac{1}{2}} \left((\log y_1)^2 + (\log y_2)^2\right)^{\frac{1}{2}}},$$

if and only if

$$(\log x_1 \log y_1 + \log x_2 \log y_2)^2 = \left((\log x_1)^2 + (\log x_2)^2\right)\left((\log y_1)^2 + (\log y_2)^2\right)$$

if and only if

$$\begin{aligned} &(\log x_1)^2(\log y_1)^2 + (\log x_2)^2(\log y_2)^2 + 2\log x_1 \log x_2 \log y_1 \log y_2 \\ &= (\log x_1)^2(\log y_1)^2 + (\log x_1)^2(\log y_2)^2 \\ &\quad + (\log x_2)^2(\log y_1)^2 + (\log x_2)^2(\log y_2)^2, \end{aligned}$$

if and only if

$$\begin{aligned} &\log x_1 \log x_2 \log y_1 \log y_2 - (\log x_2)^2(\log y_1)^2 \\ &= -\log x_1 \log x_2 \log y_1 \log y_2 + (\log x_1)^2(\log y_2)^2, \end{aligned}$$

if and only if

$$\begin{aligned} &\log x_2 \log y_1 (\log x_1 \log y_2 - \log x_2 \log y_1) \\ &= \log x_1 \log y_2 (\log x_1 \log y_2 - \log x_2 \log y_1), \end{aligned}$$

if and only if

$$\log x_2 \log y_1 = \log x_1 \log y_2,$$

if and only if

$$e^{\log x_2 \log y_1} = e^{\log x_1 \log y_2},$$

if and only if

$$x_2 \cdot_\star y_1 = x_1 \cdot_\star y_2,$$

if and only if

$$x_1 /_\star y_1 = x_2 /_\star y_2.$$

This completes the proof. □

Corollary 2.2. *For any $x, y \in E_\star^2$, we have*

$$|x +_\star y|_\star \leq |x|_\star \cdot_\star |y|_\star.$$

Proof. Using the multiplicative Cauchy–Schwartz inequality, we get

$$\begin{aligned}
|x +_\star y|_\star^{2_\star} &= \langle x +_\star y, x +_\star y \rangle_\star \\
&= \langle x, x \rangle_\star +_\star e^2 \cdot_\star \langle x, y \rangle_\star +_\star \langle y, y \rangle_\star \\
&= |x|_\star^{2_\star} +_\star e^2 \cdot_\star \langle x, y \rangle_\star +_\star |y|_\star^{2_\star} \\
&\leq |x|_\star^{2_\star} +_\star e^2 \cdot_\star |x|_\star \cdot_\star |y|_\star +_\star |y|_\star^{2_\star} \\
&= (|x|_\star +_\star |y|_\star)^{2_\star},
\end{aligned}$$

whereupon

$$|x +_\star y|_\star \leq |x|_\star +_\star |y|_\star.$$

This completes the proof. □

2.3 The Multiplicative Euclidean Plane $E_\star^2$

Suppose that $P = (x_1, x_2)$, $Q = (y_1, y_2) \in \mathbb{R}_\star^2$.

Definition 2.5. *Define the multiplicative distance between the points P and Q as follows:*

$$d_\star(P, Q) = |P -_\star Q|_\star.$$

The symbol $E_\star^2$ will be used to denote the set $\mathbb{R}_\star^2$ equipped with the multiplicative distance $d_\star$.

In fact, we have

$$\begin{aligned}
P -_\star Q &= (x_1, x_2) -_\star (y_1, y_2) \\
&= (x_1 -_\star y_1, x_2 -_\star y_2) \\
&= \left(\frac{x_1}{y_1}, \frac{x_2}{y_2} \right),
\end{aligned}$$

and

$$\begin{aligned}
d_\star(P, Q) &= |P -_\star Q|_\star \\
&= \left| \left(\frac{x_1}{y_1}, \frac{x_2}{y_2} \right) \right|_\star \\
&= e^{\left(\left(\log \frac{x_1}{y_1} \right)^2 + \left(\log \frac{x_2}{y_2} \right)^2 \right)^{\frac{1}{2}}}.
\end{aligned}$$

Example 2.5. *Let $P = (4, 16)$, $Q = (2, 4)$. Then*

$$\begin{aligned}
d_\star(P, Q) &= e^{\left(\left(\log \frac{4}{2} \right)^2 + \left(\log \frac{16}{4} \right)^2 \right)^{\frac{1}{2}}} \\
&= e^{\left((\log 2)^2 + 4 (\log 2)^2 \right)^{\frac{1}{2}}} \\
&= e^{5^{\frac{1}{2}} \log 2}.
\end{aligned}$$

Exercise 2.3. *Let* $P = (2,3)$, $Q = (4,2)$ *and* $R = (8,2)$. *Find*

1. $d_\star(P,Q)$.
2. $d_\star(P,R)$.
3. $d_\star(Q,R)$.

Answer 2.8.

1. $e^{\left((\log 2)^2+\left(\log \frac{2}{3}\right)^2\right)^{\frac{1}{2}}}$.
2. $e^{\left(4(\log 2)^2+\left(\log \frac{2}{3}\right)^2\right)^{\frac{1}{2}}}$.
3. 2.

By the properties of the multiplicative length of a vector, they follow the following properties of the multiplicative distance.

1. $d_\star(P,Q) \geq_\star 0_\star$ for any $P,Q \in E^2_\star$.
2. $d_\star(P,Q) = 0_\star$ if and only if $P = Q$.
3. $d_\star(P,Q) = d_\star(Q,P)$ for any $P,Q \in E^2_\star$.
4. $d_\star(P,Q) \leq_\star d_\star(P,R) +_\star d_\star(R,Q)$ for any $P,Q,R \in E^2_\star$.

2.4 Multiplicative Lines

For a given vector $v = (v_1, v_2)$, set

$$[v]_\star = \{t \cdot_\star v : t \in \mathbb{R}_\star\}.$$

Definition 2.6. *If P is any point and v is a vector, different than the vector* $0_\star$, *then*

$$l = \{X : X -_\star P \in [v]_\star\},$$

is called multiplicative line through P with multiplicative direction v.

In Fig. 2.1, it is shown a multiplicative line through the point $P = (4,9)$ and multiplicative direction $(2,3)$.

If l is a multiplicative line and X is a point, there are some phrases used to express the relationship $X \in l$.

1. $X \in l$.
2. l contains X.
3. X lies on l.
4. l passes through X.
5. X and l are incident.
6. X is incident with l.
7. l is incident with X.

A fundamental property of a multiplicative line is that it is uniquely determined by two points that lie on it.

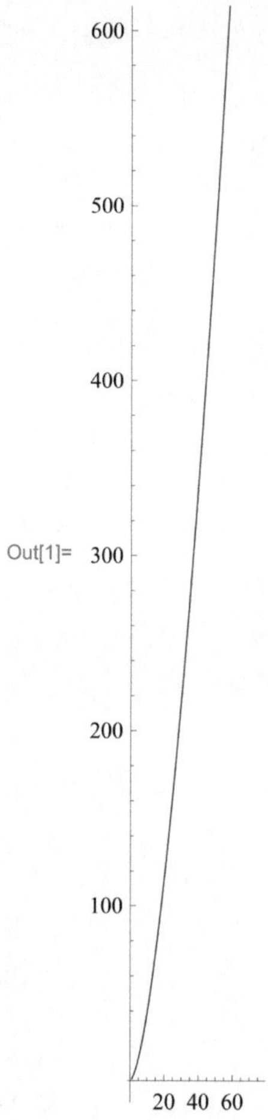

FIGURE 2.1
A multiplicative line.

Theorem 2.9. *Let P and Q be distinct points of $E_\star^2$. Then there is a unique multiplicative line containing P and Q,*

Proof. Let v be a nonzero vector in $E_\star^2$. The multiplicative line $P +_\star [v]_\star$ passes through Q if and only if $Q -_\star P \in [v]_\star$. Therefore,

$$[Q -_\star P]_\star = [v]_\star.$$

Hence, the multiplicative line

$$P +_\star [Q - P]_\star$$

is the unique multiplicative line. This completes the proof. □

Suppose that l is a multiplicative line through the points P and Q. By Theorem 2.9, it follows that a typical point X on the line l is written as

$$\begin{aligned} \alpha(t) &= P +_\star t \cdot_\star (Q -_\star P) \\ &= (1_\star -_\star t) \cdot_\star P +_\star t \cdot_\star Q. \end{aligned}$$

Note that

$$d_\star(\alpha(t_1), \alpha(t_2)) = |t_1 -_\star t_2|_\star \cdot_\star d_\star(P, Q).$$

Definition 2.7. *If*

$$X=(1_\star -_\star t)\cdot_\star P+_\star t\cdot_\star Q, \tag{2.6}$$

and $t\in(0_\star,1_\star)$ *(or* $t\in(1,e)$*), we say that X is between P and Q.*

Theorem 2.10. *If P, X and Q are distinct points of* $E_\star^2$*, then X is between P and Q if and only if*

$$d_\star(P,X)+d_\star(X,Q)=d_\star(P,Q). \tag{2.7}$$

Proof. Suppose that X is between P and Q. Then there is some $t\in(0_\star,1_\star)$ so that (2.6) holds. Then $t\in(1,e)$, $|t|_\star=t$ and

$$\begin{aligned} 1_\star -_\star t &= \frac{e}{t} \\ &> 1. \end{aligned}$$

Hence,

$$|1_\star -_\star t|_\star = 1_\star -_\star t.$$

Then

$$\begin{aligned} d_\star(P,X) &= |t|_\star\cdot_\star|P-_\star Q|_\star \\ &= t\cdot_\star|P-_\star Q|_\star. \end{aligned}$$

Next,

$$\begin{aligned} d_\star(X,Q) &= |Q-\star X|_\star \\ &= |1_\star -_\star t|_\star\cdot_\star|Q-_\star P|_\star \\ &= (1_\star -_\star t)\cdot_\star|Q-_\star P|_\star. \end{aligned}$$

Hence,

$$\begin{aligned} d_\star(P,X)+d_\star(X,Q) &= t\cdot_\star|P-_\star Q|_\star +_\star (1_\star -_\star t)\cdot_\star|P-_\star Q|_\star \\ &= 1_\star\cdot_\star|P-_\star Q|_\star \\ &= |P-_\star Q|_\star \\ &= d_\star(P,Q). \end{aligned}$$

Conversely, suppose that $X\subset E_\star^2$ is such that (2.7) holds. Then there is a $\lambda>0_\star$ so that

$$X-_\star P = \lambda\cdot_\star(Q-_\star X).$$

Hence,

$$(1_\star +_\star \lambda)\cdot_\star X = P+_\star\lambda\cdot_\star Q$$

and

$$X=(1_\star/_\star(1_\star+_\star\lambda))\cdot_\star P+_\star(\lambda/_\star(1_\star+_\star\lambda))\cdot_\star Q$$

Let

$$t=\lambda/_\star(1_\star+_\star\lambda).$$

We have

$$\begin{aligned} t &= \lambda/_\star(e+_\star\lambda) \\ &= \lambda/_\star(e\lambda) \\ &= e^{\frac{\log\lambda}{\log(e\lambda)}} \\ &= e^{\frac{\log\lambda}{1+\log\lambda}} \\ &\le e \\ &= 1_\star, \end{aligned}$$

and $t \geq 0_\star$. Moreover,

$$\begin{aligned} 1_\star -_\star t &= e -_\star e^{\frac{\log\lambda}{1+\log\lambda}} \\ &= \frac{e}{e^{\frac{\log\lambda}{1+\log\lambda}}} \\ &= e^{\frac{1}{1+\log\lambda}} \end{aligned}$$

and

$$\begin{aligned} 1_\star /_\star (1_\star +_\star \lambda) &= e /_\star (e +_\star \lambda) \\ &= e /_\star (e\lambda) \\ &= e^{\frac{\log e}{\log(e\lambda)}} \\ &= e^{\frac{1}{1+\log\lambda}} \\ &= 1_\star -_\star t. \end{aligned}$$

Thus,

$$X = (1_\star -_\star t) \cdot_\star P +_\star t \cdot_\star Q$$

and X is between P and Q. This completes the proof. □

Example 2.6. *Let $P(3,4)$, $Q(2,7)$ and*

$$R\left(e^{\log 2(1-\log 2+\log 3)}, e^{2(\log 2)^2+(1-\log 2)\log 7}\right).$$

Then

$$\begin{aligned} 2 \cdot_\star P +_\star (1_\star -_\star 2) \cdot_\star Q &= 2 \cdot_\star (3,4) +_\star \frac{e}{2} \cdot_\star (2,7) \\ &= \left(e^{\log 2 \log 3}, e^{2(\log 2)^2}\right) +_\star \left(e^{(1-\log 2)\log 2}, e^{(1-\log 2)\log 7}\right) \\ &= \left(e^{\log 2(1-\log 2+\log 3)}, e^{2(\log 2)^2+(1-\log 2)\log 7}\right) \\ &= R. \end{aligned}$$

Thus, R is between P and Q.

Exercise 2.4. *Let $P(7,9)$, $Q(3,11)$. Check if the point $R(2,3)$ is between the points P and Q.*

Answer 2.11. *No.*

Definition 2.8. *Let P and Q be two distinct points. The set consisting of P, Q and all points that are between P and Q is said to be multiplicative segment. It will be denoted by PQ.*

Definition 2.9. *Let P and Q be two distinct points. The point*

$$M = 1_\star /_\star e^2 \cdot_\star (P +_\star Q),$$

is said to be the multiplicative midpoint of the multiplicative segment PQ.

Let $P(x_1, x_2)$ and $Q(y_1, y_2)$. Then

$$\begin{aligned} P +_\star Q &= (x_1, x_2) +_\star (y_1, y_2) \\ &= (x_1 y_1, x_2 y_2), \end{aligned}$$

and

$$\begin{aligned} M &= 1_\star/_\star e^2 \cdot_\star (P +_\star Q) \\ &= e/_\star e^2 \cdot_\star (x_1y_1, x_2y_2) \\ &= e^{\frac{\log e}{\log e^2}} \cdot_\star (x_1y_1, x_2y_2) \\ &= e^{\frac{1}{2}} \cdot (x_1y_1, x_2y_2) \\ &= \left(e^{\log e^{\frac{1}{2}} \log(x_1y_1)}, e^{\log e^{\frac{1}{2}} \log(x_2y_2)}\right) \\ &= \left(e^{\frac{\log(x_1y_1)}{2}}, e^{\frac{\log(x_2y_2)}{2}}\right). \end{aligned}$$

We will show that M is between P and Q. We have

$$\begin{aligned} & 1_\star/_\star e^2 \cdot_\star P +_\star (1_\star -_\star 1_\star/_\star e^2) \cdot_\star Q \\ &= e^{\frac{1}{2}} \cdot_\star (x_1, x_2) +_\star \left(e -_\star e^{\frac{1}{2}}\right) \cdot_\star (y_1, y_2) \\ &= \left(e^{\log e^{\frac{1}{2}} \log x_1}, e^{\log e^{\frac{1}{2}} \log x_2}\right) +_\star e^{\frac{1}{2}} \cdot_\star (y_1, y_2) \\ &= \left(e^{\frac{\log x_1}{2}}, e^{\frac{\log x_2}{2}}\right) +_\star \left(e^{\log e^{\frac{1}{2}} \log y_1}, e^{\log e^{\frac{1}{2}} \log y_2}\right) \\ &= \left(e^{\frac{\log x_1}{2}}, e^{\frac{\log x_2}{2}}\right) +_\star \left(e^{\frac{\log y_1}{2}}, e^{\frac{\log y_2}{2}}\right) \\ &= \left(e^{\frac{\log x_1 + \log y_1}{2}}, e^{\frac{\log x_2 + \log y_2}{2}}\right) \\ &= \left(e^{\frac{\log(x_1y_1)}{2}}, e^{\frac{\log(x_2y_2)}{2}}\right) \\ &= M. \end{aligned}$$

Moreover, $1_\star/_\star e^2 \in [0_\star, 1_\star]$.

Definition 2.10. *If two multiplicative lines pass through a point P, we say that they intersect at P and P is their point of intersection.*

Definition 2.11. *If three or more multiplicative lines pass through a point P, we say that the multiplicative lines are multiplicative concurrent.*

Definition 2.12. *If three or more points lie on some multiplicative line, the points are said to be multiplicative collinear.*

2.5 Multiplicative Orthonormal Pairs

Definition 2.13. *Let $x, y \in E^2_\star$. We say that x and y are multiplicative orthogonal and we write $x \perp_\star y$ if*

$$\langle x, y \rangle_\star = 0_\star.$$

Let $x, y \in E^2_\star$, $x = (x_1, x_2)$, $y = (y_1, y_2)$ and $x \perp_\star y$. Then

$$\begin{aligned} 1 &= 0_\star \\ &= \langle x, y \rangle_\star \\ &= e^{\log x_1 \log y_1 + \log x_2 \log y_2}, \end{aligned}$$

whereupon

$$\log x_1 \log y_1 + \log x_2 \log y_2 = 0. \tag{2.8}$$

If (2.8) holds, then $x \perp_\star y$. For $x \in E^2_\star$, $x = (x_1, x_2)$, with $x^{\perp_\star}$ denote

$$x^{\perp_\star} = \left(\frac{1}{x_2}, x_1 \right).$$

Then

$$\begin{aligned} \langle x, x^{\perp_\star} \rangle_\star &= e^{\log x_1 \log \frac{1}{x_2} + \log x_2 \log x_1} \\ &= e^0 \\ &= 1 \\ &= 0_\star. \end{aligned}$$

Thus, $x \perp_\star x^{\perp_\star}$.

Definition 2.14. *A vector of multiplicative length $1_\star$ is said to be a multiplicative unit vector.*

Example 2.7. *Let $x = \left(e^{\frac{1}{\sqrt{2}}}, e^{\frac{1}{\sqrt{2}}} \right)$. Then*

$$\begin{aligned} |x|_\star &= e^{\left(\left(\log e^{\frac{1}{\sqrt{2}}} \right)^2 + \left(\log e^{\frac{1}{\sqrt{2}}} \right)^2 \right)^{\frac{1}{2}}} \\ &= e^{\left(\frac{1}{2} + \frac{1}{2} \right)^{\frac{1}{2}}} \\ &= e \\ &= 1_\star. \end{aligned}$$

Thus, x is a multiplicative unit vector.

Definition 2.15. *A pair $\{x, y\}$ of multiplicative unit vectors in $E^2_\star$ will be called a multiplicative orthonormal pair if $x \perp_\star y$.*

Definition 2.16. *A pair $\{x, y\}$ of vectors in $E^2_\star$ is said to be multiplicative linearly independent if*

$$\lambda_1 \cdot_\star x +_\star \lambda_2 \cdot_\star y = 0_\star \tag{2.9}$$

holds for $\lambda_1 = \lambda_2 = 0_\star$. Otherwise, if there are $\lambda_1, \lambda_2 \in \mathbb{R}_\star$, $(\lambda_1, \lambda_2) \neq (0_\star, 0_\star)$ so that (2.9) *holds, then $\{x, y\}$ are said to be multiplicative linearly dependent.*

Theorem 2.12. *Let $\{x, y\}$ be a multiplicative orthonormal pair. Then they are multiplicative linearly independent.*

Proof. Firstly, note that $x \neq 0_\star$, $y \neq 0_\star$. Let $\lambda_1, \lambda_2 \in \mathbb{R}_\star$ be such that

$$\lambda_1 \cdot_\star x +_\star \lambda_2 \cdot_\star y = 0_\star,$$

or

$$(\lambda_1 \cdot_\star x_1, \lambda_1 \cdot_\star x_2) +_\star (\lambda_2 \cdot_\star y_1, \lambda_2 \cdot_\star y_2) = (1,1),$$

or

$$\left(e^{\log\lambda_1 \log x_1}, e^{\log\lambda_1 \log x_2}\right) +_\star \left(e^{\log\lambda_2 \log y_1}, e^{\log\lambda_2 \log y_2}\right) = (1,1),$$

or

$$\left(e^{\log\lambda_1 \log x_1 + \log\lambda_2 \log y_1}, e^{\log\lambda_1 \log x_1 + \log\lambda_2 \log y_2}\right) = (1,1),$$

or

$$\begin{aligned} \log\lambda_1 \log x_1 + \log\lambda_2 \log y_1 &= 0 \\ \log\lambda_1 \log x_2 + \log\lambda_2 \log y_2 &= 0. \end{aligned} \tag{2.10}$$

Let

$$\begin{aligned} \Delta &= \begin{vmatrix} \log x_1 & \log y_1 \\ \log x_2 & \log y_1 \end{vmatrix} \\ &= \log x_1 \log y_2 - \log x_2 \log y_1. \end{aligned}$$

Since $x \perp_\star y$, we have that

$$\langle x, y\rangle_\star = 0_\star$$

or

$$e^{\log x_1 \log y_1 + \log x_2 \log y_2} = e^0,$$

or

$$\log x_1 \log y_1 + \log x_2 \log y_2 = 0.$$

Without loss of generality, suppose that $\log y_1 \neq 0$. Then

$$\log x_1 = -\frac{\log x_2 \log y_2}{\log y_1}.$$

Hence,

$$\begin{aligned} \Delta &= -\frac{\log x_2 (\log y_2)^2}{\log y_1} - \log x_2 \log y_1 \\ &= -\frac{\log x_2 \left((\log y_1)^2 + (\log y_2)^2\right)}{\log y_1}. \end{aligned}$$

If $\log x_2 = 0$, then $\log x_1 = 0$ and $x = 0_\star$. This is impossible. Therefore, $\log x_2 \neq 0$. Next, if

$$(\log y_1)^2 + (\log y_2)^2 = 0,$$

then

$$\begin{aligned} \log y_1 &= 0, \\ \log y_2 &= 0, \end{aligned}$$

and $y = 0_\star$, which is a contradiction. Consequently $\Delta \neq 0$ and the system (2.10) has unique solution

$$\log\lambda_1 = \log\lambda_2 = 0,$$

i.e.,

$$\lambda_1 = \lambda_2 = 0_\star.$$

Thus, x and y are multiplicative linearly independent. This completes the proof. □

Theorem 2.13. *Let $\{x,y\}$ be a multiplicative orthonormal pair of vectors in $E^2_\star$. Then, any $z \in E^2_\star$ can be represented in the form*

$$z = \langle z,x\rangle_\star \cdot_\star x +_\star \langle z,y\rangle_\star \cdot_\star y.$$

Proof. We will search the vector z in the form

$$z = a_1 \cdot_\star x +_\star a_2 \cdot_\star y.$$

Then

$$\begin{aligned} \langle z,x\rangle_\star &= a_1 \cdot_\star \langle x,x\rangle_\star +_\star a_2 \cdot_\star \langle y,x\rangle_\star \\ &= a_1, \end{aligned}$$

and

$$\begin{aligned} \langle z,y\rangle_\star &= a_1 \cdot_\star \langle x,y\rangle_\star +_\star a_2 \cdot_\star \langle y,y\rangle_\star \\ &= a_2. \end{aligned}$$

This completes the proof. □

2.6 Equations of a Multiplicative Line

Definition 2.17. *Let l be a multiplicative line with multiplicative direction v. The vector $v^{\perp_\star}$ will be called a multiplicative normal vector to l.*

Note that any two multiplicative normal vectors to the same multiplicative line are multiplicative proportional.

Theorem 2.14. *Let l be a multiplicative line and $P \in l$. Let also, $\{v,N\}$ be a multiplicative orthonormal pair. Then*

$$P +_\star [v]_\star = \{X : \langle X -_\star P, N\rangle_\star = 0_\star\}.$$

Proof. By Theorem 2.13, we have the identity

$$X -_\star P = \langle X -_\star P, v\rangle_\star \cdot_\star v +_\star \langle X -_\star P, N\rangle_\star \cdot_\star N, \tag{2.11}$$

for any point $X \in E^2_\star$. Suppose that

$$X = P +_\star t \cdot_\star v.$$

Then

$$\begin{aligned} \langle X -_\star P, N\rangle_\star &= \langle t \cdot_\star v, N\rangle_\star \\ &= t \cdot_\star \langle v, N\rangle_\star \\ &= 0_\star. \end{aligned}$$

Let now,

$$\langle X -_\star P, N\rangle_\star = 0_\star.$$

By (2.11), we find

$$X -_\star P = \langle X -_\star P, v\rangle_\star \cdot_\star v$$

or

$$X = P +_\star \langle X -_\star P, v\rangle_\star \cdot_\star v.$$

This completes the proof. □

Remark 2.1. *By Theorem 2.14, it follows that the set*

$$\{X : \langle X -_\star P, N\rangle_\star = 0_\star\}, \tag{2.12}$$

is the multiplicative line through P with multiplicative normal vector N and multiplicative direction $N^{\perp_\star}$.

Let l be a multiplicative line with multiplicative normal vector $N = (N_1, N_2)$. Let also, $P = (P_1, P_2) \in l$. If $X = (X_1, X_2)$ is an arbitrary point on l. Then

$$X -_\star P = \left(\frac{X_1}{P_1}, \frac{X_2}{P_2}\right),$$

and

$$\begin{aligned} 1 &= 0_\star \\ &= \langle X -_\star P, N\rangle_\star \\ &= e^{\log \frac{X_1}{P_1} \log N_1 + \log \frac{X_2}{P_2} \log N_2}, \end{aligned}$$

whereupon

$$\log \frac{X_1}{P_1} \log N_1 + \log \frac{X_2}{P_2} \log N_2 = 0,$$

or

$$\log \left(\frac{X_1}{P_1}\right)^{\log N_1} + \log \left(\frac{X_2}{P_2}\right)^{\log N_2} = 0,$$

or

$$\log \left(\left(\frac{X_1}{P_1}\right)^{\log N_1} \left(\frac{X_2}{P_2}\right)^{\log N_2} \right) = 0,$$

or

$$\left(\frac{X_1}{P_1}\right)^{\log N_1} \left(\frac{X_2}{P_2}\right)^{\log N_2} = 1. \tag{2.13}$$

If l is a multiplicative line with a multiplicative direction $v = (v_1, v_2)$, then $v^{\perp_\star} = \left(\frac{1}{v_2}, v_1\right)$ is a multiplicative normal vector to l and by (2.13), we get

$$\left(\frac{X_1}{P_1}\right)^{\log \frac{1}{v_2}} \left(\frac{X_2}{P_2}\right)^{\log v_1} = 1,$$

or

$$\left(\frac{X_2}{P_2}\right)^{\log v_1} = \left(\frac{X_1}{P_1}\right)^{\log v_2}. \tag{2.14}$$

Example 2.8. *Let l be a multiplicative line through the point* $P(2,3)$ *and multiplicative direction* $v = (4,8)$. *We find its equation. By* (2.14), *we get*

$$\left(\frac{X_2}{3}\right)^{\log 4} = \left(\frac{X_1}{2}\right)^{\log 8},$$

or

$$\left(\frac{X_2}{3}\right)^{2\log 2} = \left(\frac{X_1}{2}\right)^{3\log 2}.$$

In Fig. 2.2 it is shown the multiplicative line l.

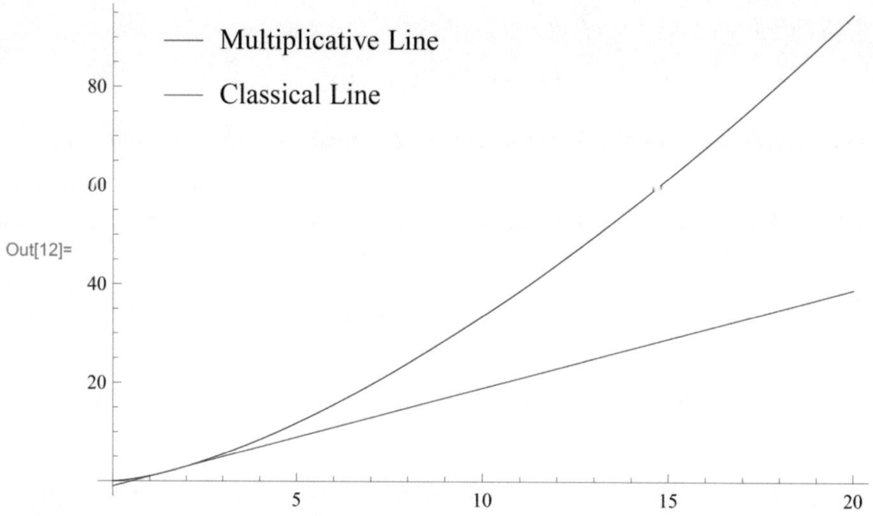

FIGURE 2.2
A multiplicative line and a classical line.

Exercise 2.5. *Find the equation of the multiplicative line through $P(4,5)$ and multiplicative direction $v = (10,15)$.*

Answer 2.15.

$$\left(\frac{X_2}{5}\right)^{\log 10} = \left(\frac{X_1}{4}\right)^{\log 15}.$$

Example 2.9. *Let l be a multiplicative line through $P(1,5)$ with multiplicative normal vector $N = (e,4)$. Its equation is*

$$X_1^{\log e}\left(\frac{X_2}{5}\right)^{\log 4} = 1,$$

or

$$X_1\left(\frac{X_2}{5}\right)^{2\log 2} = 1.$$

Exercise 2.6. *Find the equation of the multiplicative line l through $P(1,4)$ with multiplicative normal vector $N = (7,2)$.*

Answer 2.16.

$$X_1^{\log 7}\left(\frac{X_2}{4}\right)^{\log 2} = 1.$$

Definition 2.18. *For two points $P(p_1,p_2)$, $Q = (q_1,q_2) \in E_\star^2$, define a multiplicative vector $\vec{PQ}$ as follows:*

$$\begin{aligned} \vec{PQ} &= (q_1 -_\star p_1, q_2 -_\star p_2) \\ &= \left(\frac{q_1}{p_1}, \frac{q_2}{p_2}\right). \end{aligned}$$

Theorem 2.17. *Let $P(p_1,p_2)$, $Q(q_1,q_2)$ and $R(r_1,r_2) \in E_\star^2$ be given points. Then*

$$\begin{aligned} \vec{PQ} +_\star \vec{QR} &= \vec{PR}, \\ \vec{PQ} -_\star \vec{PR} &= \vec{RQ}. \end{aligned}$$

Proof. We have

$$\begin{aligned}\vec{PQ} &= \left(\frac{q_1}{p_1}, \frac{q_2}{p_2}\right),\\ \vec{QR} &= \left(\frac{r_1}{q_1}, \frac{r_2}{q_2}\right),\\ \vec{PR} &= \left(\frac{r_1}{p_1}, \frac{r_2}{p_2}\right),\\ \vec{RQ} &= \left(\frac{q_1}{r_1}, \frac{q_2}{r_2}\right).\end{aligned}$$

Then

$$\begin{aligned}\vec{PQ} +_\star \vec{QR} &= \left(\frac{q_1}{p_1}, \frac{q_2}{p_2}\right) +_\star \left(\frac{r_1}{q_1}, \frac{r_2}{q_2}\right)\\ &= \left(\frac{r_1}{p_1}, \frac{r_2}{p_2}\right)\\ &= \vec{PR},\end{aligned}$$

and

$$\begin{aligned}\vec{PQ} -_\star \vec{PR} &= \left(\frac{q_1}{p_1}, \frac{q_2}{p_2}\right) -_\star \left(\frac{r_1}{p_1}, \frac{r_2}{p_2}\right)\\ &= \left(\frac{q_1}{r_1}, \frac{q_2}{r_2}\right)\\ &= \vec{RQ}.\end{aligned}$$

This completes the proof. □

Example 2.10. *Let* $P(3,4)$ *and* $\vec{PQ} = (7,4)$. *We find the coordinates of the point* Q. *Let* $Q \in \mathbb{R}^2_\star$, $Q = Q(q_1, q_2)$. *Then*

$$\vec{PQ} = \left(\frac{q_1}{3}, \frac{q_2}{4}\right).$$

Hence,

$$\left(\frac{q_1}{3}, \frac{q_2}{4}\right) = (7,4),$$

and

$$q_1 = 21, \quad q_2 = 16.$$

Thus, $Q(21,16)$.

Exercise 2.7. *Let* $P(1,2)$ *and* $\vec{PQ} = (3,2)$. *Find the coordinates of the point* Q.

Answer 2.18. $Q(3,4)$.

Example 2.11. *Let* $P(e,4)$ *and* $v = (1,e)$. *We find the coordinates of the point* Q *so that* $|\vec{PQ}|_\star = e^2$ *and* $\vec{PQ} \perp_\star v$. *Let* $Q(q_1, q_2)$, *where* $q_1, q_2 \in \mathbb{R}_\star$. *Then*

$$\vec{PQ} = \left(\frac{q_1}{p_1}, \frac{q_2}{p_2}\right),$$

and

$$|\vec{PQ}|_\star = e^{\left(\left(\log\frac{q_1}{p_1}\right)^2+\left(\log\frac{q_2}{p_2}\right)^2\right)^{\frac{1}{2}}}$$
$$= e^2.$$

Next,

$$0_\star = e^0$$
$$= e^{\log\frac{q_1}{e}\log 1+\log\frac{q_2}{4}\log e}$$
$$= e^{\log\frac{q_2}{4}}.$$

Thus, we get the system

$$\left(\left(\log\frac{q_1}{p_1}\right)^2+\left(\log\frac{q_2}{p_2}\right)^2\right)^{\frac{1}{2}} = 2,$$
$$\log\frac{q_2}{4} = 0,$$

whereupon

$$\left(\log\frac{q_1}{p_1}\right)^2+\left(\log\frac{q_2}{p_2}\right)^2 = 4,$$
$$q_2 = 4,$$

and

$$\log\frac{q_1}{e} = 2,$$

and

$$\frac{q_1}{e} = e^2,$$

and

$$q_1 = e^3.$$

Thus, $Q\left(e^3,4\right)$.

Exercise 2.8. *Let* $P\left(e^2,e^3\right)$, $v=\left(e,e^2\right)$. *Find the coordinates of the point* Q *so that* $\vec{PQ}\perp_\star v$ *and* $|\vec{PQ}|_\star = e^9$.

Solution 2.1. *Let* $Q(q_1,q_2)$, *where* $q_1,q_2\in\mathbb{R}_\star$. *Then*

$$\vec{PQ} = \left(\frac{q_1}{p_1},\frac{q_2}{p_2}\right),$$

and

$$0_\star = e^0$$
$$= \langle\vec{PQ},v\rangle_\star$$
$$= e^{\log\frac{q_1}{e^2}\log e+\log\frac{q_2}{e^3}\log e^2}$$
$$= e^{\log\frac{q_1}{e^2}+2\log\frac{q_2}{e^3}}$$
$$= e^{\log q_1-2+2\log q_2-6}$$
$$= e^{\log q_1+2\log q_2-8},$$

i.e.,

$$\log q_1 + 2\log q_2 = 8.$$

Next,

$$\begin{aligned} |\vec{PQ}|_\star &= e^{\left(\left(\log \frac{q_1}{e^2}\right)^2 + \left(\log \frac{q_2}{e^3}\right)^2\right)^{\frac{1}{2}}} \\ &= e^9, \end{aligned}$$

whereupon

$$(\log q_1 - 2)^2 + (\log q_2 - 3)^2 = 81.$$

Therefore, we get the system

$$\begin{aligned} \log q_1 + 2\log q_2 &= 8 \\ (\log q_1 - 2)^2 + (\log q_2 - 3)^2 &= 81. \end{aligned}$$

We have

$$\begin{aligned} \log q_1 &= 8 - 2\log q_2 \\ 81 &= (\log q_1 - 2)^2 + (\log q_2 - 3)^2. \end{aligned}$$

Thus, we get the equation

$$\begin{aligned} 81 &= (6 - 2\log q_2)^2 + (\log q_2 - 3)^2 \\ &= 36 - 24\log q_2 + 4(\log q_2)^2 + (\log q_2)^2 - 6\log q_2 + 9 \\ &= 5(\log q_2)^2 - 30\log q_2 + 45, \end{aligned}$$

or

$$5(\log q_2)^2 - 30\log q_2 - 36 = 0,$$

and

$$\begin{aligned} \log q_2 &= \frac{15 \pm \sqrt{225 + 180}}{5} \\ &= \frac{15 \pm \sqrt{405}}{5} \\ &= \frac{15 \pm 9\sqrt{5}}{5}, \end{aligned}$$

and

$$\begin{aligned} \log q_1 &= 8 - \frac{30 \pm 18\sqrt{5}}{5} \\ &= \frac{10 \mp 18\sqrt{5}}{5}. \end{aligned}$$

Therefore,

$$\begin{aligned} q_1 &= e^{\frac{10 \mp 18\sqrt{5}}{5}}, \\ q_2 &= e^{\frac{15 \pm 9\sqrt{5}}{5}}. \end{aligned}$$

Consequently,

$$Q_1\left(e^{\frac{10-18\sqrt{5}}{5}}, e^{\frac{15+9\sqrt{5}}{5}}\right), \quad Q_2\left(e^{\frac{10+18\sqrt{5}}{5}}, e^{\frac{15-9\sqrt{5}}{5}}\right).$$

This completes the solution.

We can find the equation of a multiplicative line if we know two points that lie on it. Let $P(p_1, p_2)$ and $Q(q_1, q_2)$ be two points on the line l. Then the vector

$$\vec{PQ} = \left(\frac{q_1}{p_1}, \frac{q_2}{p_2}\right),$$

is a multiplicative direction of l. Hence, applying (2.14), we find

$$\left(\frac{X_2}{p_2}\right)^{\log\frac{q_1}{p_1}} = \left(\frac{X_1}{p_1}\right)^{\log\frac{q_2}{p_2}}$$

or

$$\left(\frac{X_2}{q_2}\right)^{\log\frac{q_1}{p_1}} = \left(\frac{X_1}{q_1}\right)^{\log\frac{q_2}{p_2}}.$$

Example 2.12. *Let $P(2,3)$, $Q(4,11)$, $R\left(\frac{1}{8},\frac{1}{2}\right)$, $S\left(\frac{1}{8},5\right)$ be given points and l_1, l_2, l_3, l_4 be multiplicative lines so that*

$$P,Q \in l_1; \quad P,R \in l_2; \quad Q,R \in l_3; \quad R,S \in l_4.$$

We will find the equations of l_1, l_2, l_3 and l_4. We have

$$\begin{aligned}
\vec{PQ} &= (4 -_\star 2, 11 -_\star 3)\\
&= \left(2, \frac{11}{3}\right),\\
\vec{PR} &= \left(\frac{1}{8} -_\star 2, \frac{1}{2} -_\star 3\right)\\
&= \left(\frac{1}{16}, \frac{1}{6}\right),\\
\vec{QR} &= \left(\frac{1}{8} -_\star 4, \frac{1}{2} -_\star 11\right)\\
&= \left(\frac{1}{32}, \frac{1}{22}\right),\\
\vec{PS} &= \left(\frac{1}{8} -_\star 2, 5 -_\star 3\right)\\
&= \left(\frac{1}{16}, \frac{5}{3}\right).
\end{aligned}$$

Hence, the equations of l_1, l_2, l_3 and l_4 are

$$\begin{aligned}
\left(\frac{X_2}{3}\right)^{\log 2} &= \left(\frac{X_1}{2}\right)^{\log\frac{11}{3}},\\
\left(\frac{X_2}{3}\right)^{\log\frac{1}{16}} &= \left(\frac{X_1}{2}\right)^{\log\frac{1}{6}},\\
\left(\frac{X_2}{11}\right)^{\log\frac{1}{32}} &= \left(\frac{X_1}{4}\right)^{\log\frac{1}{22}},\\
\left(\frac{X_2}{3}\right)^{\log\frac{1}{16}} &= \left(\frac{X_1}{2}\right)^{\log\frac{5}{3}},
\end{aligned}$$

respectively.

Exercise 2.9. *Let* $P\left(\frac{1}{6},\frac{1}{12}\right)$, $Q\left(\frac{1}{2},\frac{1}{8}\right)$, $R\left(\frac{1}{4},\frac{1}{21}\right)$, $S(2,4)$. *Let also,* l_1, l_2, l_3, l_4 *and* l_5 *be multiplicative lines so that*

$$P,Q\in l_1;\quad P,R\in l_2;\quad P,S\in l_3;\quad S,R\in l_4;\quad Q,S\in l_5.$$

Find the equations of the multiplicative lines l_1, l_2, l_3, l_4 *and* l_5.

Answer 2.19.

1. $l_1 : (12X_2)^{\log 3} = (6X_1)^{\log\frac{3}{2}}$.
2. $l_2 : (12X_2)^{\log\frac{3}{2}} = (6X_1)^{\log\frac{4}{7}}$.
3. $l_3 : (12X_2)^{\log 12} = (6X_1)^{\log 48}$.
4. $l_4 : \left(\frac{X_2}{4}\right)^{3\log 2} = \left(\frac{X_1}{2}\right)^{\log 84}$.
5. $l_5 : \left(\frac{X_2}{4}\right)^{2\log 2} = \left(\frac{X_1}{2}\right)^{5\log 2}$.

Example 2.13. *Let the equation of the multiplicative line* l *be*

$$\left(\frac{X_1}{2}\right)^2\left(\frac{X_2}{3}\right)^4 = 1.$$

By the equation (2.13), *it follows that the point* $P(2,3)\in l$. *We find a multiplicative normal vector* $N=(N_1,N_2)$ *for the multiplicative line* l. *By the equation* (2.13), *we find*

$$\log N_1 = 2 \quad \text{and} \quad \log N_2 = 4.$$

Therefore,

$$N_1 = e^2 \quad \text{and} \quad N_2 = e^4.$$

Consequently, $N=\left(e^2,e^4\right)$ *is a multiplicative normal vector for* l. *Next, the vector* $v=\left(e^{-4},e^2\right)$ *is a multiplicative direction vector for* l.

Exercise 2.10. *Let the equation of the multiplicative line* l *be*

$$(4X_1)^{\frac{2}{3}}(3X_2)^7 = 1.$$

Find a point $P(p_1,p_2)\in l$, *a multiplicative normal vector and a multiplicative direction vector for* l.

Answer 2.20.

1. $P\left(\frac{1}{4},\frac{1}{3}\right)$.
2. $N=\left(e^{\frac{2}{3}},e^7\right)$.
3. $v=\left(e^{-7},e^{\frac{2}{3}}\right)$.

Example 2.14. *Let* $P(p_1,p_2)$ *and* $Q(q_1,q_2)$ *be two distinct points of* $E_\star^2$. *We will find the equation of the multiplicative line* l *through the multiplicative midpoint* M *of the multiplicative segment* PQ *with a multiplicative normal vector* $\vec{PQ}$. *We have*

$$\vec{PQ} = \left(\frac{q_1}{p_1},\frac{q_2}{p_2}\right).$$

Moreover,

$$\begin{aligned} M &= 1_\star/_\star e^2\cdot_\star(P+_\star Q)\\ &= e/_\star e^2\cdot_\star(p_1q_1,p_2q_2)\\ &= e^{\frac{\log e}{\log e^2}}\cdot_\star(p_1q_1,p_2q_2)\\ &= e^{\frac{1}{2}}\cdot_\star(p_1q_1,p_2q_2)\\ &= \left(e^{\frac{\log(p_1q_1)}{2}},e^{\frac{\log(p_2q_2)}{2}}\right). \end{aligned}$$

Therefore,

$$l: \left(\frac{X_1}{e^{\frac{\log(p_1 q_1)}{2}}}\right)^{\log \frac{q_1}{p_1}} \left(\frac{X_2}{e^{\frac{\log(p_2 q_2)}{2}}}\right)^{\log \frac{q_2}{p_2}} = 1.$$

Now, suppose that $a, b, c \in \mathbb{R}_\star$, $(a, b) \neq (0_\star, 0_\star)$. Consider the equation

$$a \cdot_\star x +_\star b \cdot_\star y +_\star c = 0_\star, \tag{2.15}$$

or

$$e^{\log a \log x + \log b \log y + \log c} = e^0,$$

or

$$\log a \log x + \log b \log y + \log c = 0,$$

or

$$\log x^{\log a} + \log y^{\log b} + \log c = 0,$$

or

$$c x^{\log a} y^{\log b} = 1. \tag{2.16}$$

Let $P(x_1, x_2)$, $x_1, x_2 \in \mathbb{R}_\star$, satisfies the equation (2.15). Then

$$a \cdot_\star x_1 +_\star b \cdot_\star x_2 +_\star c = 0_\star,$$

or

$$c = -_\star a \cdot_\star x_1 -_\star b \cdot_\star x_2, \tag{2.17}$$

and

$$a \cdot_\star (x -_\star x_1) +_\star b \cdot_\star (y -_\star x_2) = 0_\star.$$

Set

$$N = (a, b).$$

Then

$$\langle X -_\star P, N \rangle_\star = 0_\star.$$

Thus, (2.15) or (2.16) is an equation of a multiplicative line with multiplicative normal vector N and if c satisfies the equation (2.17), then the point P lies on it.

2.7 Perpendicular Multiplicative Lines

Definition 2.19. *Two multiplicative lines l and m are said to be multiplicative perpendicular if they have multiplicative orthogonal direction vectors. In this case, we write $l \perp_\star m$.*

Example 2.15. *Let the multiplicative line l has the equation*

$$\left(\frac{X_1}{2}\right)^{\log 3} = \left(\frac{X_2}{3}\right)^{\log 2},$$

and the multiplicative line m has the equation

$$X_1^{\log 2} = X_2^{-\log 3}.$$

By the equation (2.14)*, it follows that the vectors*

$$v^1 = (2,3) \quad \textit{and} \quad v^2 = \left(\frac{1}{3}, 2\right),$$

are multiplicative directions of l and m, respectively. We have

$$\begin{aligned} \langle v^1, v^2 \rangle_\star &= e^{\log 2 \log \frac{1}{3} + \log 3 \log 2} \\ &= e^{-\log 2 \log 3 + \log 2 \log 3} \\ &= e^0 \\ &= 1 \\ &= 0_\star. \end{aligned}$$

Consequently, $l \perp_\star m$.

Lemma 2.1. *For any $x, y \in E^2_\star$, we have*

$$|x +_\star y|_\star^{2_\star} = |x|_\star^{2_\star} +_\star e^2 \cdot_\star \langle x, y \rangle_\star +_\star |y|_\star^{2_\star}.$$

Proof. We have

$$\begin{aligned} |x +_\star y|_\star^{2_\star} &= |(x_1 y_1, x_2 y_2)|_\star^{2_\star} \\ &= e^{(\log(x_1 y_1))^2 + (\log(x_2 y_2))^2}, \\ |x|_\star^{2_\star} &= e^{(\log x_1)^2 + (\log x_2)^2}, \\ e^2 \cdot_\star \langle x, y \rangle_\star &= e^2 \cdot_\star e^{\log x_1 \log y_1 + \log x_2 \log y_2} \\ &= e^{2(\log x_1 \log y_1 + \log x_2 \log y_2)}, \\ |y|_\star^{2_\star} &= e^{(\log y_1)^2 + (\log y_2)^2}. \end{aligned}$$

Hence,

$$\begin{aligned} & |x|_\star^{2_\star} +_\star e^2 \cdot_\star \langle x, y \rangle_\star +_\star |y|_\star^{2_\star} \\ &= e^{(\log x_1)^2 + (\log x_2)^2} e^{2(\log x_1 \log y_1 + \log x_2 \log y_2)} e^{(\log y_1)^2 + (\log y_2)^2} \\ &= e^{(\log x_1)^2 + (\log y_1)^2 + (\log x_2)^2 + (\log y_2)^2 + 2(\log x_1 \log y_1 + \log x_2 \log y_2)} \\ &= e^{(\log x_1 + \log y_1)^2 + (\log x_2 + \log y_2)^2} \\ &= e^{(\log(x_1 y_1))^2 + (\log(x_2 y_2))^2} \\ &= |x +_\star y|_\star^{2_\star}. \end{aligned}$$

This completes the proof. □

Theorem 2.21 (Multiplicative Pythagoras)**.** *Let P, Q and R be three distinct points. Then*

$$|R -_\star P|_\star^{2_\star} = |Q -_\star P|_\star^{2_\star} +_\star |R -_\star Q|_\star^{2_\star},$$

if and only if $\vec{QP}$ and $\vec{RQ}$ are multiplicative perpendicular.

Proof. By Lemma 2.1, it follows that

$$|x +_\star y|_\star^{2_\star} = |x|_\star^{2_\star} +_\star |y|_\star^{2_\star}, \quad x, y \in E^2_\star,$$

if and only if $\langle x, y \rangle_\star = 0_\star$, i.e., $x \perp_\star y$. Now, we take

$$x = \vec{QP}, \quad y = \vec{RQ}$$

and we get the desired result. This completes the proof. □

Theorem 2.22. *Let $l \perp_\star m$. Then l and m have a unique point in common.*

Proof. Let v and w be multiplicative direction vectors of l and m, respectively. Without loss of generality, suppose that v and w are multiplicative unit vectors. Thus, $\{v, w\}$ is a multiplicative orthonormal pair. Suppose that

$$\begin{aligned} l &= P +_\star [v]_\star, \\ m &= Q +_\star [w]_\star. \end{aligned}$$

Then

$$P -_\star Q = \langle P -_\star Q, v\rangle_\star \cdot_\star v +_\star \langle P -_\star Q, w\rangle_\star \cdot_\star w,$$

whereupon

$$P -_\star \langle P -_\star Q, v\rangle_\star \cdot_\star v = Q +_\star \langle P -_\star Q, v\rangle_\star \cdot_\star w.$$

Setting

$$\begin{aligned} F &= P -_\star \langle P -_\star Q, v\rangle_\star \cdot_\star v \\ &= Q +_\star \langle P -_\star Q, v\rangle_\star \cdot_\star w, \end{aligned}$$

we get that F is the common point of l and m. The point F is unique because otherwise the multiplicative lines l and m will coincide. This completes the proof. □

Example 2.16. *Let*

$$\begin{aligned} l : \left(\frac{X_1}{2}\right)^{\log 2} &= \left(\frac{X_2}{4}\right)^{\log 3}, \\ m : X_1^{\log 3} &= X_2^{\log \frac{1}{2}}. \end{aligned}$$

Set

$$\begin{aligned} v^1 &= (3, 2), \\ v^2 &= \left(\frac{1}{2}, 3\right). \end{aligned}$$

Then v^1 and v^2 are multiplicative direction vectors for the multiplicative lines l and m, respectively. Next,

$$\begin{aligned} \langle v^1, v^2\rangle_\star &= e^{\log 3 \log \frac{1}{2} + \log 2 \log 3} \\ &= e^{-\log 2 \log 3 + \log 2 \log 3} \\ &= e^0 \\ &= 1 \\ &= 0_\star. \end{aligned}$$

Thus, $v^1 \perp_\star v^2$ and $l \perp_\star m$. Let $F = l \cap m$. Note that

$$\begin{aligned} l : X_1 &= 2\left(\frac{X_2}{4}\right)^{\frac{\log 3}{\log 2}}, \\ m : X_1 &= X_2^{-\frac{\log 2}{\log 3}}. \end{aligned}$$

Then

$$X_2^{-\frac{\log 2}{\log 3}} = 2\left(\frac{X_2}{4}\right)^{\frac{\log 3}{\log 2}},$$

or

$$\begin{aligned} 1 &= \frac{2}{4^{\frac{\log 3}{\log 2}}} X_2^{\frac{\log 3}{\log 2}+\frac{\log 2}{\log 3}} \\ &= \frac{1}{2^{2\frac{\log 3}{\log 2}-1}} X_2^{\frac{(\log 3)^2+(\log 2)^2}{\log 2 \log 3}}, \end{aligned}$$

or

$$X_2^{\frac{(\log 3)^2+(\log 2)^2}{\log 2 \log 3}} = 2^{2\frac{\log 3}{\log 2}-1}.$$

Hence,

$$X_2 = 2^{\frac{(2\log 3-\log 2)\log 3}{(\log 2)^2+(\log 3)^2}}$$

and

$$X_1 = 2^{-\frac{(2\log 3-\log 2)\log 2}{(\log 2)^2+(\log 3)^2}}.$$

Consequently,

$$F\left(2^{-\frac{(2\log 3-\log 2)\log 2}{(\log 2)^2+(\log 3)^2}}, 2^{\frac{(2\log 3-\log 2)\log 3}{(\log 2)^2+(\log 3)^2}}\right),$$

is the common point of the multiplicative lines l and m.

Exercise 2.11. *Let*

$$\begin{aligned} l: \left(\frac{X_1}{4}\right)^{\log 5} &= \left(\frac{X_2}{3}\right)^{\log 7}, \\ m: \left(\frac{X_1}{11}\right)^{\log 7} &= \left(\frac{X_2}{4}\right)^{\log \frac{1}{5}}. \end{aligned}$$

Prove that $l \perp_\star m$ and find the point $F = l \cap m$.

Answer 2.23.

$$F\left(11^{\frac{(\log 7)^2}{(\log 5)^2+(\log 7)^2}} 3^{-\frac{\log 5 \log 7}{(\log 5)^2+(\log 7)^2}} 4^{\frac{\log 5(\log 5+\log 7)}{(\log 5)^2+(\log 7)^2}}, 11^{\frac{\log 5 \log 7}{(\log 5)^2+(\log 7)^2}} 4^{\frac{\log 5(\log 5-\log 7)}{(\log 5)^2+(\log 7)^2}} 3^{\frac{(\log 7)^2)}{(\log 5)^2+(\log 7)^2}}\right).$$

Theorem 2.24. *Let R be a point in $E_\star^2$ and let l be a multiplicative line. Then there is a unique multiplicative line through R multiplicative perpendicular to l. Furthermore,*

1. *$m = R +_\star [N]_\star$, where N is a multiplicative unit normal vector to l.*
2. *l and m intersect in the point*

   $$F = R -_\star \langle R -_\star P, N \rangle_\star \cdot_\star N,$$

   where P is any point on l.
3. *$d_\star(R, F) = |\langle R -_\star P, N \rangle_\star|_\star$.*

Proof. Let $Q(Q_1, Q_2) \in l$ and the equation of l is given by

$$\left(\frac{X_1}{Q_1}\right)^{\log N_1} \left(\frac{X_2}{Q_2}\right)^{\log N_2} = 1,$$

where $N = (N_1, N_2)$ is a multiplicative unit normal vector to l. Then

$$m = R +_\star [N]_\star,$$

and

$$m: \left(\frac{X_1}{P_1}\right)^{\log N_2} = \left(\frac{X_2}{P_2}\right)^{\log N_1}.$$

We have

$$F = R +_\star \mu \cdot_\star N,$$

where $\mu \in \mathbb{R}_\star$ will be determined below. Note that

$$F -_\star P = R -_\star P +_\star \mu \cdot_\star N.$$

Hence,

$$\begin{aligned} 0_\star &= \langle F -_\star P, N\rangle_\star \\ &= \langle R -_\star P, N\rangle_\star +_\star \mu \cdot_\star \langle N, N\rangle_\star \\ &= \langle R -_\star P, N\rangle_\star +_\star \mu, \end{aligned}$$

whereupon

$$\mu = -_\star \langle R -_\star P, N\rangle_\star.$$

Consequently,

$$F = R -_\star \langle R -_\star P, N\rangle_\star \cdot_\star N,$$

and hence,

$$\begin{aligned} d_\star(F,R) &= |\langle R -_\star P, N\rangle_\star \cdot_\star N|_\star \\ &= |\langle R -_\star P, N\rangle_\star|_\star \cdot_\star |N|_\star \\ &= |\langle R -_\star P, N\rangle_\star|_\star. \end{aligned}$$

This completes the proof. □

Remark 2.2. *Suppose that all conditions of Theorem 2.24 hold. Let $P(P_1, P_2)$, $R(R_1, R_2)$, $F(F_1, F_2)$. Then*

$$\begin{aligned} R -_\star P &= \left(\frac{R_1}{P_1}, \frac{R_2}{P_2}\right), \\ \langle R -_\star P, N\rangle_\star &= e^{\log \frac{R_1}{P_1} \log N_1 + \log \frac{R_2}{P_2} \log N_2}, \end{aligned}$$

and

$$\begin{aligned} \langle R -_\star P, N\rangle_\star \cdot_\star N &= \Big(e^{\left(\log \frac{R_1}{P_1} \log N_1 + \log \frac{R_2}{P_2} \log N_2\right) \log N_1}, \\ &\qquad e^{\left(\log \frac{R_1}{P_1} \log N_1 + \log \frac{R_2}{P_2} \log N_2\right) \log N_2} \Big), \end{aligned}$$

and

$$\begin{aligned} F &= (F_1, F_2) \\ &= \Big(R_1 e^{-\left(\log \frac{R_1}{P_1} \log N_1 + \log \frac{R_2}{P_2} \log N_2\right) \log N_1}, \\ &\qquad R_2 e^{-\left(\log \frac{R_1}{P_1} \log N_1 + \log \frac{R_2}{P_2} \log N_2\right) \log N_2} \Big), \end{aligned}$$

and

$$d_\star(R,F) = \left| e^{\log \frac{R_1}{P_1} \log N_1 + \log \frac{R_2}{P_2} \log N_2} \right|_\star.$$

Definition 2.20. *The number $d_\star(R,F)$ is called the multiplicative distance of the point R to the line l and it is written as $d_\star(R,l)$.*

Example 2.17. *Let*

$$l : \left(\frac{X_1}{2}\right)^{\log 3} \left(\frac{X_2}{4}\right)^{\log 7} = 1,$$

and $R(3,5)$. We will find a multiplicative line m so that $R \in m$ and $l \perp_\star m$. By the equation (2.13), it follows that $N(3,7)$ is a multiplicative normal vector to l. Then N is a multiplicative direction vector to m. Hence, applying the equation (2.14), we find

$$m : \left(\frac{X_2}{5}\right)^{\log 3} = \left(\frac{X_1}{3}\right)^{\log 7}.$$

Let $F = m \cap l$. Consider the system

$$\begin{aligned} \left(\frac{X_1}{2}\right)^{\log 3} \left(\frac{X_2}{4}\right)^{\log 7} &= 1 \\ \left(\frac{X_2}{5}\right)^{\log 3} &= \left(\frac{X_1}{3}\right)^{\log 7}. \end{aligned}$$

We have

$$\begin{aligned} X_1 &= 3\left(\frac{X_2}{5}\right)^{\frac{\log 3}{\log 7}} \\ \left(\frac{3}{2}\left(\frac{X_2}{5}\right)^{\frac{\log 3}{\log 7}}\right)^{\log 3} \left(\frac{X_2}{4}\right)^{\log 7} &= 1, \end{aligned}$$

whereupon

$$\frac{3^{\log 3}}{2^{\log 3} 5^{\frac{(\log 3)^2}{\log 7}}} X_2^{\frac{(\log 3)^2}{\log 7}} \frac{X_2^{\log 7}}{4^{\log 7}} = 1,$$

or

$$X_2^{\frac{(\log 3)^2 + (\log 7)^2}{\log 7}} = \frac{2^{\log 3} 5^{\frac{(\log 3)^2}{\log 7}} 4^{\log 7}}{3^{\log 3}},$$

or

$$X_2 = \frac{2^{\frac{\log 3 \log 7}{(\log 3)^2 + (\log 7)^2}} 5^{\frac{(\log 3)^2}{(\log 3)^2 + (\log 7)^2}} 4^{\frac{(\log 7)^2}{(\log 3)^2 + (\log 7)^2}}}{3^{\frac{\log 3 \log 7}{(\log 3)^2 + (\log 7)^2}}},$$

or

$$X_2 = \frac{2^{\frac{\log 7(\log 3 + 2\log 7)}{(\log 3)^2 + (\log 7)^2}} 5^{\frac{(\log 3)^2}{(\log 3)^2 + (\log 7)^2}}}{3^{\frac{\log 3 \log 7}{(\log 3)^2 + (\log 7)^2}}},$$

and

$$
\begin{aligned}
X_1 &= \frac{3\; 2^{\frac{\log 3(\log 3+2\log 7)}{(\log 3)^2+(\log 7)^2}}\; 5^{\frac{(\log 3)^3}{\log 7\left((\log 3)^2+(\log 7)^2\right)}}}{5^{\frac{\log 3}{\log 7}}\; 3^{\frac{(\log 3)^2}{(\log 3)^2+(\log 7)^2}}}\\
&= \frac{2^{\frac{\log 3(\log 3+2\log 7)}{(\log 3)^2+(\log 7)^2}}\; 5^{-\frac{\log 3\log 7}{\left((\log 3)^2+(\log 7)^2\right)}}}{3^{-\frac{(\log 7)^2}{(\log 3)^2+(\log 7)^2}}}\\
&= \frac{2^{\frac{\log 3(\log 3+2\log 7)}{(\log 3)^2+(\log 7)^2}}\; 3^{\frac{(\log 7)^2}{(\log 3)^2+(\log 7)^2}}}{5^{\frac{\log 3\log 7}{\left((\log 3)^2+(\log 7)^2\right)}}}.
\end{aligned}
$$

Consequently,

$$
F\left(\frac{2^{\frac{\log 3(\log 3+2\log 7)}{(\log 3)^2+(\log 7)^2}}\; 3^{\frac{(\log 7)^2}{(\log 3)^2+(\log 7)^2}}}{5^{\frac{\log 3\log 7}{\left((\log 3)^2+(\log 7)^2\right)}}}, \frac{2^{\frac{\log 7(\log 3+2\log 7)}{(\log 3)^2+(\log 7)^2}}\; 5^{\frac{(\log 3)^2}{(\log 3)^2+(\log 7)^2}}}{3^{\frac{\log 3\log 7}{(\log 3)^2+(\log 7)^2}}}\right) = m\cap l.
$$

Exercise 2.12. *Let $P(7,8)$ and*

$$
l : X_1^{\log 10}\left(\frac{X_2}{4}\right)^{\log 15} = 1.
$$

Find a multiplicative line m so that $P\in m$ and $l \perp_\star m$.

Answer 2.25.

$$
\left(\frac{X_1}{7}\right)^{\log 15} = \left(\frac{X_2}{8}\right)^{\log 10}.
$$

Definition 2.21. *Let PQ be a multiplicative segment and M be its multiplicative midpoint. The multiplicative line through M that is multiplicative perpendicular to $\vec{PQ}$ is called multiplicative perpendicular bisector of the multiplicative segment PQ.*

2.8 Multiplicative Parallel and Intersecting Multiplicative Lines

Definition 2.22. *Two distinct multiplicative lines l and m are said to be multiplicative parallel if they have no point of intersection. In this case, we will write $l \parallel_\star m$.*

Example 2.18. *Consider the multiplicative lines*

$$
\begin{aligned}
l_1 : X_1^2 X_2 &= 1,\\
l_2 : X_1^2\left(\frac{X_2}{2}\right) &= 1.
\end{aligned}
$$

These multiplicative lines have not any point of intersection. In Fig. 2.3 they are shown their graphics.

Theorem 2.26. *Two distinct multiplicative lines l and m are multiplicative parallel if and only if they have the same multiplicative direction.*

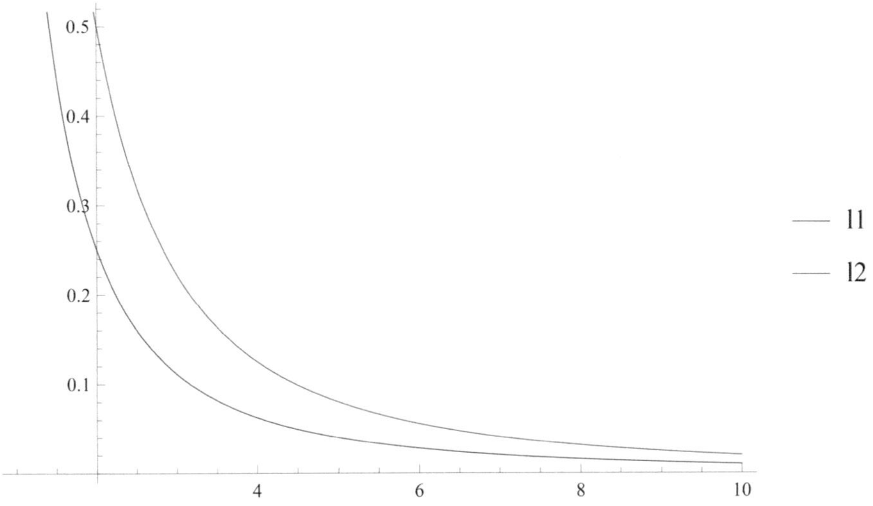

FIGURE 2.3
Parallel multiplicative lines.

Proof.

1. Let $l \parallel_\star m$. Suppose that l and m have different multiplicative direction vectors v and w, respectively. Let $P \in l$ and $Q \in m$. Because v and w are not multiplicative proportional, there exist numbers $s, t \in \mathbb{R}_\star$ so that

$$P -_\star Q = t \cdot_\star v +_\star s \cdot_\star w,$$

i.e.,

$$P -_\star t \cdot_\star v = Q +_\star s \cdot_\star w.$$

Thus, the point

$$\begin{aligned} F &= P -_\star t \cdot_\star v \\ &= Q +_\star s \cdot_\star w \end{aligned}$$

lies on l and m. This is a contradiction. Therefore, l and m have the same multiplicative direction vector.

2. Suppose that l and m have the same multiplicative direction vectors. Let l and m have a common point F. Then

$$l = F +_\star [v]_\star \quad \text{and} \quad m = F +_\star [w]_\star.$$

Because l and m are distinct, we have that

$$[v]_\star \neq [w]_\star.$$

This is a contradiction. Therefore, l and m have not any common point. Thus, $l \parallel_\star m$. This completes the proof.

□

Theorem 2.27. *Let l, m and n be multiplicative lines so that $l \parallel_\star m$ and $l \parallel_\star n$. Then $m = n$ or $m \parallel_\star n$.*

Proof. Since $l \parallel_\star m$, then l and m are distinct and have the same multiplicative direction v. Because $l \parallel_\star n$, we have that l and n are distinct and have the same multiplicative direction v. Hence, m and n have the same multiplicative direction v. Therefore, $m = n$ or $m \parallel_\star n$. This completes the proof. □

Theorem 2.28. *Let l, m and n be multiplicative lines so that $l \parallel_\star m$ and $m \perp_\star n$. Then $l \perp_\star n$.*

Proof. Since $l \parallel_\star m$, we have that l and m are distinct and have the same multiplicative direction v and the same multiplicative normal vector N. Because $m \perp_\star n$, then we have that N is a multiplicative direction vector for n. Hence, $l \perp_\star n$. This completes the proof. □

Theorem 2.29. *Let l, m and n be multiplicative lines so that $l \perp_\star m$ and $l \perp_\star n$. Then $m \parallel_\star n$ or $m = n$.*

Proof. Let v and N be a multiplicative direction vector and a multiplicative normal vector, respectively, of l. Since $l \perp_\star m$ and $l \perp_\star n$, we have that N is a multiplicative direction vector for m and n. Hence, $m \parallel_\star n$ or $m = n$. This completes the proof. □

2.9 Multiplicative Reflections

Let l be a given multiplicative line, $X \in E^2_\star$, $X \notin l$. Suppose that $F \in L$ is so that $\vec{XF} \perp_\star l$ and N is a multiplicative unit normal vector to l.

Definition 2.23. *Define $X' \in E^2_\star$ as follows*

$$e^{\frac{1}{2}} \cdot_\star (X +_\star X') = F.$$

The point X' will be called the multiplicative symmetric point of X.

Let $P \in l$. Then, we have

$$e^{\frac{1}{2}} \cdot_\star \left(X +_\star X'\right) = X -_\star \langle X -_\star P, N\rangle_\star \cdot_\star N,$$

whereupon

$$X +_\star X' = e^2 \cdot_\star X -_\star e^2 \cdot_\star \langle X -_\star P, N\rangle_\star \cdot_\star N,$$

and

$$\begin{aligned} X' &= e^2 \cdot_\star X -_\star X -_\star e^2 \cdot_\star \langle X -_\star P, N\rangle_\star \cdot_\star N \\ &= \left(e^2 -_\star 1_\star\right) \cdot_\star X -_\star e^2 \cdot_\star \langle X -_\star P, N\rangle_\star \cdot_\star N \\ &= \left(e^2 -_\star e\right) \cdot_\star X -_\star e^2 \cdot_\star \langle X -_\star P, N\rangle_\star \cdot_\star N \\ &= e \cdot_\star X -_\star e^2 \cdot_\star \langle X -_\star P, N\rangle_\star \cdot_\star N \\ &= X -_\star e^2 \cdot_\star \langle X -_\star P, N\rangle_\star \cdot_\star N. \end{aligned}$$

Define the operator $\Omega_l : E^2_\star \to E^2_\star$ as follows

$$\Omega_l X = X'.$$

Theorem 2.30. *We have $\Omega_l \Omega_l X = X$.*

Proof. We have $\Omega_l X = X'$ and

$$\begin{aligned}
\Omega_l\Omega_l X &= \Omega_l X' \\
&= X' -_\star e^2 \cdot_\star \langle X' -_\star P, N\rangle_\star \cdot_\star N \\
&= X -_\star e^2 \cdot_\star \langle X -_\star P, N\rangle_\star \cdot_\star N \\
&\quad -_\star e^2 \cdot_\star \langle X -_\star e^2 \cdot_\star \langle X -_\star P, N\rangle_\star \cdot_\star N -_\star P, N\rangle_\star \cdot_\star N \\
&= X -_\star e^2 \cdot_\star \langle X -_\star P, N\rangle_\star \cdot_\star N \\
&\quad -_\star e^2 \cdot_\star \langle X -_\star P, N\rangle_\star \cdot_\star N \\
&\quad +_\star e^2 \cdot_\star e^2 \cdot_\star \langle X -_\star P, N\rangle_\star \cdot_\star \langle N, N\rangle_\star \cdot_\star N \\
&= X -_\star \left(e^2 +_\star e^2\right) \cdot_\star \langle X -_\star P, N\rangle_\star \cdot_\star N \\
&\quad +_\star e^4 \cdot_\star \langle X -_\star P, N\rangle_\star \cdot_\star N \\
&= X -_\star e^4 \cdot_\star \langle X -_\star P, N\rangle_\star \cdot_\star N \\
&\quad +_\star e^4 \cdot_\star \langle X -_\star P, N\rangle_\star \cdot_\star N \\
&= X.
\end{aligned}$$

This completes the proof. □

Theorem 2.31. *We have $\Omega_l X = X$ if and only if $X \in l$, where l is a multiplicative line passing through the point P and having a multiplicative unit normal vector N.*

Proof. We have

$$\begin{aligned}
X &= \Omega_l X \\
&= X -_\star e^2 \cdot_\star \langle X -_\star P, N\rangle_\star,
\end{aligned}$$

if and only if

$$\langle X -_\star P, N\rangle_\star = 0_\star$$

if and only if $X \in l$. This completes the proof. □

Theorem 2.32. *We have*

$$d_\star(\Omega_l X, \Omega_l Y) = d_\star(X, Y),$$

for any two points $X, Y \in E^2_\star$.

Proof. We have

$$\begin{aligned}
\Omega_l X -_\star \Omega_l Y &= X -_\star e^2 \cdot_\star \langle X -_\star P, N\rangle_\star \cdot_\star N \\
&\quad -_\star Y +_\star e^2 \cdot_\star \langle Y -_\star P, N\rangle_\star \cdot_\star N \\
&= X -_\star Y -_\star e^2 \cdot_\star \langle X -_\star Y, N\rangle_\star \cdot_\star N,
\end{aligned}$$

and

$$\begin{aligned}
|\Omega_l X -_\star \Omega_l Y|_\star^{2_\star} &= |X -_\star Y -_\star e^2 \cdot_\star \langle X -_\star Y, N\rangle_\star \cdot_\star N|_\star^{2_\star} \\
&= |X -_\star Y|_\star^{2_\star} -_\star e^2 \cdot_\star \langle X -_\star Y, e^2 \cdot_\star \langle X -_\star Y, N\rangle_\star \cdot_\star N\rangle_\star \\
&\quad +_\star e^4 \cdot_\star |\langle X -_\star Y, N\rangle_\star \cdot_\star N|_\star^{2_\star} \\
&= |X -_\star Y|_\star^{2_\star} -_\star e^4 \cdot_\star |\langle X -_\star Y, N\rangle_\star|_\star^{2_\star} \\
&\quad +_\star e^4 \cdot_\star |\langle X -_\star Y, N\rangle_\star|_\star^{2_\star} \\
&= |X -_\star Y|_\star^{2_\star},
\end{aligned}$$

whereupon

$$d_\star(\Omega_l X, \Omega_l Y) = d_\star(X,Y).$$

This completes the proof. □

Theorem 2.33. *We have $\Omega_l : E_\star^2 \to E_\star^2$ is a bijection.*

Proof.

1. Let Y be any point of $E_\star^2$. Set $X = \Omega_l Y$. Then, applying Theorem 2.30, we find

$$\begin{aligned}\Omega_l X &= \Omega_l \Omega_l Y\\ &= Y.\end{aligned}$$

Thus, $\Omega_l : E_\star^2 \to E_\star^2$ is surjective.

2. Let $X, Y \in E_\star^2$ be arbitrarily chosen. Assume that

$$\Omega_l X = \Omega_l Y.$$

Then

$$\Omega_l \Omega_l X = \Omega_l \Omega_l Y$$

or

$$X = Y.$$

Therefore, $\Omega_l : E_\star^2 \to E_\star^2$ is injective, and hence, it is bijective. This completes the proof.

□

Example 2.19. *Let $P(2,4) \in l$ and $N = \left(e^{\frac{1}{3}}, e^{\frac{2\sqrt{2}}{3}}\right)$ be multiplicative normal vector to l. We have*

$$\begin{aligned}|N|_\star &= e^{\left(\left(\log e^{\frac{1}{3}}\right)^2 + \left(\log e^{\frac{2\sqrt{2}}{3}}\right)^2\right)}\\ &= e^{\left(\frac{1}{9}+\frac{8}{9}\right)^2}\\ &= e\\ &= 1_\star,\end{aligned}$$

i.e., N is a multiplicative unit normal vector to l. Take $X(4,6) \in E_\star^2$. Then

$$\begin{aligned}X -_\star P &= (2,2),\\ \langle X -_\star P, N\rangle_\star &= e^{\log 2 \log e^{\frac{1}{3}} + \log 2 \log e^{\frac{2\sqrt{2}}{3}}}\\ &= e^{\frac{2\sqrt{2}+1}{3}\log 2},\\ e^2 \cdot_\star \langle X -_\star P, N\rangle_\star &= e^{\log e^2 \log e^{\frac{2\sqrt{2}+1}{3}\log 2}}\\ &= e^{\frac{4\sqrt{2}+2}{3}\log 2},\end{aligned}$$

and

$$e^2 \cdot_\star \langle X -_\star P, N\rangle_\star \cdot_\star N = \left(e^{\frac{1}{3}+\frac{4\sqrt{2}+2}{3}\log 2}, e^{\frac{2\sqrt{2}}{3}+\frac{4\sqrt{2}+2}{3}\log 2}\right),$$

and

$$\begin{aligned}X' &= X -_\star e^2 \cdot_\star \langle X -_\star P, N\rangle_\star \cdot_\star N\\ &= \left(4e^{-\frac{1}{3}-\frac{4\sqrt{2}+2}{3}\log 2}, 6e^{-\frac{2\sqrt{2}}{3}-\frac{4\sqrt{2}+2}{3}\log 2}\right).\end{aligned}$$

Exercise 2.13. *Let $P(7,8)$, $N = \left(e^{\frac{\sqrt{2}}{4}}, e^{\frac{\sqrt{14}}{4}}\right)$.*

1. *Find the equation of the multiplicative line l through P and with multiplicative unit normal vector N.*
2. *Let $X(4,5)$. Find $\Omega_l X$.*

Answer 2.34.

1.
$$l : \left(\frac{X_1}{7}\right)^{\frac{\sqrt{2}}{4}} \left(\frac{X_2}{8}\right)^{\frac{\sqrt{14}}{4}} = 1.$$

2.
$$\Omega_l X = \left(4e^{-\frac{1}{4}\log\frac{4}{7} - \frac{\sqrt{7}}{4}\log\frac{5}{8}}, 5e^{-\frac{\sqrt{7}}{4}\log\frac{4}{7} - \frac{7}{4}\log\frac{5}{8}}\right).$$

2.10 Multiplicative Congruence and Multiplicative Isometries

Definition 2.24. *A mapping $T : E_\star^2 \to E_\star^2$ is said to be a multiplicative isometry if*

$$d_\star(TX, TY) = d_\star(X, Y)$$

for any points X, $Y \in E_\star^2$.

Definition 2.25. *Two figures $\mathscr{F}_1$ and $\mathscr{F}_2$ are said to be multiplicative congruent if there exists a multiplicative isometry $T : E_\star^2 \to E_\star^2$ so that*

$$T\mathscr{F}_1 = \mathscr{F}_2.$$

Theorem 2.35. *Every multiplicative isometry $T : E_\star^2 \to E_\star^2$ is a bijection.*

Proof. Let $X, Y \in E_\star^2$ be arbitrarily chosen. If $X = Y$, then

$$\begin{aligned} 0_\star &= d_\star(X,Y) \\ &= d_\star(TX,TY), \end{aligned}$$

whereupon $TX = TY$. Thus, $T^{-1} : E_\star^2 \to E_\star^2$ exists. Next,

$$\begin{aligned} d_\star(T^{-1}X, T^{-1}Y) &= d_\star(TT^{-1}X, TT^{-1}Y) \\ &= d_\star(X,Y), \end{aligned}$$

i.e., $T^{-1} : E_\star^2 \to E_\star^2$ is a multiplicative isometry. This completes the proof. □

Theorem 2.36. *If T and S are multiplicative isometries, then TS is a multiplicative isometry.*

Proof. Let $X, Y \in E_\star^2$ be arbitrarily chosen. Then

$$\begin{aligned} d_\star(TX,TY) &= d_\star(X,Y), \\ d_\star(SX,SY) &= d_\star(X,Y). \end{aligned}$$

Hence,

$$\begin{aligned} d_\star(TSX, TSY) &= d_\star(SX, SY) \\ &= d_\star(X, Y). \end{aligned}$$

This completes the proof. □

Definition 2.26. *The set of all multiplicative isometries is a group called multiplicative isometry group, and it is denoted by $\mathscr{S}_\star(E^2_\star)$.*

Definition 2.27. *Let $\mathscr{F}$ be a figure in $E^2_\star$. The set*

$$\mathscr{S}_\star(\mathscr{F}) = \{T \in \mathscr{S}_\star(E^2_\star) : T\mathscr{F} = \mathscr{F}\}$$

is a subgroup of $\mathscr{S}_\star(E^2_\star)$ and it is called multiplicative symmetry group of $\mathscr{F}$.

2.11 Multiplicative Translations

Let m and n be two multiplicative parallel lines in $E^2_\star$. Take $P \in m$ arbitrarily. Let $Q \in n$ be such that $\vec{PQ} \perp_\star n$. Denote with N a multiplicative unit normal vector to m. Then N is a multiplicative unit normal vector to n. Therefore, for $X \in E^2_\star$, we get

$$\begin{aligned} \Omega_m\Omega_n X = \Omega_n X -_\star e^2 \cdot_\star \langle \Omega_n X -_\star P, N\rangle_\star \cdot_\star N \\ = \; & X -_\star e^2 \cdot_\star \langle X -_\star Q, N\rangle_\star \cdot_\star N \\ & -_\star e^2 \cdot_\star \langle X -_\star e^2 \cdot_\star \langle X -_\star Q, N\rangle_\star \cdot_\star N -_\star P, N\rangle_\star \cdot_\star N \\ = \; & X -_\star e^2 \cdot_\star \langle X -_\star Q, N\rangle_\star \cdot_\star N \\ & -_\star e^2 \cdot_\star \langle X -_\star P, N\rangle_\star \cdot_\star N \\ & +_\star e^4 \cdot_\star \langle X -_\star Q, N\rangle_\star \cdot_\star \langle N, N\rangle_\star \cdot_\star N \\ = \; & X +_\star e^2 \cdot_\star \langle X -_\star Q, N\rangle_\star \cdot_\star N \\ & -_\star e^2 \cdot_\star \langle X -_\star P, N\rangle_\star \cdot_\star N \\ = \; & X +_\star e^2 \cdot_\star \langle P -_\star Q, N\rangle_\star \cdot_\star N \\ = \; & X +_\star e^2_\star \cdot_\star (P -_\star Q). \end{aligned}$$

Definition 2.28. *Let l be any multiplicative line and m, n be multiplicative lines so that $m, n \perp_\star l$. Then the multiplicative transformation $\Omega_m\Omega_n$ is called a multiplicative translation along l. If $m \neq n$, the multiplicative translation is said to be nontrivial.*

Definition 2.29. *The set of all multiplicative lines multiplicative perpendicular to a given multiplicative line l will be called a multiplicative pencil of multiplicative parallels and will be denoted by $\mathscr{P}_l$. The multiplicative line l is a common multiplicative perpendicular for the multiplicative pencil.*

Fig. 2.4 shows a multiplicative pencil. Let

$$\lambda = e^2 \cdot_\star \langle P -_\star Q, N\rangle_\star.$$

We introduce the notation

$$T_\lambda X = X +_\star \lambda \cdot_\star N.$$

With $T_\star(l)$ we will denote the set of all multiplicative translations along l.

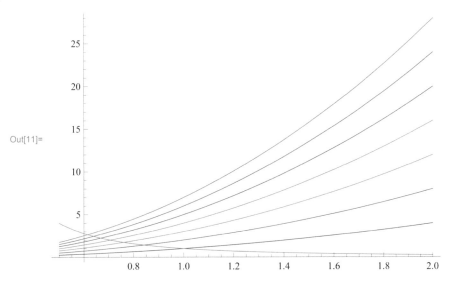

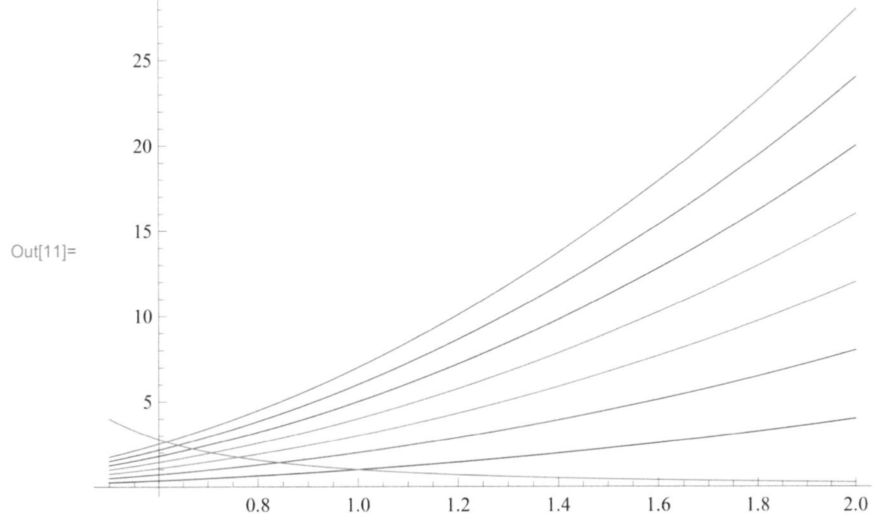

FIGURE 2.4
Multiplicative pencil.

Theorem 2.37. *$T_\star(l)$ is an Abelian group isomorphic to the multiplicative additive group of multiplicative real numbers.*

Proof. Let $\lambda, \mu \in \mathbb{R}_\star$ and $X \in E_\star^2$ be arbitrarily chosen. Then

$$\begin{aligned} T_\lambda T_\mu X &= T_\lambda (X +_\star \mu \cdot_\star N) \\ &= X +_\star \mu \cdot_\star N +_\star \lambda \cdot_\star N \\ &= X +_\star (\mu +_\star \lambda) \cdot_\star N \\ &= T_{\mu +_\star \lambda} X. \end{aligned}$$

As above,

$$T_\mu T_\lambda X = T_{\mu +_\star \lambda} X.$$

Since

$$\lambda +_\star \mu = \mu +_\star \lambda,$$

we get that the multiplicative translations commute. Note that $T_{0_\star} = Id$ and

$$T_\lambda T_{-_\star \lambda} = T_{0_\star}.$$

Therefore,

$$(T_\lambda)^{-1} = T_{-_\star \lambda}.$$

Consequently, $T(l)$ is an Abelian group and it is an isomorphism. This completes the proof. □

Theorem 2.38. *Let T be a nontrivial multiplicative translation along l. Then l has a multiplicative direction vector v such that*

$$Tx = x +_\star v, \tag{2.18}$$

for any $x \in E_\star^2$. Conversely, if v is any multiplicative nonzero vector and l be any multiplicative line with a multiplicative direction vector v, then the transformation T, defined by (2.18), *is a multiplicative translation along l.*

Proof.

1. Let N be a multiplicative direction vector for the multiplicative line l. Take $P \in E_\star^2$ arbitrarily. Let also, α and β be multiplicative lines so that $\alpha \perp_\star l$ and $\beta \perp_\star l$. Suppose that $a, b \in \mathbb{R}_\star$ are the unique numbers so that

$$P +_\star a \cdot_\star N \in \alpha, \quad P +_\star b \cdot_\star N \in \beta.$$

Then

$$\begin{aligned} \Omega_\alpha \Omega_\beta x &= x +_\star e^2 \cdot_\star (P +_\star a \cdot_\star N -_\star P -_\star b \cdot_\star N) \\ &= x +_\star e^2 \cdot_\star (a -_\star b) \cdot_\star N. \end{aligned}$$

Hence,

$$v = e^2 \cdot_\star (a -_\star b) \cdot_\star N.$$

Note that v is a multiplicative direction vector for the multiplicative line l.

2. For each $\lambda \in \mathbb{R}_\star$, define the mapping

$$T_\lambda x = x +_\star \lambda \cdot_\star N.$$

Take $a, b \in \mathbb{R}_\star$ so that

$$\lambda = e^2 \cdot_\star (a -_\star b).$$

Construct

$$\begin{aligned} \alpha &= P +_\star a \cdot_\star N +_\star [N^{\perp_\star}]_\star, \\ \beta &= P +_\star b \cdot_\star N +_\star [N^{\perp_\star}]_\star. \end{aligned}$$

Then

$$\begin{aligned} \Omega_\alpha \Omega_\beta x &= x +_\star e^2 \cdot_\star (a -_\star b) \cdot_\star N \\ &= T_\lambda x. \end{aligned}$$

We have that $\alpha \perp_\star l$ and $\beta \perp_\star l$. Thus, T_λ is a multiplicative translation along l. This completes the proof.

□

Theorem 2.39. *Let α, β and γ be three multiplicative lines of a multiplicative pencil $\mathscr{P}$ with common multiplicative perpendicular l. Then there exists a unique multiplicative line δ of this multiplicative pencil such that*

$$\Omega_\alpha \Omega_\beta \Omega_\gamma = \Omega_\delta.$$

Proof. Let $\alpha, \beta, \gamma \in \mathscr{P}_l$ with corresponding numbers a, b and c, respectively. Then

$$\Omega_\alpha \Omega_\beta \Omega_\gamma = \Omega_\alpha T_{e^2 \cdot_\star (b -_\star c)},$$

and

$$\begin{aligned} \Omega_\alpha \Omega_\beta \Omega_\gamma x &= \Omega_\alpha T_{e^2 \cdot_\star (b -_\star c)} x \\ &= \Omega_\alpha (x +_\star e^2 \cdot_\star (b -_\star c) \cdot_\star N) \\ &= x +_\star e^2 \cdot_\star (b -_\star c) \cdot_\star N \\ &\quad -_\star e^2 \cdot_\star \langle x +_\star e^2 \cdot_\star (b -_\star c) \cdot_\star N -_\star P -_\star a \cdot_\star N, N \rangle_\star \cdot_\star N \\ &= x +_\star e^2 \cdot_\star (b -_\star c) \cdot_\star N \end{aligned}$$

$$
\begin{aligned}
& -_\star e^2 \cdot_\star \langle x -_\star P, N\rangle_\star \cdot_\star N \\
& -_\star e^4 \cdot_\star (b -_\star c) \cdot_\star N +_\star e^2 \cdot_\star a \cdot_\star N \\
= \; & x -_\star e^2 \cdot_\star (b -_\star c) \cdot_\star N +_\star e^2 \cdot_\star a \cdot_\star N \\
& -_\star e^2 \cdot_\star \langle x -_\star P, N\rangle_\star \cdot_\star N \\
= \; & x -_\star e^2 \cdot_\star (b -_\star c -_\star a) \cdot_\star N \\
& -_\star e^2 \cdot_\star \langle X -_\star P, N\rangle_\star \cdot_\star N \\
= \; & x -_\star e^2 \cdot_\star \langle x -_\star (P +_\star (a -_\star b +_\star c) \cdot_\star N, N\rangle_\star \cdot_\star N.
\end{aligned}
$$

Note that the right-hand side of the last equation is a representation of a multiplicative reflection in a multiplicative line $\delta \in \mathscr{P}_l$ through the point $P +_\star d \cdot_\star N$, where

$$d = a -_\star b +_\star c.$$

This completes the proof. □

Theorem 2.40. *Let $T = \Omega_\alpha \Omega_\beta \in T(l)$. If m and n are arbitrary multiplicative lines that are multiplicative perpendicular to l, then there are unique multiplicative lines m' and n' so that*

$$\Omega_m \Omega_{m'} = \Omega_n \Omega_{n'}.$$

Proof. By Theorem 2.39, it follows that there exists a unique multiplicative line m' so that

$$\Omega_m \Omega_\alpha \Omega_\beta = \Omega_{m'}.$$

Now, using that $\Omega_m \Omega_m = Id$, we get

$$
\begin{aligned}
\Omega_\alpha \Omega_\beta &= \Omega_m \Omega_m \Omega_\alpha \Omega_\beta \\
&= \Omega_m \Omega_{m'}.
\end{aligned}
$$

As above, there exists a unique multiplicative line n' such that

$$\Omega_\alpha \Omega_\beta = \Omega_n \Omega_{n'}.$$

Consequently,

$$\Omega_m \Omega_{m'} = \Omega_n \Omega_{n'}.$$

This completes the proof. □

Now, suppose that Ω_α and Ω_β are multiplicative reflections with corresponding numbers a and b, respectively. For convenience, we use Ω_a instead of Ω_α and Ω_b instead of Ω_β. Let $\lambda, \mu \in \mathbb{R}_\star$. Then

1. $\Omega_a \Omega_b = T_{e^2 \cdot_\star (a -_\star b)}$.
2. $\Omega_a T_\mu = \Omega_{a -_\star (1_\star /_\star e^2) \cdot_\star \mu}$.

   Proof. We have

$$
\begin{aligned}
\Omega_\alpha T_\mu x &= \Omega_\alpha (x +_\star \mu \cdot_\star N) \\
&= x +_\star \mu \cdot_\star N -_\star e^2 \cdot_\star \langle x +_\star \mu \cdot_\star N -_\star P -_\star a \cdot_\star N, N\rangle_\star \cdot_\star N \\
&= x +_\star \mu \cdot_\star N -_\star e^2 \cdot_\star \langle x -_\star P -_\star (a -_\star \mu) \cdot_\star N, N\rangle_\star \cdot_\star N \\
&= x -_\star e^2 \cdot_\star \langle x -_\star P -_\star (a -_\star \mu +_\star 1_\star /_\star e^2 \cdot_\star \mu) \cdot_\star N, N\rangle_\star \cdot_\star N \\
&= \Omega_{a -_\star (1_\star /_\star e^2) \cdot_\star \mu} x, \quad x \in E_\star^2.
\end{aligned}
$$

   This completes the proof. □

3. $T_\lambda \Omega_\beta = \Omega_{b +_\star (1_\star /_\star e^2) \cdot_\star \lambda}$.

Proof. We have

$$\begin{aligned} T_\lambda \Omega_\beta x &= T_\lambda (x -_\star e^2 \cdot_\star \langle x -_\star P -_\star b \cdot_\star N, N\rangle_\star \cdot_\star N) \\ &= x -_\star e^2 \cdot_\star \langle x -_\star P -_\star b \cdot_\star N, N\rangle_\star \cdot_\star N \\ &\quad +_\star \lambda \cdot_\star N \\ &= x -_\star e^2 \cdot_\star \langle x -_\star P -_\star b \cdot_\star N, N\rangle_\star \cdot_\star N \\ &\quad -_\star e^2 \cdot_\star \langle -_\star (1_\star /_\star e^2) \cdot_\star \lambda \cdot_\star N, N\rangle_\star \cdot_\star N \\ &= x -_\star e^2 \cdot_\star \langle x -_\star P -_\star (b +_\star (1_\star /_\star e^2) \cdot_\star \lambda) \cdot_\star N, N\rangle_\star \cdot_\star N \\ &= \Omega_{b +_\star (1_\star /_\star e^2) \cdot_\star \lambda} x, \quad x \in E_\star^2. \end{aligned}$$

This completes the proof. □

4. $T_\lambda T_\mu = T_{\lambda +_\star \mu}$.

Thus, the elements of $\mathscr{S}_\star(\mathscr{P}_l)$ are either multiplicative translations along l or multiplicative reflections in multiplicative lines of $\mathscr{P}_l$.

Definition 2.30. *Let v be any multiplicative vector of $E_\star^2$. Define a multiplicative translation by v as follows*

$$\tau_v x = x +_\star v, \quad x \in E_\star^2.$$

With $\mathscr{T}(E_\star^2)$, we denote the set

$$\mathscr{T}(E_\star^2) = \{\tau_v : v \in E_\star^2\}.$$

Theorem 2.41. *The set $\mathscr{T}(E_\star^2)$ is an Abelian group of $S_\star(E_\star^2)$.*

Proof. Let $v, w \in E_\star^2$ be arbitrarily chosen. Then

$$\begin{aligned} \tau_v x &= x +_\star v, \\ \tau_w x &= x +_\star w, \quad x \in E_\star^2, \end{aligned}$$

and

$$\begin{aligned} \tau_v \tau_w x &= \tau_v (x +_\star w) \\ &= x +_\star v +_\star w \\ &= \tau_{v +_\star w} x, \quad x \in E_\star^2. \end{aligned}$$

Next,

$$\begin{aligned} \tau_{-_\star v} \tau_v x &= \tau_{-_\star v} (x +_\star v) \\ &= x +_\star v -_\star v \\ &= x, \quad x \in E_\star^2, \end{aligned}$$

and

$$\begin{aligned} \tau_v \tau_{-_\star v} x &= \tau_v (x -_\star v) \\ &= x -_\star v +_\star v \\ &= x, \quad x \in E_\star^2. \end{aligned}$$

Moreover,

$$\begin{aligned}\tau_{0_\star}x &= x+_\star 0_\star \\ &= x, \quad x\in E_\star^2.\end{aligned}$$

This completes the proof. □

Note that $\mathscr{T}(E_\star^2)$ is isomorphic to $\mathbb{R}_\star^2$ with multiplicative vector multiplicative addition.

2.12 Multiplicative Rotations

Let $l = P +_\star [v]_\star$ be a multiplicative line with multiplicative unit direction vector v. Then there is a $\theta \in (e^{-\pi}, e^{\pi}]$ so that

$$v = (\cos_\star \theta, \sin_\star \theta).$$

Consider the vector

$$N = (-_\star \sin_\star \theta, \cos_\star \theta).$$

We have

$$\begin{aligned}\langle N, v\rangle_\star &= e^{\log\cos_\star\theta\log(-_\star\sin_\star\theta)+\log\sin_\star\theta\log\cos_\star\theta} \\ &= e^{\log\left(e^{\cos(\log\theta)}\right)\log\left(-_\star e^{\sin(\log\theta)}\right)} \\ &\quad e^{\log\left(e^{\sin(\log\theta)}\right)\log\left(e^{\cos(\log\theta)}\right)} \\ &= e^{\cos(\log\theta)\log\left(e^{-\sin(\log\theta)}\right)+\sin(\log\theta)\cos(\log\theta)} \\ &= e^{-\cos(\log\theta)\sin(\log\theta)+\sin(\log\theta)\cos(\log\theta)} \\ &= e^0 \\ &= 1 \\ &= 0_\star.\end{aligned}$$

Thus, N is a multiplicative unit normal vector to l. For $X \in E_\star^2$, we have

$$\Omega_l X = X -_\star e^2 \cdot_\star \langle X -_\star P, N\rangle_\star \cdot_\star N.$$

Hence,

$$\Omega_l X -_\star P = X -_\star P -_\star e^2 \cdot_\star \langle X -_\star P, N\rangle_\star \cdot_\star N$$

for any $P \in E_\star^2$. Let $l_{O_\star}$ be the multiplicative line through $O_\star = (0_\star, 0_\star)$ with multiplicative direction vector v. Then

$$\Omega_{l_{O_\star}} X = X -_\star e^2 \cdot_\star \langle X, N\rangle_\star \cdot_\star N.$$

Therefore,

$$\Omega_l X -_\star P = \Omega_{l_{O_\star}}(X -_\star P)$$

or

$$\Omega_l X = T_P \Omega_{l_{O_\star}} T_{-_\star P} X.$$

Let $X = \begin{pmatrix} X_1 \\ X_2 \end{pmatrix}$. Then

$$\begin{aligned} \Omega_{l_{O_\star}} \begin{pmatrix} X_1 \\ X_2 \end{pmatrix} &= \begin{pmatrix} X_1 \\ X_2 \end{pmatrix} \\ &\quad -_\star e^2 \cdot_\star (-_\star X_1 \cdot_\star \sin_\star \theta +_\star X_2 \cdot_\star \cos_\star \theta) \cdot_\star \begin{pmatrix} {}_\star \sin_\star \theta \\ \cos_\star \theta \end{pmatrix} \\ &= \begin{pmatrix} X_1 \\ X_2 \end{pmatrix} \\ &\quad -_\star (-_\star e^2 \cdot_\star X_1 \cdot_\star \sin_\star \theta +_\star e^2 \cdot_\star X_2 \cdot_\star \cos_\star \theta) \cdot_\star \begin{pmatrix} -_\star \sin_\star \theta \\ \cos_\star \theta \end{pmatrix} \\ &= \begin{pmatrix} (e -_\star e^2 \cdot_\star (\sin_\star \theta)^{2_\star}) \cdot_\star X_1 +_\star (e^2 \cdot_\star \sin_\star \theta \cdot_\star \cos_\star \theta) \cdot_\star X_2 \\ (e^2 \cdot_\star \sin_\star \theta \cdot_\star \cos_\star \theta) \cdot_\star X_1 +_\star (e -_\star e^2 \cdot_\star (\cos_\star \theta)^{2_\star}) \cdot_\star X_2 \end{pmatrix} \\ &= \begin{pmatrix} \cos_\star (e^2 \cdot \star\theta) & \sin_\star (e^2 \cdot_\star \theta) \\ \sin_\star (e^2 \cdot_\star \theta) & -_\star \cos_\star (e^2 \cdot_\star \theta) \end{pmatrix} \cdot_\star \begin{pmatrix} X_1 \\ X_2 \end{pmatrix}. \end{aligned}$$

Set

$$\mathrm{Ref}_\star \theta = \begin{pmatrix} \cos_\star (e^2 \cdot \star\theta) & \sin_\star (e^2 \cdot_\star \theta) \\ \sin_\star (e^2 \cdot_\star \theta) & -_\star \cos_\star (e^2 \cdot_\star \theta) \end{pmatrix}.$$

Then

$$\Omega_{l_{O_\star}} \begin{pmatrix} X_1 \\ X_2 \end{pmatrix} = \mathrm{Ref}_\star \theta \cdot_\star \begin{pmatrix} X_1 \\ X_2 \end{pmatrix}.$$

Now, we consider another multiplicative line m through P and the associated multiplicative line $m_{O_\star}$. If

$$(\cos_\star \phi, \sin_\star \phi)$$

is a multiplicative direction vector of m, then

$$\begin{aligned} \mathrm{Ref}_\star \theta \cdot_\star \mathrm{Ref}_\star \phi &= \begin{pmatrix} \cos_\star (e^2 \cdot \star\theta) & \sin_\star (e^2 \cdot_\star \theta) \\ \sin_\star (e^2 \cdot_\star \theta) & -_\star \cos_\star (e^2 \cdot_\star \theta) \end{pmatrix} \\ &\quad \cdot_\star \begin{pmatrix} \cos_\star (e^2 \cdot \star\phi) & \sin_\star (e^2 \cdot_\star \phi) \\ \sin_\star (e^2 \cdot_\star \phi) & -_\star \cos_\star (e^2 \cdot_\star \phi) \end{pmatrix} \\ &= \begin{pmatrix} \cos_\star (e^2 \cdot \star(\theta - \star\phi)) & \sin_\star (e^2 \cdot_\star (\theta -_\star \phi)) \\ \sin_\star (e^2 \cdot_\star (\theta -_\star \phi)) & -_\star \cos_\star (e^2 \cdot_\star (\theta -_\star \phi)) \end{pmatrix} \\ &= \mathrm{Ref}_\star (\theta -_\star \phi). \end{aligned} \tag{2.19}$$

Set

$$\mathrm{Rot}_\star \theta = \begin{pmatrix} \cos_\star \theta & -_\star \sin_\star \theta \\ \sin_\star \theta & \cos_\star \theta \end{pmatrix}.$$

Definition 2.31. *If α and β are multiplicative lines passing through a point $P \in E_\star^2$, the multiplicative isometry $\Omega_\alpha \Omega_\beta$ is called a multiplicative rotation about P. If $\alpha \perp_\star \beta$, the rotation $\Omega_\alpha \Omega_\beta$ is called a multiplicative half-turn.*

Let l_1 and l_2 be two multiplicative lines such that $l_1 \perp_\star l_2$, $l_1 \cap l_2 = P$. Let also, N_1 and N_2 be multiplicative unit normal vectors to l_1 and l_2, respectively. Then $N_1 \perp_\star N_2$. Take $X \in E_\star^2$ arbitrarily. Then

$$\Omega_{l_1} X = X -_\star e^2 \cdot_\star \langle X -_\star P, N_1 \rangle_\star \cdot_\star N_1,$$

and

$$\begin{aligned}\Omega_{l_2}\Omega_{l_1}X &= \Omega_{l_1}X -_\star e^2 \cdot_\star \langle \Omega_{l_1}X -_\star P, N_2\rangle_\star \cdot_\star N_2\\ &= X -_\star e^2 \cdot_\star \langle X -_\star P, N_1\rangle_\star \cdot_\star N_1\\ &\quad -_\star e^2 \cdot_\star \langle X -_\star e^2 \cdot_\star \langle X -_\star P, N_1\rangle_\star \cdot \star N_1 -_\star P, N_2\rangle_\star \cdot_\star N_2\\ &= X -_\star e^2 \cdot_\star \langle X -_\star P, N_1\rangle_\star \cdot_\star N_1\\ &\quad -_\star e^2 \cdot_\star \langle X -_\star P, N_2\rangle_\star \cdot_\star N_2\\ &= X -_\star e^2 \cdot_\star (\langle X -_\star P, N_1\rangle_\star \cdot_\star N_1 +_\star \langle X -_\star P, N_2\rangle_\star \cdot_\star N_2).\end{aligned}$$

Let $N_1 = (y_1, y_2)$ and $N_2 = (-_\star y_2, y_1)$, and $|N_1|_\star = 1_\star$. Let also,

$$X -_\star P = (z_1, z_2).$$

Then

$$\begin{aligned}\langle X -_\star P, N_1\rangle_\star &= e^{\log z_1 \log y_1 + \log z_2 \log y_2},\\ \langle X -_\star P, N_2\rangle_\star &= e^{-\log z_1 \log y_2 + \log z_2 \log y_1},\end{aligned}$$

and

$$\begin{aligned}\langle X -_\star P, N_1\rangle_\star \cdot_\star N_1 &= \left(e^{(\log z_1 \log y_1 + \log z_2 \log y_2)\log y_1}, e^{(\log z_1 \log y_1 + \log z_2 \log y_2)\log y_2}\right),\\ \langle X -_\star P, N_2\rangle_\star \cdot_\star N_2 &= \left(e^{-(-\log z_1 \log y_2 + \log z_2 \log y_1)\log y_2}, e^{(-\log z_1 \log y_2 + \log z_2 \log y_1)\log y_1}\right).\end{aligned}$$

Thus,

$$\begin{aligned}&\langle X -_\star P, N_1\rangle_\star \cdot_\star N_1 +_\star \langle X -_\star P, N_2\rangle_\star \cdot_\star N_2\\ &= \left(e^{(\log z_1 \log y_1 + \log z_2 \log y_2)\log y_1}, e^{(\log z_1 \log y_1 + \log z_2 \log y_2)\log y_2}\right)\\ &\quad +_\star \left(e^{-(-\log z_1 \log y_2 + \log z_2 \log y_1)\log y_2}, e^{(-\log z_1 \log y_2 + \log z_2 \log y_1)\log y_1}\right)\\ &= \left(e^{\log z_1((\log y_1)^2 + (\log y_2)^2)}, e^{\log z_2((\log y_1)^2 + (\log y_2)^2)}\right)\\ &= (z_1, z_2) \cdot_\star |N_1|_\star^{2_\star}\\ &= (z_1, z_2)\\ &= X -_\star P.\end{aligned}$$

Hence,

$$\begin{aligned}\Omega_{l_2}\Omega_{l_1}X &= X -_\star e^2 \cdot_\star (\langle X -_\star P, N_1\rangle_\star \cdot_\star N_1 +_\star \langle X -_\star P, N_2\rangle_\star \cdot_\star N_2)\\ &= X -_\star e^2 \cdot_\star (X -_\star P)\\ &= X -_\star e^2 \cdot_\star X +_\star e^2 \cdot_\star P\\ &= -_\star X +_\star e^2 \cdot_\star P.\end{aligned}$$

The multiplicative half-turn about P is denoted by H_P. Consequently, for any $X \in E_\star^2$, we have

$$H_P X = -_\star X +_\star e^2 \cdot_\star P.$$

Example 2.20. *Let $P(2,3)$, $X(e^3, e^4)$. Then*

$$\begin{aligned}e^2 \cdot_\star P &= e^2 \cdot_\star (2,3)\\ &= \left(e^{2\log 2}, e^{2\log 3}\right)\end{aligned}$$

and

$$-_\star X = \left(e^{-3}, e^{-4}\right),$$

and

$$\begin{aligned} H_P X &= \left(e^{-3}, e^{-4}\right) +_\star \left(e^{2\log 2}, e^{2\log 3}\right) \\ &= \left(e^{-3+2\log 2}, e^{-4+2\log 3}\right). \end{aligned}$$

Exercise 2.14. *Let $P(e^{-1}, 4)$, $X(3, e^7)$. Find $H_P X$.*

Answer 2.42.

$$H_P X = \left(\frac{1}{3e^2}, e^{-7+4\log 2}\right).$$

Theorem 2.43. *The set of all multiplicative rotations is an Abelian group.*

Proof. We have

$$\begin{aligned} \mathrm{Rot}_\star 0_\star &= \begin{pmatrix} \cos_\star 0_\star & -_\star \sin_\star 0_\star \\ \sin_\star 0_\star & \cos_\star 0_\star \end{pmatrix} \\ &= \begin{pmatrix} \cos_\star 1 & -_\star \sin_\star 1 \\ \sin_\star 1 & \cos_\star 1 \end{pmatrix} \\ &= \begin{pmatrix} e^{\cos(\log 1)} & e^{-\sin(\log 1)} \\ e^{\sin(\log 1)} & e^{\cos(\log 1)} \end{pmatrix} \\ &= \begin{pmatrix} e^{\cos 0} & e^{\sin 0} \\ e^{\sin 0} & e^{\cos 0} \end{pmatrix} \\ &= \begin{pmatrix} e & 1 \\ 1 & e \end{pmatrix} \\ &= \begin{pmatrix} 1_\star & 0_\star \\ 0_\star & 1_\star \end{pmatrix} \\ &= I_\star. \end{aligned}$$

Next,

$$\begin{aligned} \mathrm{Rot}_\star\theta \cdot_\star \mathrm{Rot}_\star\phi &= \begin{pmatrix} \cos_\star \theta & -_\star \sin_\star \theta \\ \sin_\star \theta & \cos_\star \theta \end{pmatrix} \cdot_\star \begin{pmatrix} \cos_\star \phi & -_\star \sin_\star \phi \\ \sin_\star \phi & \cos_\star \phi \end{pmatrix} \\ &= \begin{pmatrix} \cos_\star(\theta +_\star \phi) & -_\star \sin_\star(\theta +_\star \phi) \\ \sin_\star(\theta +_\star \phi) & \cos_\star(\theta +_\star \phi) \end{pmatrix} \\ &= \mathrm{Rot}_\star(\theta +_\star \phi). \end{aligned}$$

Now, we find $(\mathrm{Rot}_\star\theta)^{-1_\star}$. We have

$$\begin{aligned} \det{}_\star \mathrm{Rot}_\star\theta &= e^{\log(\cos_\star\theta)\log(\cos_\star\theta) -_\star \log(-_\star \sin_\star\theta)\log(\sin_\star\theta)} \\ &= e^{(\cos(\log\theta))^2 + (\sin(\log\theta))^2} \\ &= e \\ &= 1_\star. \end{aligned}$$

Hence,

$$\begin{aligned}(\mathrm{Rot}_\star)^{-1_\star} &= 1_\star \cdot_\star \begin{pmatrix} \cos_\star\theta & \sin_\star\theta \\ \frac{1}{\sin_\star\theta} & \cos_\star\theta \end{pmatrix} \\ &= \begin{pmatrix} \cos_\star\theta & \sin_\star\theta \\ \frac{1}{\sin_\star\theta} & \cos_\star\theta \end{pmatrix} \\ &= \begin{pmatrix} e^{\cos(\log\theta)} & e^{\sin(\log\theta)} \\ e^{-\sin(\log\theta)} & e^{\cos(\log\theta)} \end{pmatrix}\end{aligned}$$

and

$$\begin{aligned}\mathrm{Rot}_\star(-_\star\theta) &= \begin{pmatrix} \cos_\star(-_\star\theta) & -_\star\sin_\star(-_\star\theta) \\ \sin_\star(-_\star\theta) & \cos_\star(-_\star\theta) \end{pmatrix} \\ &= \begin{pmatrix} \cos_\star\left(\frac{1}{\theta}\right) & -_\star\sin_\star\left(\frac{1}{\theta}\right) \\ \\ \sin_\star\left(\frac{1}{\theta}\right) & \cos_\star\left(\frac{1}{\theta}\right) \end{pmatrix} \\ &= \begin{pmatrix} e^{\cos\left(\log\frac{1}{\theta}\right)} & e^{-\sin\left(\log\frac{1}{\theta}\right)} \\ e^{\sin\left(\log\frac{1}{\theta}\right)} & e^{\cos\left(\log\frac{1}{\theta}\right)} \end{pmatrix} \\ &= \begin{pmatrix} e^{\cos(\log\theta)} & e^{\sin(\log\theta)} \\ e^{-\sin(\log\theta)} & e^{\cos(\log\theta)} \end{pmatrix}.\end{aligned}$$

Therefore,

$$(\mathrm{Rot}_\star\theta)^{-1_\star} = \mathrm{Rot}_\star(-_\star\theta).$$

This completes the proof. □

Definition 2.32. *The set of all multiplicative rotations about the multiplicative origin* $O_\star = (0_\star, 0_\star)$ *will be called* $SO_\star\left(e^2\right)$.

By Theorem 2.43, it follows that $SO_\star(e^2)$ is an Abelian group.

Theorem 2.44. *We have*

$$Ref_\star\phi_1 \cdot_\star Ref_\star\phi_2 \cdot_\star Ref_\star\phi_3 = Ref_\star(\phi_1 -_\star \phi_2 +_\star \phi_3).$$

Proof. By (2.19), we have

$$\mathrm{Ref}_\star\phi_2 \cdot_\star \mathrm{Ref}_\star\phi_3 = \mathrm{Ref}_\star(\phi_2 -_\star \phi_3).$$

Again we apply (2.19) and we arrive at

$$\begin{aligned}\mathrm{Ref}_\star\phi_1 \cdot_\star \mathrm{Ref}_\star\phi_2 \cdot_\star \mathrm{Ref}_\star\phi_3 &= \mathrm{Ref}_\star\phi_1 \cdot_\star \mathrm{Ref}_\star(\phi_2 -_\star \phi_3) \\ &= \mathrm{Ref}_\star(\phi_1 -_\star \phi_2 +_\star \phi_3).\end{aligned}$$

This completes the proof. □

Theorem 2.45. *We have*

$$Ref_\star\theta \cdot_\star Rot_\star\phi = Ref_\star\left(\theta -_\star \left(1_\star /_\star e^2\right) \cdot_\star \phi\right).$$

Proof. We have

$$\begin{aligned}
&\cos_\star(e^2 \cdot_\star \theta) \cdot_\star \cos_\star \phi +_\star \sin_\star(e^2 \cdot_\star \theta) \cdot \sin_\star \phi \\
&= \cos_\star\left(e^2 \cdot_\star \theta -_\star \phi\right) \\
&= \cos_\star\left(e^2 \cdot_\star \left(\theta -_\star \left(1_\star /_\star e^2\right) \cdot_\star \phi\right)\right)
\end{aligned}$$

and

$$\begin{aligned}
&\cos_\star(e^2 \cdot_\star \theta) \cdot_\star (-_\star \sin_\star \phi) +_\star \sin_\star(e^2 \cdot_\star \theta) \cdot_\star \cos_\star \phi \\
&= \sin_\star(e^2 \cdot_\star \theta) \cdot_\star \cos_\star \phi -_\star \cos_\star(e^2 \cdot_\star \theta) \cdot_\star \sin_\star \phi \\
&= \sin_\star\left(e^2 \cdot_\star \theta -_\star \phi\right) \\
&= \sin_\star\left(e^2 \cdot_\star \left(\theta -_\star \left(1_\star /_\star e^2\right) \cdot_\star \phi\right)\right),
\end{aligned}$$

and

$$\begin{aligned}
&\sin_\star(e^2 \cdot_\star \theta) \cdot_\star \cos_\star \phi -_\star \sin_\star \phi \cdot_\star \cos_\star(e^2 \cdot_\star \theta) \\
&= \sin_\star\left(e^2 \cdot_\star \theta -_\star \phi\right) \\
&= \sin_\star\left(e^2 \cdot \left(\theta -_\star \left(1_\star /_\star e^2\right) \cdot_\star \phi\right)\right),
\end{aligned}$$

and

$$\begin{aligned}
&-_\star(e^2 \cdot_\star \theta) \cdot_\star \sin_\star \phi -_\star \cos_\star(e^2 \cdot_\star \theta) \cdot_\star \cos_\star \phi \\
&= -_\star\left(\sin_\star(e^2 \cdot_\star \theta) \cdot_\star \sin_\star \phi +_\star \cos_\star(e^2 \cdot_\star \theta) \cdot_\star \cos_\star \phi\right) \\
&= -_\star \cos_\star\left(\theta -_\star \left(1_\star /_\star e^2\right) \cdot_\star \phi\right).
\end{aligned}$$

Thus,

$$\begin{aligned}
\mathrm{Ref}_\star \theta \cdot_\star \mathrm{Rot}_\star \phi &= \begin{pmatrix} \cos_\star\left(e^2 \cdot \star \theta\right) & \sin_\star\left(e^2 \cdot_\star \theta\right) \\ \sin_\star\left(e^2 \cdot_\star \theta\right) & -_\star \cos_\star\left(e^2 \cdot_\star \theta\right) \end{pmatrix} \\
&\quad \cdot_\star \begin{pmatrix} \cos_\star \phi & -_\star \sin_\star \phi \\ \sin_\star \phi & \cos_\star \phi \end{pmatrix} \\
&= \begin{pmatrix} \cos_\star\left(e^2 \cdot_\star \left(\theta -_\star \left(1_\star /_\star e^2\right) \cdot_\star \phi\right)\right) & \sin_\star\left(e^2 \cdot_\star \left(\theta -_\star \left(1_\star /_\star e^2\right) \cdot_\star \phi\right)\right) \\ \sin_\star\left(e^2 \cdot_\star \left(\theta -_\star \left(1_\star /_\star e^2\right) \cdot_\star \phi\right)\right) & -_\star \cos_\star\left(e^2 \cdot_\star \left(\theta -_\star \left(1_\star /_\star e^2\right) \cdot_\star \phi\right)\right) \end{pmatrix} \\
&= \mathrm{Ref}\left(e^2 \cdot_\star \left(\theta -_\star \left(1_\star /_\star e^2\right) \cdot_\star \phi\right)\right).
\end{aligned}$$

This completes the proof. □

Exercise 2.15. *Prove that*

$$Rot_\star \theta \cdot_\star Ref_\star \phi = Ref_\star\left(\phi +_\star \left(1_\star /_\star e^2\right) \cdot_\star \theta\right).$$

Definition 2.33. *The set of all multiplicative rotations about the multiplicative origin $O_\star$ and all multiplicative reflections in multiplicative lines through $O_\star$ is denoted by $O_\star(e^2)$.*

Note that $\mathrm{SO}_\star(e^2)$ is a subgroup of $O_\star(e^2)$. Let $\mathscr{P}(P)$ be a multiplicative pencil of all multiplicative lines through the point P. We denote by $\mathrm{Ref}_\star(\mathscr{P}(P))$ the multiplicative symmetry group of multiplicative isometries containing all Ω_l, where $l \in \mathscr{P}(P)$. With $\mathrm{Rot}_\star(P)$, we denote the set of all multiplicative rotations about P. Then $\tau_P \mathrm{Ref}_\star \theta \tau_{-_\star P}$ is a multiplicative reflection and $\tau_P \mathrm{Rot}_\star \theta \tau_{-_\star P}$ is a multiplicative rotation. Therefore, $O_\star(e^2)$ is isomorphic onto $\mathrm{Ref}_\star(P)$ and $\mathrm{SO}_\star(e^2)$ is isomorphic onto $\mathrm{Rot}_\star(P)$.

Theorem 2.46. *Let α, β and γ be multiplicative lines through a point $P \in E_\star^2$. Then there is a unique multiplicative line δ through P so that*

$$\Omega_\alpha \Omega_\beta \Omega_\gamma = \Omega_\delta.$$

Proof. Let

$$\begin{aligned} \Omega_\alpha &= \tau_P \mathrm{Ref}_\star \theta \tau_{-_\star P}, \\ \Omega_\beta &= \tau_P \mathrm{Ref}_\star \phi \tau_{-_\star P}, \\ \Omega_\gamma &= \tau_P \mathrm{Ref}_\star \psi \tau_{-_\star P}. \end{aligned}$$

Then

$$\begin{aligned} \Omega_\alpha \Omega_\beta \Omega_\gamma &= \tau_P \mathrm{Ref}_\star \theta \tau_{-_\star P} \tau_P \mathrm{Ref}_\star \phi \tau_{-_\star P} \tau_P \mathrm{Ref}_\star \psi \tau_{-_\star P} \\ &= \tau_P \mathrm{Ref}_\star \theta \mathrm{Ref}_\star \phi \mathrm{Ref}_\star \psi \tau_{-_\star P} \\ &= \tau_P \mathrm{Ref}_\star (\theta -_\star \phi +_\star \psi) \tau_{-_\star P}. \end{aligned}$$

Take δ to be the multiplicative line through the point P and a multiplicative direction vector

$$(\cos_\star(\theta -_\star \phi +_\star \psi), \sin_\star(\theta -_\star \phi +_\star \psi)).$$

Hence,

$$\Omega_\alpha \Omega_\beta \Omega_\gamma = \Omega_\delta.$$

This completes the proof. □

Theorem 2.47. *Let $T = \Omega_\alpha \Omega_\beta \in Rot_\star(P)$ and l be any multiplicative line through P. Then there exist unique multiplicative lines m and m' through P so that*

$$T = \Omega_l \Omega_m = \Omega_{m'} \Omega_l.$$

Proof. By Theorem 2.46, it follows that there exists a unique multiplicative line m' through the point P such that

$$\Omega_{m'} = \Omega_\alpha \Omega_\beta \Omega_l.$$

Since $\Omega_l \Omega_l = \mathrm{Id}$, we get

$$\begin{aligned} \Omega_\alpha \Omega_\beta &= \Omega_\alpha \Omega_\beta \Omega_l \Omega_l \\ &= \Omega_{m'} \Omega_l. \end{aligned}$$

By Theorem 2.46, it follows that there exists a unique multiplicative line m through the point P such that

$$\Omega_m = \Omega_l \Omega_\alpha \Omega_\beta.$$

Then

$$\begin{aligned} \Omega_\alpha \Omega_\beta &= \Omega_l \Omega_l \Omega_\alpha \Omega_\beta \\ &= \Omega_l \Omega_m. \end{aligned}$$

This completes the proof. □

2.13 Multiplicative Glide Reflections

Definition 2.34. *A multiplicative glide reflection is defined to be a multiplicative reflection followed by a multiplicative translation along the mirror.*

If $l = P +_\star [v]_\star$, the multiplicative glide reflection is given by

$$\tau_v \Omega_l X = X -_\star e^2 \cdot_\star \langle X -_\star P, N\rangle_\star \cdot_\star N +_\star v.$$

Here N is the multiplicative unit normal vector to l. Note that

$$\begin{aligned} \Omega_l \tau_v X &= X -_\star e^2 \cdot_\star \langle X +_\star v -_\star P, N\rangle_\star \cdot_\star N +_\star v \\ &= X -_\star e^2 \cdot_\star \langle X -_\star P, N\rangle_\star \cdot_\star N +_\star v \end{aligned}$$

because $N \perp_\star v$.

Theorem 2.48. *Let α, β, γ be three distinct multiplicative lines that are not multiplicative concurrent and not all multiplicative parallel. Then $\Omega_\alpha \Omega_\beta \Omega_\gamma$ is a multiplicative glide reflection.*

Proof. Let $\alpha \cap \beta = P$. Let also, l be a multiplicative line through P that is multiplicative perpendicular to γ. Set $F = l \cap \gamma$. There is a unique multiplicative line m through P such that

$$\Omega_\alpha \Omega_\beta = \Omega_m \Omega_l.$$

Hence,

$$\Omega_\alpha \Omega_\beta \Omega_\gamma = \Omega_m \Omega_l \Omega_\gamma.$$

Let n be the multiplicative line through F that is multiplicative perpendicular to m and n' be the multiplicative line through F that is perpendicular to n. Now,

$$\begin{aligned} \Omega_l \Omega_\gamma &= \Omega_{n'} \Omega_n \\ &= H_F. \end{aligned}$$

Hence,

$$\Omega_\alpha \Omega_\beta \Omega_\gamma = \Omega_m \Omega_{n'} \Omega_n.$$

Note that $\Omega_m \Omega_{n'}$ is a multiplicative translation along n. Since $F \in m$, we have that m and n' are distinct. Thus, $\Omega_\alpha \Omega_\beta \Omega_\gamma$ is a multiplicative glide reflection. Let $\alpha \not\cap \beta$. Then $\gamma \cap \beta = P_1$ and as above, $\Omega_\gamma \Omega_\beta \Omega_\alpha$ is a multiplicative glide reflection and

$$\Omega_\gamma \Omega_\beta \Omega_\alpha = \tau_v \Omega_l.$$

Hence,

$$\begin{aligned} \Omega_\alpha \Omega_\beta \Omega_\gamma &= (\Omega_\gamma \Omega_\beta \Omega_\alpha)^{-1} \\ &= (\tau_v \Omega_l)^{-1} \\ &= \Omega_l \tau_{-_\star v} \\ &= \tau_{-_\star v} \Omega_l. \end{aligned}$$

This completes the proof. □

Theorem 2.49. *Let T be a multiplicative glide reflection and let Ω_α be any multiplicative reflection. Then $\Omega_\alpha T$ is a multiplicative translation or a multiplicative rotation.*

Proof. Let l be the multiplicative axis of the multiplicative glide reflection T. We have the following cases.

1. Let $l \cap \alpha = P$. Then there is a multiplicative line through P and a multiplicative line β so that $a \perp_\star l$ and $\beta \perp_\star l$ and
$$T = \Omega_l \Omega_a \Omega_\beta.$$
Hence,
$$\Omega_\alpha T = \Omega_\alpha \Omega_l \Omega_a \Omega_\beta.$$
Now, l, a and α pass through P. Then there is a unique multiplicative line c through P so that
$$\Omega_\alpha \Omega_l \Omega_a = \Omega_c.$$
Hence,
$$\Omega_\alpha T = \Omega_c \Omega_\beta.$$
Thus, $\Omega_\alpha T$ is either a multiplicative translation or a multiplicative rotation.
2. Let $l \parallel_\star \alpha$. Then
$$\begin{aligned}\Omega_\alpha T &= \Omega_\alpha \Omega_l \Omega_a \Omega_\beta \\ &= \Omega_\alpha \Omega_a \Omega_l \Omega_\beta.\end{aligned}$$
Note that $\beta \perp_\star l$ and $a \perp_\star \alpha$. Thus, $\Omega_\alpha \Omega_a$ and $\Omega_l \Omega_\beta$ are distinct multiplicative half-turns. Hence, $\Omega_\alpha T$ is a multiplicative translation. This completes the proof.

□

Definition 2.35. *A multiplicative isometry that is the product of a finite number of multiplicative reflections is called a multiplicative motion.*

Exercise 2.16. *Prove that the group of multiplicative motions consists of all multiplicative translations, multiplicative rotations, multiplicative reflections and multiplicative glide reflections.*

2.14 Structure of the Multiplicative Isometry Group

We start this section with the following useful multiplicative polarization identity.

Lemma 2.2 (The Multiplicative Polarization Identity). *Let* $x = (x_1, x_2)$, $y = (y_1, y_2) \in E_\star^2$. *Then*
$$\langle x, y\rangle_\star = 1_\star /_\star e^2 \cdot_\star \left(|x|_\star^{2_\star} +_\star |y|_\star^{2_\star} -_\star |x -_\star y|_\star^{2_\star}\right). \tag{2.20}$$

Proof. We have
$$\langle x, y\rangle_\star = e^{\log x_1 \log y_1 + \log x_2 \log y_2} \tag{2.21}$$
and
$$\begin{aligned}|x|_\star^{2_\star} &= e^{(\log x_1)^2 + (\log x_2)^2}, \\ |y|_\star^{2_\star} &= e^{(\log y_1)^2 + (\log y_2)^2}, \\ x -_\star y &= \left(\frac{x_1}{y_1}, \frac{x_2}{y_2}\right),\end{aligned}$$

and

$$\begin{aligned}
|x -_\star y|_\star^{2_\star} &= e^{\left(\log \frac{x_1}{y_1}\right)^2 + \left(\log \frac{x_2}{y_2}\right)^2} \\
&= e^{(\log x_1 - \log y_1)^2 + (\log x_2 - \log y_2)^2} \\
&= e^{(\log x_1)^2 + (\log x_2)^2 + (\log y_1)^2 + (\log y_2)^2 - 2\log x_1 \log y_1 - 2\log x_2 \log y_2}
\end{aligned}$$

Hence,

$$\begin{aligned}
&|x|_\star^{2_\star} +_\star |y|_\star^{2_\star} -_\star |x -_\star y|_\star^{2_\star} \\
&= e^{(\log x_1)^2 + (\log x_2)^2} +_\star e^{(\log y_1)^2 + (\log y_2)^2} \\
&\quad -_\star e^{(\log x_1)^2 + (\log x_2)^2 + (\log y_1)^2 + (\log y_2)^2 - 2\log x_1 \log y_1 - 2\log x_2 \log y_2} \\
&= e^{(\log x_1)^2 + (\log x_2)^2 + (\log y_1)^2 + (\log y_2)^2} \\
&\quad -_\star e^{(\log x_1)^2 + (\log x_2)^2 + (\log y_1)^2 + (\log y_2)^2 - 2\log x_1 \log y_1 - 2\log x_2 \log y_2} \\
&= e^{2\log x_1 \log y_1 + 2\log x_2 \log y_2},
\end{aligned}$$

and

$$\begin{aligned}
&1_\star /_\star e^2 \cdot_\star \left(|x|_\star^{2_\star} +_\star |y|_\star^{2_\star} -_\star |x -_\star y|_\star^{2_\star}\right) \\
&= e^{\frac{1}{2}} \cdot_\star e^{2\log x_1 \log y_1 + 2\log x_2 \log y_2} \\
&= e^{\log x_1 \log y_1 + \log x_2 \log y_2}.
\end{aligned}$$

By the last equality and (2.21), we get the equation (2.20). This completes the proof. □

Lemma 2.3. *Let T be a multiplicative isometry with $T(0_\star) = 0_\star$. Then*

$$\langle Tx, Ty\rangle_\star = \langle x, y\rangle_\star$$

for any $x, y \in E_\star^2$.

Proof. Let $x, y \in E_\star^2$. By the multiplicative polarization identity, we have

$$\langle x, y\rangle_\star = 1_\star /_\star e^2 \cdot_\star \left(|x|_\star^{2_\star} +_\star |y|_\star^{2_\star} -_\star |x -_\star y|_\star^{2_\star}\right), \tag{2.22}$$

and

$$\langle Tx, Ty\rangle_\star = 1_\star /_\star e^2 \cdot_\star \left(|Tx|_\star^{2_\star} +_\star |Ty|_\star^{2_\star} -_\star |Tx -_\star Ty|_\star^{2_\star}\right). \tag{2.23}$$

Note that

$$\begin{aligned}
|Tx|_\star &= d_\star(0_\star, Tx) \\
&= d_\star(T(0_\star), Tx) \\
&= d_\star(0_\star, x) \\
&= |x|_\star.
\end{aligned}$$

As above,

$$|Ty|_\star = |y|_\star.$$

Hence and (2.22), (2.23), we get

$$\begin{aligned}
\langle Tx, Ty\rangle_\star &= 1_\star /_\star e^2 \cdot_\star \left(|x|_\star^{2_\star} +_\star |y|_\star^{2_\star} -_\star |x -_\star y|_\star^{2_\star}\right) \\
&= \langle x, y\rangle_\star.
\end{aligned}$$

This completes the proof. □

Theorem 2.50. *If T is a multiplicative isometry with $T(0_\star) = 0_\star$, then*

$$T = Ref_\star\theta \quad or \quad T = Rot_\star(\theta /_\star e^2),$$

for some θ.

Proof. Let

$$e_1 = (1_\star, 0_\star) \quad \text{and} \quad e_2 = (0_\star, 1_\star).$$

Then

$$e_1 = (e, 1) \quad \text{and} \quad e_2 = (1, e).$$

Moreover,

$$\begin{aligned} \langle e_1, e_2 \rangle_\star &= e^{\log e \log 1 + \log 1 \log e} \\ &= e^0 \\ &= 1 \\ &= 0_\star. \end{aligned}$$

Next,

$$\begin{aligned} |e_1|_\star &= e^{\left((\log e)^2 + (\log 1)^2\right)^{\frac{1}{2}}} \\ &= e \\ &= 1_\star, \end{aligned}$$

and as above, $|e_2|_\star = 1_\star$. Thus, $\{e_1, e_2\}$ is a multiplicative orthonormal pair in $E_\star^2$. Since T is a multiplicative isometry, we find

$$\begin{aligned} \langle Te_1, Te_2 \rangle_\star &= \langle e_1, e_2 \rangle_\star \\ &= 0_\star. \end{aligned}$$

Hence, $\{Te_1, Te_2\}$ is a multiplicative orthonormal pair. Let

$$Tx = x_1 \cdot_\star e_1 +_\star x_2 \cdot_\star e_2.$$

Then, we get

$$\begin{aligned} Tx &= \langle Tx, Te_1 \rangle_\star \cdot_\star Te_1 +_\star \langle Tx, Te_2 \rangle_\star \cdot_\star Te_2 \\ &= \langle x, e_1 \rangle_\star \cdot_\star Te_1 +_\star \langle Tx, e_2 \rangle_\star \cdot_\star Te_2 \\ &= x_1 \cdot_\star Te_1 +_\star x_2 \cdot_\star Te_2. \end{aligned}$$

Write

$$Te_1 = \lambda_1 \cdot_\star e_1 +_\star \lambda_2 \cdot_\star e_2.$$

By the multiplicative Cauchy–Schwartz inequality, we find

$$\begin{aligned} |\lambda_1|_\star &= |\langle Te_1, e_1 \rangle_\star|_\star \\ &\leq |Te_1|_\star \cdot_\star |e_1|_\star \\ &= 1_\star. \end{aligned}$$

As above, $|\lambda_2|_\star \leq 1_\star$. Next,

$$\begin{aligned} |Te_1|_\star^{2_\star} &= e^{(\log \lambda_1)^2 + (\log \lambda_2)^2} \\ &= e \\ &= 1_\star. \end{aligned}$$

So,

$$(\log \lambda_1)^2 + (\log \lambda_2)^2 = 1$$

or

$$\lambda_1^{2_\star} +_\star \lambda_2^{2_\star} = 1_\star.$$

Then, there is a $\theta \in (e^{-\pi}, e^{\pi}]$ such that

$$\lambda_1 = \cos_\star \theta, \quad \lambda_2 = \sin_\star \theta.$$

Thus,

$$Te_1 = \cos_\star \theta \cdot_\star e_1 +_\star \sin_\star \theta \cdot_\star e_2.$$

Since $Te_1 \perp_\star Te_2$, we have

$$Te_2 = \pm_\star ((-_\star \sin_\star \theta) \cdot_\star e_1 +_\star \cos_\star \theta \cdot_\star e_2).$$

Therefore,

$$Tx = \mathrm{Rot}_\star \theta \quad \text{or} \quad Tx = \mathrm{Ref}_\star (\theta /_\star e^2).$$

This completes the proof. □

2.15 Fixed Points and Fixed Multiplicative Lines

Theorem 2.51. *A nontrivial multiplicative isometry has no fixed points if and only if it is a nontrivial multiplicative translation or a nontrivial multiplicative reflection.*

Proof. By the previous section, we have that

$$Tx = (\mathrm{Rot}_\star \theta) \cdot_\star x +_\star P$$

or

$$Tx = (\mathrm{Ref}_\star \theta) \cdot_\star x +_\star P$$

for some θ and $P \in E_\star^2$. In the first case, $Tx = x$ if and only if

$$(I_\star -_\star \mathrm{Rot}_\star \theta) \cdot_\star x = P. \tag{2.24}$$

Note that

$$I_\star -_\star \mathrm{Rot}_\star \theta = \begin{pmatrix} 1_\star -_\star \cos_\star \theta & -_\star \sin_\star \theta \\ \sin_\star \theta & 1_\star -_\star \cos_\star \theta \end{pmatrix}$$

and

$$\begin{aligned} \det{}_\star (I_\star -_\star \mathrm{Rot}_\star \theta) &= (1_\star -_\star \cos_\star \theta) \cdot_\star (1_\star -_\star \cos_\star \theta) +_\star \sin_\star \theta \cdot_\star \sin_\star \theta \\ &= (1_\star -_\star \cos_\star \theta)^{2_\star} +_\star (\sin_\star \theta)^{2_\star} \\ &= \left(e -_\star e^{\cos(\log \theta)}\right)^{2_\star} +_\star \left(e^{\sin(\log \theta)}\right)^{2_\star} \\ &= \left(e^{1-\cos(\log \theta)}\right)^{2_\star} +_\star \left(e^{\sin(\log \theta)}\right)^{2_\star} \\ &= e^{(1-\cos(\log \theta))^2} +_\star e^{(\sin(\log \theta))^2} \\ &= e^{(1-\cos(\log \theta))^2 + (\sin(\log \theta))^2} \\ &= e^{1-2\cos(\log \theta) + (\cos(\log \theta))^2 + (\sin(\log \theta))^2} \\ &= e^{2(1-\cos(\log \theta))} \\ &= e^{4\left(\sin\left(\frac{\log \theta}{2}\right)\right)^2}. \end{aligned}$$

Note that

$$e^{4\left(\sin\left(\frac{\log \theta}{2}\right)\right)^2} = e^0$$

if and only if

$$\sin\left(\frac{\log\theta}{2}\right) = 0.$$

Hence,

$$\cos\left(\frac{\log\theta}{2}\right) = 1.$$

Then

$$\cos(\log\theta) = 1 \quad \text{and} \quad \sin(\log\theta) = 0.$$

Consequently,

$$\begin{aligned} \mathrm{Rot}_\star\theta &= \begin{pmatrix} 1_\star & 0_\star \\ 0_\star & 1_\star \end{pmatrix} \\ &= I_\star. \end{aligned}$$

Therefore, unless $\mathrm{Rot}_\star\theta = I_\star$, the equation (2.24) has a solution. Conversely, any nontrivial multiplicative translation has no fixed points. Let now,

$$Tx = (\mathrm{Ref}_\star\theta)\cdot_\star x +_\star P.$$

Observe that T is a multiplicative product of three reflections and from here, T is a multiplicative glide reflection. Conversely, a nontrivial multiplicative glide reflection has no fixed points. This completes the proof. □

Theorem 2.52. *A nontrivial multiplicative isometry has exactly one fixed point if and only if it is a nontrivial multiplicative rotation.*

Proof. Let T be a nontrivial multiplicative isometry that has exactly one fixed point. By Lemma 2.50, it is a nontrivial multiplicative rotation or a nontrivial multiplicative reflection. By Theorem 2.31, it follows that T must be a nontrivial multiplicative rotation. Let now, T be a nontrivial multiplicative rotation. Then

$$TX = P +_\star (\mathrm{Rot}_\star\theta)\cdot_\star (X -_\star P),$$

and hence, $X = TX$ if and only if

$$X = P +_\star (\mathrm{Rot}_\star\theta)\cdot_\star (X -_\star P),$$

if and only if

$$X -_\star P = (\mathrm{Rot}_\star\theta)\cdot_\star (X -_\star P),$$

if and only if $X = P$ or $\mathrm{Rot}_\star = I_\star$. This completes the proof. □

Remark 2.3. *By Theorem 2.31, it follows that a nontrivial multiplicative reflection has a multiplicative line of fixed points.*

Remark 2.4. *The multiplicative identity has a multiplicative plane of fixed points.*

2.16 Advanced Practical Problems

Problem 2.1. *Let* $x = (3,7)$, $y = (4,1) \in \mathbb{R}^2_\star$. *Find*

1. $x +_\star y$.

2. $x -_\star y$.
3. $3 \cdot_\star (x -_\star y)$.

Answer 2.53.

1. $(12,7)$.
2. $(\frac{3}{4},7)$.
3. $\left(e^{\log 3 \log \frac{3}{4}}, e^{\log 3 \log 7}\right)$.

Problem 2.2. *Let*

$$x = (2,3), \quad y = (4,5).$$

Find

1. $x +_\star y$.
2. $x -_\star y$.
3. $\langle x, y \rangle_\star$.

Answer 2.54.

1. $(8,15)$.
2. $(\frac{1}{2},\frac{3}{5})$.
3. $e^{2(\log 2)^2 + \log 3 \log 5}$.

Problem 2.3. *Let $P = (1,4)$, $Q = (3,8)$ and $R = (2,9)$. Find*

1. $d_\star(P,Q)$.
2. $d_\star(P,R)$.
3. $d_\star(R,Q)$.

Answer 2.55.

1. $e^{((\log 3)^2 + (\log 2)^2)^{\frac{1}{2}}}$.
2. $e^{\left((\log 2)^2 + \left(\log \frac{9}{4}\right)^2\right)^{\frac{1}{2}}}$.
3. $e^{\left(\left(\log \frac{3}{2}\right)^2 + \left(\log \frac{8}{9}\right)^2\right)^{\frac{1}{2}}}$.

Problem 2.4. *Find the equation of the multiplicative line through $P(7,6)$ and multiplicative direction $v = (8,5)$.*

Answer 2.56.

$$\left(\frac{X_1}{7}\right)^{\log 5} = \left(\frac{X_2}{6}\right)^{\log 8}.$$

Problem 2.5. *Find the equation of the multiplicative line l through $P(1,4)$ with multiplicative normal vector $N = (7,2)$.*

Answer 2.57.

$$X_1^{\log 7} \left(\frac{X_2}{4}\right)^{\log 2} = 1.$$

Problem 2.6. *Let* $P(6,7)$, $Q(8,9)$, $R(2,e)$, $S(1,4)$, $T(2,11)$. *Let also,* l_1, l_2, l_3, l_4, l_5 *and* l_6 *be multiplicative lines so that*

$$P,Q \in l_1; \quad P,R \in l_2; \quad P,S \in l_3; \quad P,T \in l_4; \quad R,S \in l_5; \quad S,T \in l_6.$$

Find the equations of the multiplicative lines l_1, l_2, l_3, l_4, l_5 *and* l_6.

Answer 2.58.

1. $$l_1: \quad \left(\frac{X_1}{6}\right)^{\log\frac{9}{7}} = \left(\frac{X_2}{7}\right)^{\log\frac{4}{3}}.$$

2. $$l_2: \quad \left(\frac{X_1}{6}\right)^{1-\log 7} = \left(\frac{X_2}{7}\right)^{-\log 3}.$$

3. $$l_3: \quad \left(\frac{X_1}{6}\right)^{\log\frac{4}{7}} = \left(\frac{X_2}{7}\right)^{-\log 6}.$$

4. $$l_4: \quad \left(\frac{X_1}{6}\right)^{\log\frac{11}{7}} = \left(\frac{X_2}{7}\right)^{-\log 3}.$$

5. $$l_5: \quad \left(\frac{X_1}{2}\right)^{2\log 2-1} = \left(\frac{X_2}{e}\right)^{-\log 2}.$$

6. $$l_6: \quad (X_1)^{\log\frac{11}{4}} = \left(\frac{X_2}{4}\right)^{\log 2}.$$

Problem 2.7. *Let the equation of the multiplicative line* l *be*

$$X_1^7\left(\frac{X_2}{5}\right)^8 = 1.$$

Find a point $P(p_1,p_2) \in l$, *a multiplicative normal vector and a multiplicative direction vector for the multiplicative line* l.

Answer 2.59.

1. $$P(1,5).$$

2. $$N = \left(e^7, e^8\right).$$

3. $$v = \left(e^{-8}, e^7\right).$$

Problem 2.8. *Let*

$$\begin{aligned} l: \left(\frac{X_1}{13}\right)^{\log 4} &= X_2^{\log 8}, \\ m: \left(\frac{X_2}{5}\right)^{\log 8} &= X_1^{\log\frac{3}{4}}. \end{aligned}$$

Check if $l \perp_\star m$ *and find the point* $F = l \cap m$.

Answer 2.60.

$$F\left(13^{\frac{3\log 2}{3\log 2-\frac{3}{2}\log\frac{3}{4}}}5^{\frac{\frac{9}{2}\log 2}{3\log 2-\frac{3}{2}\log\frac{3}{4}}},13^{\frac{\log\frac{3}{4}}{3\log 2-\frac{3}{2}\log\frac{3}{4}}}5^{\frac{3\log 2\log\frac{3}{4}}{3\log 2-\frac{3}{2}\log\frac{3}{4}}}\right).$$

Problem 2.9. *Check if the multiplicative lines*

$$\begin{aligned} l : X_1^{\log 18} &= \left(\frac{X_2}{3}\right)^{\log 11}, \\ m : X_1^{\log 4} &= X_2^{\log 15} \end{aligned}$$

are perpendicular.

Answer 2.61. *No.*

Problem 2.10. *Let $P(7,3)$ and*

$$l : \left(\frac{X_1}{7}\right)^{\log 2}\left(\frac{X_2}{3}\right)^{\log 5} = 1.$$

Find a multiplicative line m so that $P \in m$ and $m \perp_\star l$.

Answer 2.62.

$$\left(\frac{X_1}{7}\right)^{\log 5} = \left(\frac{X_2}{3}\right)^{\log 2}.$$

Problem 2.11. *Let*

$$\begin{aligned} l_1 : X_1X_2^4 &= 1, \\ l_2 : \left(\frac{X_1}{2}\right)^{\log 2} X_2^3 &= 1, \\ l_3 : \left(\frac{X_1}{3}\right)^{\log 4}\left(\frac{X_2}{2}\right)^{\log\frac{1}{3}} &= 1. \end{aligned}$$

Check if $l_1 \perp_\star l_2$, $l_2 \perp_\star l_3$, $l_1 \perp_\star l_3$.

Answer 2.63. *No.*

Problem 2.12. *Let $P(6,9)$, $N = \left(e^{\frac{1}{2}}, e^{\frac{\sqrt{3}}{2}}\right)$.*

1. *Find the equation of the multiplicative line l through P and with multiplicative unit normal vector N.*
2. *Let $X(2,11)$. Find $\Omega_l X$.*

Answer 2.64.

1.

$$\left(\frac{X_1}{6}\right)^{\frac{1}{2}}\left(\frac{X_2}{9}\right)^{\frac{\sqrt{3}}{2}} = 1.$$

2.

$$\left(2e^{-\frac{1}{2}\log 3-\frac{\sqrt{3}}{2}\log\frac{9}{11}}, 11e^{-\frac{\sqrt{3}}{2}\log 3-\frac{3}{2}\log\frac{9}{11}}\right).$$

3

Multiplicative Affine Transformations in the Multiplicative Euclidean Plane

In this chapter, multiplicative affine transformations, multiplicative affine reflections, multiplicative affine symmetries, multiplicative shears, multiplicative dilatations and multiplicative similarities are defined and some of their properties are deduced. Multiplicative segments, multiplicative angles and multiplicative rectilinear figures are defined. Some criteria for existence of multiplicative affine transformations that leave multiplicative lines and multiplicative points fixed are obtained. A multiplicative barycentric coordinate system is introduced and some of its applications are given. In the chapter, some criteria for congruence of multiplicative angles and triangles are deduced.

3.1 Multiplicative Affine Transformations

Let $\{e_1, e_2\}$ be a multiplicative orthonormal pair in $\mathbb{R}^2_\star$.

Definition 3.1. *A multiplicative collineation is a bijection $T : E^2_\star \to E^2_\star$ that satisfies the condition: all distinct points P, Q and R are multiplicative collinear if and only if TP, TQ and TR are multiplicative collinear.*

Definition 3.2. *A mapping $T : E^2_\star \to E^2_\star$ is called a multiplicative affine transformation if there is a multiplicative invertible 2×2 matrix A and a multiplicative vector $b \in \mathbb{R}^2_\star$ such that*

$$Tx = A \cdot_\star x +_\star b,$$

for all $x \in \mathbb{R}^2_\star$.

Remark 3.1. *Note that*

$$T0_\star = b,$$

and the columns of the multiplicative matrix A are the multiplicative vectors

$$Te_j -_\star b, \quad j = 1, 2.$$

Thus, the multiplicative matrix A and the multiplicative vector b are uniquely determined by T.

Definition 3.3. *We call the multiplicative matrix A the multiplicative linear part of T and the multiplicative vector b the multiplicative translation part of T.*

Theorem 3.1. *Every multiplicative affine transformation is a multiplicative collineation.*

Proof. Let

$$Tx = A \cdot_\star x +_\star b,$$

DOI: 10.1201/9781003325284-3

where $A \in \mathscr{M}_{\star 2\times 2}$, $\det_\star A \neq 0_\star$, $b \in \mathbb{R}^2_\star$. Note that

$$\begin{aligned} A \cdot_\star A^{-1_\star} \cdot_\star (x -_\star b) +_\star b &= x -_\star b +_\star b \\ &= x, \quad x \in \mathbb{R}^2_\star. \end{aligned}$$

Consequently, any multiplicative affine transformation is surjective. Let now, $x, y \in E^2_\star$ and $Tx = Ty$. Then

$$A \cdot_\star x +_\star b = A \cdot_\star y +_\star b,$$

whereupon

$$A \cdot_\star x = A \cdot_\star y.$$

Since $\det_\star A \neq 0_\star$, we conclude that $x = y$. Thus, any multiplicative affine transformation is injective and from here, any multiplicative affine transformation is bijective. Let $R, P, Q \in E^2_\star$ be three distinct points that are multiplicative collinear. Hence, there is a $\lambda \in \mathbb{R}_\star$ so that

$$R = (1_\star -_\star \lambda) \cdot_\star P +_\star \lambda \cdot_\star Q.$$

We have

$$\begin{aligned} TR &= T((1_\star -_\star \lambda) \cdot_\star P +_\star \lambda \cdot_\star Q) \\ &= (1_\star -_\star \lambda) \cdot_\star TP +_\star \lambda \cdot_\star TQ, \end{aligned}$$

and TR, TP, TQ are multiplicative collinear. Let now, $S_1 \in E^2_\star$ be multiplicative collinear with TP and TQ. Since T is surjective, there is a point $S \in E^2_\star$ so that $TS = S_1$. Because S_1, TP and TQ are multiplicative collinear, there is a $\mu \in \mathbb{R}_\star$ so that

$$S_1 = (1_\star -_\star \mu) \cdot_\star TP +_\star \mu \cdot_\star TQ.$$

Hence,

$$TS = (1_\star -_\star \mu) \cdot_\star TP +_\star \mu \cdot_\star TQ.$$

Because T is injective, we get

$$S = (1_\star -_\star \mu) \cdot_\star P +_\star \mu \cdot_\star Q,$$

i.e., S, P and Q are multiplicative collinear. This completes the proof. □

Corollary 3.1. *Any multiplicative isometry is a multiplicative affine transformation.*

Lemma 3.1. *Let f be a multiplicative collineation with $f(0_\star) = 0_\star$. If v and w are multiplicative nonproportional vectors, then*

$$f(v +_\star w) = f(v) +_\star f(w).$$

Proof. Note that $v +_\star w$ is the intersection of the multiplicative lines

$$l = v +_\star [w]_\star,$$

and

$$m = w +_\star [v]_\star.$$

Then $f(v +_\star w)$ is the intersection of the multiplicative lines $f(l)$ and $f(m)$. Next, $f(l)$ passes through $f(v)$ and it is multiplicative parallel to

$$f([w]_\star) = [f(w)]_\star.$$

Also, $f(m)$ passes through $f(w)$ and it is multiplicative parallel to

$$f([v]_\star) = [f(v)]_\star.$$

Since $f(v)+_\star f(w)$ satisfies these both conditions, it must be the unique point of intersection of $f(l)$ and $f(m)$. Therefore,

$$f(v+_\star w)=f(v)+_\star f(w).$$

This completes the proof. □

Lemma 3.2. *Let $\phi:\mathbb{R}_\star\to\mathbb{R}_\star$ be a bijection that satisfies the conditions*

$$\begin{aligned}\phi(s+_\star t) &= \phi(s)+_\star\phi(t),\\ \phi(s\cdot_\star t) &= \phi(s)\cdot_\star\phi(t),\end{aligned}$$

for any $s,t\in\mathbb{R}_\star$. Then ϕ is identity.

Proof. By the first condition, we get

$$\begin{aligned}\phi(0_\star) &= \phi(0_\star)+_\star\phi(0_\star)\\ &= e^2\cdot_\star\phi(0_\star),\end{aligned}$$

whereupon

$$\phi(0_\star)=0_\star.$$

Next, by the second given condition, we find

$$\begin{aligned}\phi(1_\star) &= \phi(e)\\ &= \phi(e\cdot_\star e)\\ &= \phi(e)\cdot_\star\phi(e)\\ &= (\phi(e))^{2_\star}\\ &= (\phi(1_\star))^{2_\star},\end{aligned}$$

whereupon

$$\phi(1_\star)=1_\star$$

or

$$\phi(e)=e.$$

Assume that

$$\phi\left(e^n\right)=e^n, \tag{3.1}$$

for some $n\in\mathbb{N}$. Then

$$e^{n+1}=e^n\cdot_\star e$$

and

$$\begin{aligned}\phi\left(e^{n+1}\right) &= \phi\left(e^n\cdot_\star e\right)\\ &= \phi\left(e^n\right)\cdot_\star\phi(e)\\ &= e^n\cdot_\star e\\ &= e^{n+1}.\end{aligned}$$

Therefore, (3.1) holds for any $n\in\mathbb{N}$. Next, by the first given condition, we find

$$\begin{aligned}0_\star &= \phi(0_\star)\\ &= \phi(1_\star-_\star 1_\star)\\ &= \phi(1_\star)+_\star\phi(-_\star 1_\star)\\ &= e+_\star\phi(-_\star 1_\star)\\ &= e+_\star\phi\left(e^{-1}\right),\end{aligned}$$

and

$$\begin{aligned}\phi\left(e^{-1}\right) &= 0_\star -_\star e\\ &= 1 -_\star e\\ &= e^{-1}.\end{aligned}$$

Suppose that

$$\phi\left(e^{-n}\right) = e^{-n}, \tag{3.2}$$

for some $n \in \mathbb{N}$. Then

$$\begin{aligned}\phi\left(e^{-(n+1)}\right) &= \phi\left(e^{-n} \cdot_\star e^{-1}\right)\\ &= \phi\left(e^{-n}\right) \cdot_\star \phi\left(e^{-1}\right)\\ &= e^{-n} \cdot_\star e^{-1}\\ &= e^{-(n+1)}.\end{aligned}$$

Consequently (3.2) holds for any $n \in \mathbb{N}$. Let now, $m, n \in \mathbb{N}$ and

$$q_1 = \frac{m}{n}, \quad q_2 = -\frac{m}{n}.$$

Then

$$m = nq_1 \quad \text{and} \quad m = -nq_2.$$

Hence,

$$\begin{aligned}\phi\left(e^{m}\right) &= \phi\left(e^{nq_1}\right)\\ &= e^{nq_1}\\ &= e^{m},\end{aligned}$$

and

$$\begin{aligned}\phi\left(e^{m}\right) &= \phi\left(e^{-nq_2}\right)\\ &= e^{-nq_2}\\ &= e^{m}.\end{aligned}$$

Thus, ϕ is identity on

$$\left\{e^{\frac{m}{n}} : m, n \in \mathbb{Z}, \quad n \neq 0_\star\right\}.$$

Let now, $a > 0_\star$. Then $a = e^{b^2}$, $b \in \mathbb{R}$, $b \neq 0$. Hence,

$$\begin{aligned}\phi(a) &= \phi\left(e^{b^2}\right)\\ &= e^{b^2}\\ &> e^{0}\\ &= 0_\star.\end{aligned}$$

Now, take $t > s$, $t, s \in \mathbb{R}_\star$. Then

$$t -_\star s > 0_\star$$

and

$$\begin{aligned}0_\star &< \phi(t -_\star s)\\ &= \phi(t) -_\star \phi(s),\end{aligned}$$

i.e.,

$$\phi(t) > \phi(s).$$

Consequently, ϕ preserves the order in $\mathbb{R}_\star$. Let now, $t \in \mathbb{R}_\star$ be such that

$$\phi(t) > t.$$

Then there is a $q = e^{\frac{m}{n}}$, $m, n \in \mathbb{Z}$, $n \neq 0$, with

$$\phi(t) > q > t.$$

From here,

$$\begin{aligned} \phi(t) &< \phi(q) \\ &= \phi\left(e^{\frac{m}{n}}\right) \\ &= e^{\frac{m}{n}} \\ &= q. \end{aligned}$$

This is a contradiction. Let $t_1 \in \mathbb{R}_\star$ be such that

$$\phi(t_1) < t_1.$$

Then there is a $q_1 = e^{\frac{m_1}{n_1}}$, $m_1, n_1 \in \mathbb{Z}$, $n_1 \neq 0$, with

$$\phi(t_1) < q_1 < t_1.$$

From here,

$$\begin{aligned} \phi(t_1) &> \phi(q_1) \\ &= \phi\left(e^{\frac{m_1}{n_1}}\right) \\ &= e^{\frac{m_1}{n_1}} \\ &= q_1. \end{aligned}$$

This is a contradiction. Therefore, ϕ is identity on $\mathbb{R}_\star$. This completes the proof. □

Theorem 3.2. *Let f be a multiplicative collineation so that $f(0_\star) = 0_\star$, $f(e_1) = e_1$, $f(e_2) = e_2$. Then f is identity.*

Proof. Since

$$f(0_\star) = 0_\star \quad \text{and} \quad f(e_1) = e_1,$$

there is a function ϕ so that

$$f(t \cdot_\star e_1) = \phi(t) \cdot_\star e_1, \quad t \in \mathbb{R}_\star.$$

Because

$$f(0_\star) = 0_\star \quad \text{and} \quad f(e_2) = e_2,$$

there exists a function ψ such that

$$f(s \cdot_\star e_2) = \psi(s) \cdot_\star e_2, \quad s \in \mathbb{R}_\star.$$

Since the multiplicative line $x_1 = x_2$ is fixed, we have that $\phi = \psi$. Hence,

$$\begin{aligned} f(t \cdot_\star e_1 +_\star s \cdot_\star e_2) &= f(t \cdot_\star e_1) +_\star f(s \cdot_\star e_2) \\ &= \phi(t) \cdot_\star e_1 +_\star \phi(s) \cdot_\star e_2. \end{aligned}$$

In particular, for $m \in \mathbb{R}_\star$, we have that

$$\begin{aligned} f(e_1 +_\star m \cdot_\star e_2) &= f(e_1) +_\star f(m \cdot_\star e_2) \\ &= e_1 +_\star \phi(m) \cdot_\star e_2, \end{aligned}$$

and

$$\begin{aligned} f(x \cdot_\star e_1 +_\star x \cdot_\star m \cdot_\star e_2) &= f(x \cdot_\star e_1) +_\star f(x \cdot_\star m \cdot_\star e_2) \\ &= \phi(x) \cdot_\star e_1 +_\star \phi(x \cdot_\star m) \cdot_\star e_2. \end{aligned}$$

Since

$$(1_\star, m), \quad (x, x \cdot_\star m), \quad (0_\star, 0_\star)$$

are multiplicative collinear, so are

$$(\phi(1_\star), \phi(m)), \quad (\phi(x), \phi(x \cdot_\star m)), \quad (\phi(0_\star), \phi(0)\star)) = (0_\star, 0_\star).$$

Thus,

$$\phi(x \cdot_\star m) = \phi(x) \cdot_\star \phi(m),$$

and in particular,

$$\begin{aligned} (\phi(-_\star 1_\star))^{2_\star} &= \phi(1_\star) \\ &= 1_\star. \end{aligned}$$

So,

$$\phi(-_\star 1_\star) = -_\star 1_\star.$$

Note that

$$f((t +_\star s) \cdot_\star e_1) = \phi(t +_\star s) \cdot_\star e_1,$$

and if $t \neq -_\star s$, we get

$$\begin{aligned} f((t +_\star s) \cdot_\star e_1) &= f(t \cdot_\star e_1 +_\star s \cdot_\star e_2 +_\star s \cdot_\star e_1 -_\star s \cdot_\star e_2) \\ &= f(t \cdot_\star e_1 +_\star s \cdot_\star e_2) +_\star f(t \cdot_\star e_1 -_\star s \cdot_\star e_2) \\ &= f(t \cdot_\star e_1) +_\star f(s \cdot_\star e_2) +_\star f(s \cdot_\star e_1) +_\star f(-_\star s \cdot_\star e_2) \\ &= \phi(t) \cdot_\star e_1 +_\star \phi(s) \cdot_\star e_2 +_\star \phi(s) \cdot_\star e_1 +_\star \phi(-_\star s) \cdot_\star e_2 \\ &= \phi(t) \cdot_\star e_1 +_\star \phi(s) \cdot_\star e_2. \end{aligned}$$

Therefore,

$$\phi(t +_\star s) = \phi(t) +_\star \phi(s).$$

Hence and Lemma 3.1, it follows that ϕ is identity. Thus, f is identity. This completes the proof. □

Corollary 3.2. *Any multiplicative collineation of $E_\star^2$ is a multiplicative affine transformation.*

3.2 Fixed Multiplicative Lines

Firstly, we will note that if T is a multiplicative affine transformation and l is a multiplicative line, then Tl is a multiplicative line.

Theorem 3.3. *Let $T: E^2_\star \to E^2_\star$ be a multiplicative affine transformation given by*

$$Tx = A \cdot_\star x +_\star b, \quad A \in \mathscr{M}_{\star 2\times 2}, \quad \det_\star A \neq 0_\star, \quad x, b \in \mathbb{R}^2_\star.$$

Let also, $l = P +_\star [v]_\star$ be a multiplicative line. Then Tl is the multiplicative line given by $TP +_\star [A \cdot_\star v]_\star$.

Proof. For any $\lambda \in \mathbb{R}_\star$, we have

$$\begin{aligned} T(P +_\star \lambda \cdot_\star v) &= A \cdot_\star (P +_\star \lambda \cdot_\star v) +_\star b \\ &= A \cdot_\star P +_\star b +_\star \lambda \cdot_\star A \cdot_\star v \\ &= TP +_\star \lambda \cdot_\star A \cdot_\star v. \end{aligned}$$

Thus, any point of Tl lies on $TP +_\star [A \cdot_\star v]_\star$ and conversely. This completes the proof. □

Corollary 3.3. *Let $T: E^2_\star \to E^2_\star$ be a multiplicative affine transformation given by*

$$Tx = A \cdot_\star x +_\star b, \quad A \in \mathscr{M}_{\star 2\times 2}, \quad \det_\star A \neq 0_\star, \quad x, b \in \mathbb{R}^2_\star.$$

Then a multiplicative line $l = P +_\star [v]_\star$ is a fixed line, i.e.,

$$T(P +_\star [v]_\star) = P +_\star [v]_\star,$$

if and only if v is a multiplicative eigenvector, i.e., there is a $\mu \in \mathbb{R}_\star$ so that

$$A \cdot_\star v = \mu \cdot_\star v,$$

and

$$(A -_\star I_\star) \cdot_\star P +_\star b \in [v]_\star.$$

Note that the set of all multiplicative affine transformations is a group and it will be denoted by $AF_\star(e^2)$.

Let

$$A = \begin{pmatrix} a_{11} & a_{12} \\ a_{21} & a_{22} \end{pmatrix}, \quad b = \begin{pmatrix} b_1 \\ b_2 \end{pmatrix}, \quad x = \begin{pmatrix} x_1 \\ x_2 \end{pmatrix}, \quad y = \begin{pmatrix} y_1 \\ y_2 \end{pmatrix}.$$

Then

$$\begin{aligned} y &= A \cdot_\star x +_\star b \\ &= \begin{pmatrix} a_{11} & a_{12} \\ a_{21} & a_{22} \end{pmatrix} \cdot_\star \begin{pmatrix} x_1 \\ x_2 \end{pmatrix} +_\star \begin{pmatrix} b_1 \\ b_2 \end{pmatrix} \\ &= \begin{pmatrix} e^{\log a_{11} \log x_1 + \log a_{12} \log x_2} \\ e^{\log a_{21} \log x_1 + \log a_{22} \log x_2} \end{pmatrix} +_\star \begin{pmatrix} b_1 \\ b_2 \end{pmatrix} \\ &= \begin{pmatrix} b_1 e^{\log a_{11} \log x_1 + \log a_{12} \log x_2} \\ b_2 e^{\log a_{21} \log x_1 + \log a_{22} \log x_2} \end{pmatrix}. \end{aligned}$$

Observe that

$$\begin{aligned} &\begin{pmatrix} a_{11} & a_{12} & b_1 \\ a_{21} & a_{22} & b_2 \\ 0_\star & 0_\star & 1_\star \end{pmatrix} \cdot_\star \begin{pmatrix} x_1 \\ x_2 \\ 1_\star \end{pmatrix} \\ &= \begin{pmatrix} a_{11} & a_{12} & b_1 \\ a_{21} & a_{22} & b_2 \\ 1 & 1 & e \end{pmatrix} \cdot_\star \begin{pmatrix} x_1 \\ x_2 \\ e \end{pmatrix} \end{aligned}$$

$$
\begin{aligned}
&= \begin{pmatrix} e^{\log a_{11}\log x_1+\log a_{12}\log x_2+\log b_1 \log e} \\ e^{\log a_{21}\log x_1+\log a_{22}\log x_2+\log b_2 \log e} \\ e^{\log 1 \log x_1+\log 1 \log x_2+(\log e)^2} \end{pmatrix} \\
&= \begin{pmatrix} b_1 e^{\log a_{11}\log x_1+\log a_{12}\log x_2} \\ b_2 e^{\log a_{21}\log x_1+\log a_{22}\log x_2} \\ e \end{pmatrix} \\
&= \begin{pmatrix} b_1 e^{\log a_{11}\log x_1+\log a_{12}\log x_2} \\ b_2 e^{\log a_{21}\log x_1+\log a_{22}\log x_2} \\ 1_\star \end{pmatrix}.
\end{aligned}
$$

Thus,

$$
\begin{pmatrix} y_1 \\ y_2 \\ 1_\star \end{pmatrix} = \begin{pmatrix} a_{11} & a_{12} & b_1 \\ a_{21} & a_{22} & b_2 \\ 0_\star & 0_\star & 1_\star \end{pmatrix} \cdot_\star \begin{pmatrix} x_1 \\ x_2 \\ 1_\star \end{pmatrix}.
$$

3.3 The Fundamental Theorem of the Multiplicative Affine Geometry

The multiplicative affine geometry consists of those facts about $E^2_\star$ that depend only on incidence properties and not on multiplicative perpendicularity or multiplicative distance. The fundamental theorem gives a clear and simple criterion for existence and uniqueness of multiplicative affine transformations, namely, that any two multiplicative triangles can be related by a unique multiplicative affine transformation.

Theorem 3.4. *Let P and Q be points, and T be a multiplicative affine transformation. Then for any $\lambda \in \mathbb{R}_\star$, we have*

$$
T((1_\star -_\star \lambda)\cdot_\star P +_\star \lambda \cdot_\star Q) = (1_\star -_\star \lambda)\cdot_\star TP +_\star \lambda \cdot_\star TQ.
$$

Proof. Suppose that

$$
Tx = A\cdot_\star x +_\star b, \quad x, b \in \mathbb{R}^2_\star, \quad \det_\star A \neq 0_\star.
$$

Then

$$
\begin{aligned}
T((1_\star -_\star \lambda)\cdot_\star P +_\star \lambda \cdot_\star Q) &= A\cdot_\star ((1_\star -_\star \lambda)\cdot_\star P +_\star \lambda \cdot_\star Q) +_\star b \\
&= (1_\star -_\star \lambda)\cdot_\star A \cdot_\star P +_\star (1_\star -_\star \lambda)\cdot_\star b \\
&\quad +_\star \lambda \cdot_\star A \cdot_\star Q +_\star \lambda \cdot_\star b \\
&= (1_\star -_\star \lambda)\cdot_\star (A\cdot_\star P +_\star b) \\
&\quad +_\star \lambda \cdot_\star (A \cdot_\star Q +_\star b) \\
&= (1_\star -_\star \lambda)\cdot_\star TP +_\star \lambda \cdot_\star TQ.
\end{aligned}
$$

This completes the proof. □

Theorem 3.5. *A point X lies between two points P and Q if and only if for any multiplicative affine transformation T the point TX lies between TP and TQ. Moreover,*

$$
d_\star(P,X)/_\star d_\star(P,Q) = d_\star(TP,TX)/_\star d_\star(TP,TQ). \tag{3.3}
$$

Proof. By Theorem 3.4, it follows the first part of the assertion. Now, we will prove (3.3). Let

$$
X = (1_\star -_\star \lambda)\cdot_\star P +_\star \lambda \cdot_\star Q
$$

for some $\lambda \in \mathbb{R}_\star$. Then

$$\begin{aligned} d_\star(P,X) &= |\lambda \cdot_\star P -_\star \lambda \cdot_\star Q|_\star \\ &= |\lambda|_\star \cdot_\star |P -_\star Q|_\star \\ &= |\lambda|_\star \cdot_\star d_\star(P,Q), \end{aligned}$$

and

$$d_\star(P,X)/_\star d_\star(P,Q) = |\lambda|_\star.$$

Next,

$$\begin{aligned} d_\star(TP,TX) &= |\lambda \cdot_\star TP -_\star \lambda \cdot_\star TQ|_\star \\ &= |\lambda|_\star \cdot_\star |TP -_\star TQ|_\star \\ &= |\lambda|_\star \cdot_\star d_\star(TP,TQ). \end{aligned}$$

Therefore,

$$d_\star(TP,TX)/_\star d_\star(TP,TQ) = |\lambda|_\star.$$

From here, we find (3.3). This completes the proof. □

Theorem 3.6. *If a multiplicative affine transformation leaves two distinct points, then it leaves fixed every point on the multiplicative line joining these points.*

Proof. Let l be a multiplicative line and T be a multiplicative affine transformation and $P,Q \in l$ be such that

$$TP = P \quad \text{and} \quad TQ = Q.$$

Take $X \in l$ arbitrarily. Hence and Theorem 3.5, we find

$$\begin{aligned} d_\star(P,X)/_\star d_\star(P,Q) &= d_\star(TP,TX)/_\star d_\star(TP,TQ) \\ &= d_\star(P,TX)/_\star d_\star(P,Q). \end{aligned}$$

Thus,

$$d_\star(P,X) = d_\star(P,TX).$$

From here, $X = TX$. This completes the proof. □

Theorem 3.7. *If a multiplicative affine transformation leaves three multiplicative non collinear points, it must be identity.*

Proof. Let P, Q and R be three multiplicative non collinear points and T be a multiplicative affine transformation so that

$$TP = P, \quad TQ = Q \quad \text{and} \quad TR = R.$$

Let l be the multiplicative line through P and Q, m be the multiplicative line through Q and R, n be the multiplicative line through R and P. Let also, $X \notin l$, $X \notin m$, $X \notin n$ be arbitrarily chosen and A be the multiplicative midpoint of the multiplicative segment PQ. With s we denote the multiplicative line through X and A. Then $s \nparallel_\star m$ and $s \nparallel_\star n$ simultaneously. Hence, $s \cap m$ or $s \cap n$. Without loss of generality, assume that $s \cap m = B$. Note that

$$TA = A \quad \text{and} \quad TB = B.$$

Hence and Theorem 3.6, it follows that $TX = X$. Because X was arbitrarily chosen, we conclude that T is identity. This completes the proof. □

Theorem 3.8 (Fundamental Theorem of the Multiplicative Affine Geometry). *Given two multiplicative nonlinear triples of points P, Q, R and P_1, Q_1, R_1, there is a unique multiplicative affine transformation T so that*

$$TP = P_1, \quad TQ = Q_1 \quad \text{and} \quad TR = R_1.$$

Proof. Since $\{Q -_\star P, R -_\star P\}$ and $\{Q_1 -_\star P_1, R_1 -_\star P_1\}$ are bases of $E^2_\star$, there is a multiplicative matrix $A \in \mathscr{M}_{\star 2\times 2}$ so that $\det_\star A \neq 0_\star$ and

$$\begin{aligned} A \cdot_\star (Q -_\star P) &= Q_1 -_\star P_1, \\ A \cdot_\star (R -_\star P) &= R_1 -_\star P_1. \end{aligned}$$

Let

$$T = \tau_{P_1} A \tau_{-_\star P}.$$

Then

$$\begin{aligned} TP &= \tau_{P_1} A \tau_{-_\star P} P \\ &= P_1, \end{aligned}$$

and

$$\begin{aligned} TQ &= \tau_{P_1} A \tau_{-_\star P} Q \\ &= \tau_{P_1} A (Q -_\star P) \\ &= \tau_{P_1} (Q_1 -_\star P_1) \\ &= Q_1, \end{aligned}$$

and

$$\begin{aligned} TR &= \tau_{P_1} A \tau_{-_\star P} R \\ &= \tau_{P_1} A (R -_\star P) \\ &= \tau_{P_1} (R_1 -_\star P_1) \\ &= R_1. \end{aligned}$$

Assume that the multiplicative affine transformation T_1 agrees with T on P, Q and R. Then

$$\begin{aligned} T_1^{-1_\star} TP &= T_1^{-1_\star} P_1 \\ &= P, \\ T_1^{-1_\star} TQ &= T_1^{-1_\star} Q_1 \\ &= Q, \\ T_1^{-1_\star} TR &= T_1^{-1_\star} R_1 \\ &= R. \end{aligned}$$

Hence and Theorem 3.7, it follows that $T_1^{-1_\star} T$ must be identity. This completes the proof. □

3.4 Multiplicative Affine Reflections

Definition 3.4. *Let P, Q and R be multiplicative nonlinear points of $E^2_\star$. The unique multiplicative affine transformation that leaves P fixed while interchanging Q and R is called multiplicative affine reflection. It is denoted by*

$$[P, Q \longleftrightarrow R]_\star.$$

Example 3.1. *Every ordinary multiplicative reflection is a multiplicative affine reflection.*

Theorem 3.9. *Let Q and R be distinct points, M be the multiplicative midpoint of the multiplicative segment QR and P be a point that is not multiplicative collinear with Q and R. Then the multiplicative affine transformation $[P, Q \longleftrightarrow R]_\star$ leaves fixed every point of the multiplicative line through P and M but not other points.*

Proof. Firstly, we prove that M is a multiplicative fixed point. Let

$$Tx = A \cdot_\star x +_\star b, \quad A \in \mathscr{M}_{\star 2\times 2}, \quad \det_\star A \neq 0_\star, \quad b \in \mathbb{R}^2_\star.$$

We have

$$\begin{aligned} Q &= TR \\ &= A \cdot_\star R +_\star b, \\ R &= TQ \\ &= A \cdot_\star Q +_\star b. \end{aligned}$$

Thus,

$$\begin{aligned} M &= 1_\star /_\star e^2 \cdot_\star (Q +_\star R) \\ &= 1_\star /_\star e^2 \cdot_\star (A \cdot_\star R +_\star b +_\star A \cdot_\star Q +_\star b) \\ &= 1_\star /_\star e^2 \cdot_\star (A \cdot_\star (R +_\star Q) +_\star e^2 \cdot_\star b) \\ &= A \cdot_\star M +_\star b \\ &= TM. \end{aligned}$$

Hence and Theorem 3.6, the multiplicative line through P and M consists entirely of fixed points. Since the multiplicative affine transformation is not the multiplicative identity, so it cannot leave additional fixed points. This completes the proof. □

Theorem 3.10. *The multiplicative affine reflection $[P, Q \longleftrightarrow R]_\star$ leaves fixed the multiplicative line through P and M and all multiplicative lines multiplicative parallel to the multiplicative line through Q and R and no other multiplicative lines. Here M is the multiplicative midpoint of the multiplicative segment QR.*

Proof. Let l be the multiplicative line through Q and R, m be the multiplicative line through P and M, and

$$T = [P, Q \longleftrightarrow R]_\star.$$

Since TQ and TR determine the same multiplicative line as the multiplicative line through Q and R, we conclude that $Tl = l$. Let l_1 be a multiplicative line that is multiplicative parallel to l. Let also,

$$l_1 \cap PM = M_1.$$

We have

$$TM_1 = M_1.$$

Moreover, $M_1 \in Tl_1$ and

$$Tl_1 \parallel_\star l = Tl.$$

Therefore $Tl_1 = l_1$. Let now, l_2 be a multiplicative line so that

$$Tl_2 = l_2, \quad l_2 \not\parallel_\star l, \quad l_2 \neq m.$$

Then T leaves $l_2 \cap l_1$ and $l_2 \cap l$. This is a contradiction because the multiplicative fixed points are on m. This completes the proof. □

Exercise 3.1. *Prove that the multiplicative affine reflection $[P,Q \longleftrightarrow R]_\star$ is multiplicative isometry if and only if the multiplicative line through P and M is multiplicative perpendicular to the multiplicative line through Q and R. Here M is the multiplicative midpoint of the multiplicative segment QR.*

3.5 Multiplicative Shears

Definition 3.5. *Suppose that P, Q and R are multiplicative nonlinear points. The unique multiplicative affine transformation that leaves fixed every point of the multiplicative line through P multiplicative parallel to QR and takes Q to R is said to be multiplicative shear and it will be denoted by*

$$[P,Q \longrightarrow R]_\star.$$

The multiplicative shear $[P,Q \longrightarrow R]_\star$ has the multiplicative line through P parallel to QR as its set of fixed points. The fixed multiplicative lines are those belonging to the pencil of multiplicative parallels determined by QR. Note that every matrix

$$\begin{pmatrix} 1_\star & \lambda & 0_\star \\ 0_\star & 1_\star & 0_\star \\ 0_\star & 0_\star & 1_\star \end{pmatrix},$$

where $\lambda \in \mathbb{R}_\star$, determines a multiplicative shear. If T is a multiplicative shear with axis l and ρ is any multiplicative affine transformation, then $\rho T \rho^{-1}$ is a multiplicative shear with axis ρl.

3.6 Multiplicative Dilatations

Definition 3.6. *A multiplicative dilatation is a multiplicative affine transformation T with the property that for any multiplicative line l we have $Tl = l$ or $Tl \parallel_\star l$. The identity is the trivial multiplicative dilation.*

Theorem 3.11. *A multiplicative dilatation that leaves two distinct points fixed must be identity.*

Proof. Let T be a multiplicative dilatation and $P, Q \in E^2_\star$ be two distinct points for which $TP = P$ and $TQ = Q$. Let l be the multiplicative line through P and Q. Then $Tl = l$. Take $X \notin l$ arbitrarily. Let m be the multiplicative line through the points P and X. Then $Tm = m$ or $Tm \parallel_\star m$. Because $m \cap l = P$, we have that $Tm = m$. Thus, T is a multiplicative affine transformation that leaves three multiplicative non collinear points P, Q and X fixed. Therefore, T is identity. This completes the proof. □

Remark 3.2. *By Theorem 3.11, it follows that any multiplicative dilatation that is not identity can have at most one fixed point.*

Definition 3.7. *A multiplicative dilatation that has exactly one fixed point is said to be central multiplicative dilatation, and the fixed point is said to be its center.*

Theorem 3.12. *A central multiplicative dilatation T with center C can be written as*

$$Tx = C +_\star \kappa \cdot_\star (x -_\star C).$$

Definition 3.8. *The number $|\kappa|_\star$ is said to be multiplicative magnification factor of T.*

Proof. Let

$$Tx = A \cdot_\star x +_\star b, \quad A \in \mathscr{M}_{\star 2\times 2}, \quad \det_\star A \neq 0_\star, \quad b \in \mathbb{R}^2_\star.$$

Because T is a multiplicative dilatation, every multiplicative vector v must be a multiplicative eigenvector of A. Therefore

$$A = \kappa \cdot_\star I_\star,$$

where $\kappa \in \mathbb{R}_\star$. Hence,

$$\begin{aligned} Tx &= \kappa \cdot_\star I_\star \cdot_\star x +_\star b \\ &= \kappa_\star \cdot_\star x +_\star b. \end{aligned}$$

Since $TC = C$, we get

$$\begin{aligned} C &= TC \\ &= \kappa \cdot_\star C +_\star b, \end{aligned}$$

and then

$$b = C \cdot_\star (I -_\star \kappa).$$

Therefore,

$$\begin{aligned} Tx &= \kappa \cdot_\star x +_\star C \cdot_\star (I -_\star \kappa) \\ &= \kappa \cdot_\star x +_\star C -_\star \kappa \cdot_\star C \\ &= \kappa \cdot_\star (x -_\star C) +_\star C. \end{aligned}$$

This completes the proof. □

Theorem 3.13. *A multiplicative dilatation T that has no fixed points must be a multiplicative translation.*

Proof. If $\kappa \neq 1_\star$, then the equation

$$\kappa \cdot_\star x +_\star b = x,$$

has a unique solution for any $b \in \mathbb{R}^2_\star$. Therefore, T has a fixed point. This is a contradiction and $\kappa = 1_\star$. Hence, T is a multiplicative translation. This completes the proof. □

Theorem 3.14. *The fixed multiplicative lines of a central multiplicative dilatation T with a center C are precisely those passing through its center.*

Proof. For any multiplicative vector $v \in \mathbb{R}^2_\star$, we have

$$\begin{aligned} T(C +_\star [v]_\star) &= TC +_\star T[v]_\star \\ &= C +_\star [v]_\star. \end{aligned}$$

Thus, all multiplicative lines through the center C are fixed. Let l be a multiplicative line so that $C \notin l$. Take $X \in l$ arbitrarily. With m we denote the multiplicative line through C and X so that $Tm = m$. Then $m \cap l = X$ and $TX = X$, which is a contradiction because $X \neq C$. This completes the proof. □

3.7 Multiplicative Similarities

Definition 3.9. *A mapping $T : E^2_\star \to E^2_\star$ is said to be a multiplicative similarity if there is a $\kappa \in \mathbb{R}_\star$, $\kappa > 0_\star$, called multiplicative magnification factor, so that*

$$d_\star(TX, TY) = \kappa \cdot_\star d_\star(X,Y),$$

for any $X, Y \in E^2_\star$.

Theorem 3.15. *Every multiplicative similarity T with multiplicative magnification factor κ is a central multiplicative dilatation followed by a multiplicative isometry.*

Proof. Let S be a central multiplicative dilatation defined by

$$SX = 1_\star /_\star \kappa \cdot_\star X, \quad X \in E^2_\star.$$

Then

$$\begin{aligned} d_\star(TSX, TSY) &= \kappa \cdot_\star d_\star(SX, SY) \\ &= 1_\star /_\star \kappa \cdot_\star \kappa \cdot_\star d_\star(X,Y) \\ &= d_\star(X,Y), \quad X, Y \in E^2_\star. \end{aligned}$$

Thus, TS is a multiplicative isometry and

$$T = TSS^{-1_\star}.$$

This completes the proof. □

Theorem 3.16. *Let T_1 and T_2 be two multiplicative similarities with multiplicative magnification factors κ_1 and κ_2, respectively. Then*

1. *T_1T_2 is a multiplicative similarity with a multiplicative magnification factor $\kappa_1\kappa_2$.*
2. *$T_1^{-1_\star}$ is a multiplicative similarity with a multiplicative magnification factor $\kappa_1^{-1_\star}$.*

Proof.

1. We have

$$\begin{aligned} d_\star(T_1T_2X, T_1T_2Y) &= \kappa_1 \cdot_\star d_\star(T_2X, T_2Y) \\ &= \kappa_1 \cdot_\star \kappa_2 \cdot_\star d_\star(X,Y), \end{aligned}$$

for any X, $Y \in E^2_\star$. Thus, T_1T_2 is a multiplicative similarity with a magnification factor $\kappa_1\kappa_2$.

2. Let $X, Y \in E^2_\star$ and $T_1X = X_1$, $T_1Y = Y_1$. Then

$$T_1^{-1_\star}X_1 = X, \quad T_1^{-1_\star}Y_1 = Y.$$

Hence,

$$\begin{aligned} d_\star(X_1, Y_1) &= d_\star(T_1X, T_1Y) \\ &= \kappa_1 \cdot_\star d_\star(X,Y) \\ &= \kappa_1 \cdot_\star d_\star(T_1^{-1_\star}X_1, T_1^{-1_\star}Y_1), \end{aligned}$$

whereupon

$$d_\star(T_1^{-1_\star}X_1, T_1^{-1_\star}Y_1) = 1_\star /_\star \kappa_1 \cdot_\star d_\star(X_1, Y_1).$$

This completes the proof.

□

Exercise 3.2. *Prove that the set of all multiplicative similarities of* $E^2_\star$ *is a group, which will be denoted by* $Sim_\star(E^2_\star)$.

Definition 3.10. *Two figures* $\mathscr{F}_1$ *and* $\mathscr{F}_2$ *of* $E^2_\star$ *are said to be multiplicative similar if there is a multiplicative similarity* T *so that*

$$T(\mathscr{F}_1) = \mathscr{F}_2.$$

3.8 Multiplicative Affine Symmetries

Definition 3.11. *Let* $\mathscr{F}$ *be any figure. A multiplicative affine transformation leaving* $\mathscr{F}$ *fixed is called a multiplicative affine symmetry of* $\mathscr{F}$. *The set of all multiplicative affine symmetries is a group, called the multiplicative affine symmetry group of* $\mathscr{F}$. *We write*

$$\mathscr{A}\mathscr{S}(\mathscr{F}) = \{T \in AF_\star; R\mathscr{F} = \mathscr{F}\}.$$

We have the inclusions

$$\mathscr{S}(\mathscr{F}) \subset \mathscr{A}\mathscr{S}(\mathscr{F}) \subset AF_\star(E^2_\star).$$

Theorem 3.17. *Let* $\mathscr{F}$ *be a set of a single point* P. *Then* $\mathscr{A}\mathscr{S}(\mathscr{F})$ *is isomorphic to the group of* 2×2 *multiplicative invertible matrices.*

Proof. Take $T \in \mathscr{A}\mathscr{S}(\mathscr{F})$ arbitrarily. Let

$$A = T_{-_\star P} T T_P.$$

Then

$$\begin{aligned} T0_\star &= T_{-_\star P} T T_P 0_\star \\ &= T_{-_\star P} T P \\ &= T_{-_\star P} P \\ &= 0_\star. \end{aligned}$$

Next, we have that

$$T = T_P A T_{-_\star P}.$$

Note that the map that maps T to A is an isomorphism. This completes the proof. □

Exercise 3.3. *Let* $\mathscr{F}$ *be a set of two distinct points. Prove that* $\mathscr{A}\mathscr{S}(\mathscr{F})$ *is isomorphic to the group of* 2×2 *matrices of the form*

$$\begin{pmatrix} \pm_\star 1_\star & \lambda \\ 0_\star & \mu \end{pmatrix}, \quad \mu \neq 0_\star.$$

3.9 Multiplicative Rays and Multiplicative Angles

Definition 3.12. *Let* P *be a point of* $E^2_\star$ *and* v *be a multiplicative nonzero vector. Then*

$$r = \{P +_\star t \cdot_\star v : t \geq 0_\star\}$$

is called a multiplicative ray with origin P *and a multiplicative direction* v.

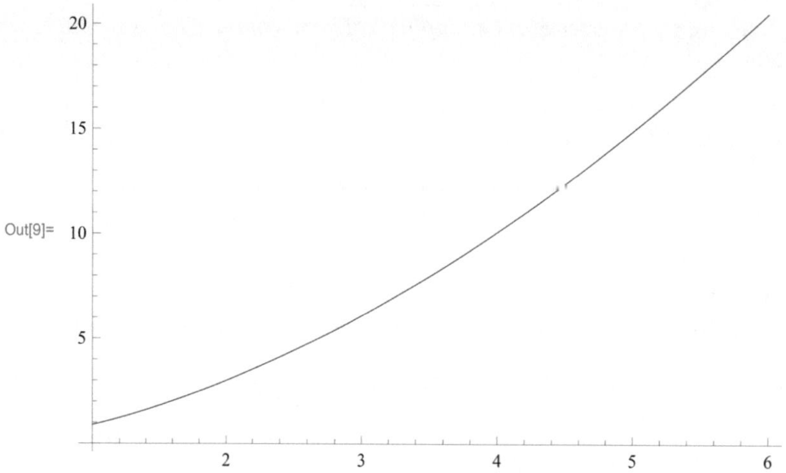

FIGURE 3.1
Multiplicative ray.

Clearly, every multiplicative line through P is the union of two multiplicative rays with origin P. Their directions are multiplicative negative of each other. In Fig. 3.1, it is shown a multiplicative ray through $P(2,3)$ with direction $v = (e^4, e^7)$.

Definition 3.13. *The union of two multiplicative rays r_1 and r_2 with common origin P is called a multiplicative angle with vertex P and multiplicative arms r_1 and r_2. When $r_1 = r_2$, then the multiplicative angle is said to be the multiplicative zero angle. If r_1 and r_2 are two halves of the same multiplicative line, we say that they are multiplicative opposite rays and the multiplicative angle is a multiplicative straight angle. If $r_1 \perp_\star r_2$, we call the multiplicative angle a multiplicative right angle.*

Let

$$\begin{aligned} r_1 &= \{(2,3) +_\star t \cdot_\star (e^4, e^7), \quad t \geq 0_\star\}, \\ r_2 &= \{(2,3) +_\star t \cdot_\star (e, e^{\frac{1}{2}}), \quad t \geq 0_\star\}. \end{aligned}$$

In Fig. 3.2, it is shown a multiplicative angle with vertex $(2,3)$ and multiplicative arms r_1 and r_2. For a given two distinct points P and Q, there is a unique multiplicative ray with origin the point P that passes through the point Q. We denote this ray by $\vec{PQ}$. The multiplicative angle with vertex the point Q and multiplicative arms $\vec{QR}$ and $\vec{QP}$ will be denoted by $\angle_\star PQR$ and $\angle_\star RQP$.

Definition 3.14. *Let $\mathscr{A}$ be a multiplicative angle whose multiplicative arms have multiplicative unit direction vectors u and v. The multiplicative radian measure of $\mathscr{A}$ is defined by*

$$\alpha = \arccos_\star(\langle u, v\rangle_\star).$$

Example 3.2. *Let* $u = \left(e^{\frac{1}{\sqrt{2}}}, e^{\frac{1}{\sqrt{2}}}\right)$ *and* $v = \left(e^{\sqrt{\frac{2}{3}}}, e^{\sqrt{\frac{1}{3}}}\right)$. *Then*

$$\begin{aligned} |u|_\star &= e^{\left(\left(\log e^{\frac{1}{\sqrt{2}}}\right)^2 + \left(\log e^{\frac{1}{\sqrt{2}}}\right)^2\right)^{\frac{1}{2}}} \\ &= e^{\left(\frac{1}{2}+\frac{1}{2}\right)^{\frac{1}{2}}} \\ &= e \\ &= 1_\star, \end{aligned}$$

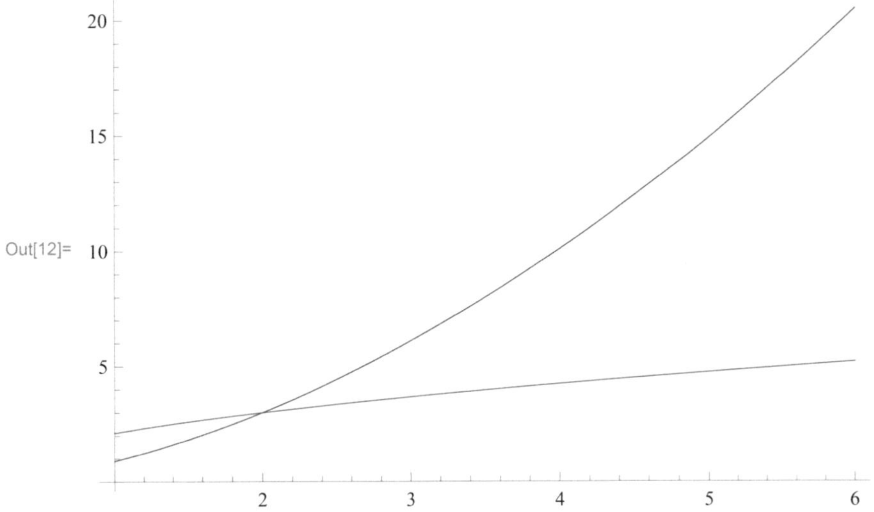

FIGURE 3.2
Multiplicative angle.

and

$$\begin{aligned} |v|_\star &= e^{\left(\left(\log e^{\sqrt{\frac{2}{3}}}\right)^2+\left(\log e^{\sqrt{\frac{1}{3}}}\right)^2\right)^{\frac{1}{2}}} \\ &= e^{\left(\frac{2}{3}+\frac{1}{3}\right)^{\frac{1}{2}}} \\ &= e \\ &= 1_\star. \end{aligned}$$

We have

$$\begin{aligned} \langle u,v\rangle_\star &= e^{\log e^{\frac{1}{\sqrt{2}}}\log e^{\sqrt{\frac{2}{3}}}+\log e^{\frac{1}{\sqrt{2}}}\log e^{\sqrt{\frac{1}{3}}}} \\ &= e^{\frac{1}{\sqrt{2}}\cdot\frac{\sqrt{2}}{\sqrt{3}}+\frac{1}{\sqrt{2}}\cdot\frac{1}{\sqrt{3}}} \\ &= e^{\frac{1+\sqrt{2}}{\sqrt{6}}} \end{aligned}$$

and

$$\begin{aligned} \alpha &= \arccos_\star(\langle u,v\rangle_\star) \\ &= e^{\arccos(\log(\langle u,v\rangle_\star))} \\ &= e^{\arccos\left(\log e^{\frac{1+\sqrt{2}}{\sqrt{6}}}\right)} \\ &= e^{\arccos\left(\frac{1+\sqrt{2}}{\sqrt{6}}\right)}. \end{aligned}$$

Let

$$u=(\cos_\star\theta,\sin_\star\theta),\quad v=(\cos_\star\phi,\sin_\star\phi).$$

Then

$$\begin{aligned} u &= \left(e^{\cos(\log\theta)},e^{\sin(\log\theta)}\right), \\ v &= \left(e^{\cos(\log\phi)},e^{\sin(\log\phi)}\right) \end{aligned}$$

and

$$\begin{aligned}\langle u,v\rangle_\star &= e^{\cos(\log\theta)\cos(\log\phi)+\sin(\log\theta)\sin(\log\phi)}\\ &= e^{\cos(\log\theta-\log\phi)}\\ &= e^{\cos\left(\log\frac{\theta}{\phi}\right)},\end{aligned}$$

and the multiplicative radian measure is

$$\begin{aligned}\alpha &= \arccos_\star(\langle u,v\rangle_\star)\\ &= e^{\arccos(\log(\langle u,v\rangle_\star)}\\ &= e^{\arccos\left(\cos\left(\log\frac{\theta}{\phi}\right)\right)}\\ &= e^{\log\frac{\theta}{\phi}}\\ &= \frac{\theta}{\phi}\\ &= \theta -_\star \phi.\end{aligned}$$

Observe that

$$\begin{aligned}(\mathrm{Rot}_\star\alpha)\cdot_\star v &= \begin{pmatrix}\cos_\star\alpha & -_\star\sin_\star\alpha\\ \sin_\star\alpha & \cos_\star\alpha\end{pmatrix}\cdot_\star\begin{pmatrix}\cos_\star\phi\\ \sin_\star\phi\end{pmatrix}\\ &= \begin{pmatrix}\cos_\star(\alpha+_\star\phi)\\ \sin_\star(\alpha+_\star\phi)\end{pmatrix}\\ &= \begin{pmatrix}\cos_\star\theta\\ \sin_\star\theta\end{pmatrix}\\ &= u.\end{aligned}$$

Thus, there is a multiplicative rotation by the multiplicative radian measure that interchanges one multiplicative arm of $\mathscr{A}$ to other.

Exercise 3.4. *Let* $u=(2,8)$ *and* $v=(3,7)$. *Find the multiplicative radian measure.*

Theorem 3.18. *Let* $\mathscr{A}$ *be a multiplicative angle. Its multiplicative radian measure* α *is* $0_\star$ *if* $\mathscr{A}$ *is the multiplicative zero angle.*

Proof. Let u, v be multiplicative unit vectors that are the multiplicative arms of $\mathscr{A}$. Then

$$\begin{aligned}\alpha &= \arccos_\star(\langle u,v\rangle_\star)\\ &= 0_\star,\end{aligned}$$

if and only if

$$e^{\arccos(\log(\langle u,v\rangle_\star))}=e^0,$$

if and only if

$$\arccos(\log(\langle u,v\rangle_\star))=0,$$

if and only if

$$\log(\langle u,v\rangle_\star)=1,$$

if and only if

$$\langle u,v\rangle_\star=e,$$

if and only if $u=v$ or $u\parallel_\star v$.This completes the proof. □

Theorem 3.19. *Let $\mathscr{A}$ be a multiplicative angle. Its multiplicative radian measure α is e^{π} if and only if $\mathscr{A}$ is a multiplicative straight angle.*

Proof. By the proof of Theorem 3.18, it follows that

$$\alpha = e^{\pi},$$

if and only if

$$\arccos(\log(\langle u, v\rangle_\star)) = \pi,$$

if and only if

$$\log(\langle u, v\rangle_\star) = -1,$$

if and only if

$$\langle u, v\rangle_\star = e^{-1},$$

if and only if

$$e^{\log u_1 \log v_1 + \log u_2 \log v_2} = e^{-1},$$

if and only if

$$\log u_1 \log v_1 + \log u_2 \log v_2 = -1,$$

if and only if

$$v = \left(\frac{1}{u_1}, \frac{1}{u_2}\right),$$

if and only if $\mathscr{A}$ is a multiplicative straight angle. This completes the proof. □

Theorem 3.20. *Let $\mathscr{A}$ be a multiplicative angle. Then its multiplicative radian measure is in the interval $[1, e^{\pi}]$.*

Proof. We have that

$$\alpha = e^{\arccos(\log(\langle u, v\rangle_\star))}.$$

Hence, $\alpha \in [1, e^{\pi}]$. This completes the proof. □

Definition 3.15. *A multiplicative angle $\mathscr{A}$ is said to be multiplicative acute if its multiplicative radiation measure is $< e^{\frac{\pi}{2}}$. If its multiplicative radian measure is $> e^{\frac{\pi}{2}}$, then it is said to be multiplicative obtuse.*

Let $\mathscr{A} = \angle_\star PQR$. By the definition of multiplicative radian measure, it follows that $\mathscr{A}$ is multiplicative acute(obtuse) if and only if

$$\langle P -_\star Q, R -_\star Q\rangle_\star > (<) 0_\star.$$

Definition 3.16. *Let $\mathscr{A}$ be a multiplicative angle with vertex P and multiplicative direction vectors u and v so that*

$$(Rot_\star \alpha) \cdot_\star u = v.$$

Then the multiplicative ray with origin P and multiplicative direction vector

$$Rot_\star(\alpha /_\star e^2) \cdot_\star u,$$

is said to be multiplicative bisector of $\mathscr{A}$.

Theorem 3.21. *For any multiplicative angle $\mathscr{A}$, there is a unique multiplicative reflection that interchanges its multiplicative arms, the multiplicative reflection in the multiplicative line containing the multiplicative bisector of $\mathscr{A}$.*

Proof. Let u and v be the multiplicative unit direction vectors of the multiplicative arms of the multiplicative angle $\mathscr{A}$ and

$$(\mathrm{Rot}_\star \alpha) \cdot_\star u = v.$$

With $P \in E_\star^2$ we denote the vertex of $\mathscr{A}$. Set

$$T = T_P \left(\mathrm{Ref}_\star((\theta +_\star \phi)/_\star e^2)\right) T_{-_\star P},$$

where

$$\begin{aligned} u &= (\cos_\star \theta, \sin_\star \theta), \\ v &= (\cos_\star \phi, \sin_\star \phi). \end{aligned}$$

Let

$$e_1 = (1_\star, 0_\star), \quad e_2 = (0_\star, 1_\star).$$

Then, for $t \in \mathbb{R}_\star$, we have

$$\begin{aligned} T(P +_\star t \cdot_\star u) &= P +_\star \left(\mathrm{Ref}_\star((\theta +_\star \phi)/_\star e^2)\right) \cdot_\star t \cdot_\star u \\ &= P +_\star t \cdot_\star \left(\mathrm{Ref}_\star((\theta +_\star \phi)/_\star e^2)\right) \cdot_\star \mathrm{Rot}_\star \theta \cdot_\star e_1 \\ &= P +_\star t \cdot_\star \left(\mathrm{Ref}_\star \left(\phi/_\star e^2\right)\right) \cdot_\star e_1 \\ &= P +_\star t \cdot_\star v. \end{aligned}$$

Therefore, $Tr_1 = r_2$. As above, $Tr_2 = r_1$. Assume that there is another multiplicative reflection T_1 with the same properties as T. Then

$$\begin{aligned} T_1 T r_1 &= T_1 r_2 \\ &= r_1, \end{aligned}$$

and

$$\begin{aligned} T_1 T r_2 &= T_1 r_1 \\ &= r_2. \end{aligned}$$

Hence, $T_1 T$ leaves P fixed. Since $TP = P$, we get that $T_1 P = P$. Thus, the axis of the multiplicative reflection T_1 passes through P and $T_1 T$ is a multiplicative rotation about P. Since the only multiplicative rotation leaves a multiplicative ray fixed is the identity, we conclude that $T_1 T$ must be the identity and $T = T_1$. This completes the proof. □

3.10 Multiplicative Rectilinear Figures

Definition 3.17. *A union of finitely many multiplicative segments, multiplicative rays and multiplicative lines is said to be a multiplicative rectilinear figure.*

Definition 3.18. *Let $\mathscr{F}$ be any multiplicative rectilinear figure. The figure $\widetilde{\mathscr{F}}$ of all multiplicative lines that contain all multiplicative lines, multiplicative rays and multiplicative segments of $\mathscr{F}$ is called the multiplicative rectilinear completion of $\mathscr{F}$.*

Definition 3.19. *A multiplicative rectilinear figure $\mathscr{F}$ is said to be multiplicative complete, if whenever a multiplicative segment is in $\mathscr{F}$, the multiplicative line containing it is in $\mathscr{F}$.*

Note that $\widetilde{\mathscr{F}}$ is the smallest multiplicative rectilinear figure containing $\mathscr{F}$.

Theorem 3.22. *Let T be a multiplicative affine transformation and let $\mathscr{F}$ be a multiplicative rectilinear figure. Then T maps the set of all multiplicative lines of $\widetilde{\mathscr{F}}$ bijectively to the set of multiplicative lines of $\widetilde{T\mathscr{F}}$.*

Proof. Let Tl be a multiplicative line of $\widetilde{T\mathscr{F}}$. Suppose that $l \notin \widetilde{\mathscr{F}}$. Then for any $m \in \widetilde{\mathscr{F}}$, Tm meets Tl in at most one point. Since $T\mathscr{F} \cap Tl$ is contained in the union of all the Tm, it can contain only finitely many points. This is impossible, because Tl contains at least one multiplicative segment of $T\mathscr{F}$. Let now, m is any multiplicative line of $\widetilde{\mathscr{F}}$. Then m contains a multiplicative segment m_0 that is contained in $\mathscr{F}$. Moreover, Tm_0 is a multiplicative segment of $T\mathscr{F}$. Thus, $\widetilde{T\mathscr{F}}$ contains the multiplicative line determined by Tm_0, i.e., $Tm \in \widetilde{T\mathscr{F}}$. This completes the proof. □

Definition 3.20. *Suppose that $\mathscr{F}$ is a multiplicative rectilinear figure. A point of $\mathscr{F}$ where two multiplicative lines of $\widetilde{\mathscr{F}}$ intersect is called a vertex of $\mathscr{F}$.*

Theorem 3.23. *Let $\mathscr{F}$ be a multiplicative rectilinear figure and T be a multiplicative affine transformation. Then T maps the set of vertices of $\mathscr{F}$ bijectively to the set of vertices of $T\mathscr{F}$.*

Proof. Let P be a vertex of $\mathscr{F}$. Then TP is a point of $T\mathscr{F}$. Since P is the intersection of two multiplicative lines of $\widetilde{\mathscr{F}}$, TP is the intersection of their images, which, by Theorem 3.22, are in $\widetilde{T\mathscr{F}}$. Thus, TP is a vertex of $T\mathscr{F}$. This completes the proof. □

Exercise 3.5. *Prove that any multiplicative affine symmetry T of a multiplicative rectilinear figure $\mathscr{F}$ is also a multiplicative affine symmetry of $\widetilde{\mathscr{F}}$.*

3.11 The Multiplicative Centroid

Suppose that $\mathscr{F}$ is a finite set of points of $E^2_\star$. Define

$$f(x) = \sum_{\star\, P \in \mathscr{F}} d_\star(x, P)^{2_\star}.$$

Theorem 3.24. *There is a unique point of $E^2_\star$ where the function f achieves its minimum value. This point will be called the multiplicative centroid of $\mathscr{F}$.*

Proof. Firstly, note that

$$\begin{aligned} d_\star(x,P)^{2_\star} &= \langle x -_\star P, x -_\star P\rangle_\star \\ &= |x|_\star^{2_\star} -_\star e^2 \cdot_\star \langle x, P\rangle_\star +_\star |P|_\star^{2_\star}. \end{aligned}$$

Without loss of generality, suppose that $\mathscr{F}$ has n points. Then

$$\begin{aligned} f(x) &= \sum_{\star\, P \in \mathscr{F}} \left(|x|_\star^{2_\star} -_\star e^2 \cdot_\star \langle x, P\rangle_\star +_\star |P|_\star^{2_\star} \right) \\ &= e^n \cdot_\star |x|_\star^{2_\star} -_\star e^2 \cdot_\star \langle x, \sum_{\star\, P \in \mathscr{F}} P\rangle_\star +_\star \sum_{\star\, P \in \mathscr{F}} |P|_\star^{2_\star}. \end{aligned}$$

Set

$$\begin{aligned} C &= (1_\star /_\star e^n) \cdot_\star \sum_{\star\, P \in \mathscr{F}} P, \\ b &= (1_\star /_\star e^n) \cdot_\star \sum_{\star\, P \in \mathscr{F}} |P|_\star^{2_\star}. \end{aligned}$$

Then

$$\begin{aligned} f(x) &= e^n \cdot_\star |x|_\star^{2_\star} -_\star e^2 \cdot_\star e^n \cdot_\star \langle x, C\rangle_\star +_\star e^n \cdot_\star b \\ &= e^n \cdot_\star \left(|x|_\star^{2_\star} -_\star e^2 \cdot_\star \langle x, C\rangle_\star +_\star b\right) \\ &= e^n \cdot_\star \left(|x|_\star^{2_\star} -_\star e^2 \cdot_\star \langle x, C\rangle_\star +_\star |C|_\star^{2_\star} +_\star b -_\star |C|_\star^{2_\star}\right) \\ &= e^n \cdot_\star \left(|x -_\star C|_\star^{2_\star} +_\star \left(b -_\star |C|_\star^{2_\star}\right)\right). \end{aligned}$$

Thus,

$$\min_{x \in \mathbb{E}_\star^2} f(x) = f(C).$$

This completes the proof. □

Example 3.3. *Let*

$$\mathscr{F} = \{P_1(1,3), \quad P_2(e,e^2), \quad P_3(2,1)\}.$$

Then

$$\begin{aligned} C &= (1_\star /_\star e^3) +_\star (P_1 +_\star P_2 +_\star P_3) \\ &= (1_\star /_\star e^3) \cdot_\star ((1,3) +_\star (e,e^2) +_\star (2,1)) \\ &= e^{\frac{1}{3}} \cdot_\star (2e, 3e^2) \\ &= \left(e^{\frac{1}{3}\log(2e)}, e^{\frac{1}{3}\log(3e^2)}\right) \\ &= \left(e^{\frac{1}{3}(1+\log 2)}, e^{\frac{1}{3}(2+\log 3)}\right). \end{aligned}$$

Exercise 3.6. *Find the multiplicative centroid of*

$$\mathscr{F} = \{P_1(e^7, e^8), \quad P_2(3,4), \quad P_3(1,4), \quad P_4(2,8)\}.$$

Theorem 3.25. *Suppose that $\mathscr{F}$ is a multiplicative rectilinear figure having a finite nonzero number of multiplicative vertices. Let C be the multiplicative centroid of $\mathscr{F}$. If T is any multiplicative isometry, then TC is the multiplicative centroid of $T\mathscr{F}$.*

Proof. Let P be a multiplicative vertex of $\mathscr{F}$. Then TP is a multiplicative vertex of $T\mathscr{F}$. Next, using that T is a multiplicative isometry, we have

$$\begin{aligned} {\sum_\star}_{P \in \mathscr{F}} |x -_\star TP|_\star^{2_\star} &= {\sum_\star}_{P \in \mathscr{F}} |TT^{-1_\star} x -_\star TP|_\star^{2_\star} \\ &= {\sum_\star}_{P \in \mathscr{F}} |T^{-1_\star} x -_\star P|_\star^{2_\star}. \end{aligned}$$

Thus, $T^{-1_\star}x$ is the multiplicative centroid C of $\mathscr{F}$, i.e., $x = TC$ is the multiplicative centroid of $T\mathscr{F}$. This completes the proof. □

Exercise 3.7. *If T is a multiplicative symmetry of a multiplicative rectilinear figure $\mathscr{F}$ with a finite number of multiplicative vertices, prove that T leaves the multiplicative centroid of $\mathscr{F}$ fixed.*

3.12 Multiplicative Symmetries of a Multiplicative Segment

Suppose that PQ is a multiplicative segment, $P, Q \in E_\star^2$. Take T to be a multiplicative affine transformation. Let $TP = P_1$ and $TQ = Q_1$.

Lemma 3.3. *Let PQ and P_2Q_2 denote the same multiplicative segment. Then $\{P,Q\} = \{P_2,Q_2\}$.*

Proof. Note that there are $\lambda, \mu \in [0_\star, 1_\star]$, $\lambda < \mu$, so that

$$\begin{aligned} P_2 &= (1_\star -_\star \lambda) \cdot_\star P +_\star \lambda \cdot_\star Q, \\ Q_2 &= (1_\star -_\star \mu) \cdot_\star P +_\star \mu \cdot_\star Q. \end{aligned}$$

Also, there are $\gamma, \delta \in [0_\star, 1_\star]$ so that

$$\begin{aligned} P &= (1_\star -_\star \gamma) \cdot_\star P_2 +_\star \gamma \cdot_\star Q_2, \\ Q &= (1_\star -_\star \delta) \cdot_\star P_2 +_\star \delta \cdot_\star Q_2. \end{aligned}$$

Then

$$\begin{aligned} P &= (1_\star -_\star \gamma) \cdot_\star ((1_\star -_\star \lambda) \cdot_\star P +_\star \lambda \cdot_\star Q) \\ &\quad +_\star \gamma \cdot_\star ((1_\star -_\star \mu) \cdot_\star P +_\star \mu \cdot_\star Q) \\ &= ((1_\star -_\star \gamma) \cdot_\star (1_\star -_\star \lambda) +_\star \gamma \cdot_\star (1_\star -_\star \mu)) \cdot_\star P \\ &\quad +_\star ((1_\star -_\star \gamma) \cdot_\star \lambda +_\star \gamma \cdot_\star \mu) \cdot_\star Q \\ &= (1_\star -_\star \gamma +_\star \gamma -_\star ((1_\star -_\star \gamma) \cdot_\star \lambda +_\star \gamma \cdot_\star \mu)) \cdot_\star P \\ &\quad +_\star ((1_\star -_\star \gamma) \cdot_\star \lambda +_\star \gamma \cdot_\star \mu) \cdot_\star Q \\ &= P +_\star ((1_\star -_\star \gamma) \cdot_\star \lambda +_\star \gamma \cdot_\star \mu) \cdot_\star (Q -_\star P). \end{aligned}$$

Because $P \neq Q$, we conclude that

$$\begin{aligned} 0_\star &= (1_\star -_\star \gamma) \cdot_\star \lambda +_\star \gamma \cdot_\star \mu \\ &= \lambda -_\star \lambda \cdot_\star \gamma +_\star \gamma \cdot_\star \mu \\ &= \lambda +_\star (\mu -_\star \lambda) \cdot_\star \gamma. \end{aligned}$$

Since $\lambda \geq 0_\star$, $\gamma \geq 0_\star$, $\mu -_\star \lambda > 0_\star$, by the last equation, we obtain

$$\lambda = 0_\star \quad \text{and} \quad \gamma = 0_\star.$$

Thus, $P = P_2$. As above, $Q = Q_2$. This completes the proof. □

Now, we will describe the multiplicative symmetry group $\mathscr{S}_\star(PQ)$. There are two multiplicative isometries that leave $\{P,Q\}$ fixed. One of them is the multiplicative reflection Ω_l with axis $l = PQ$ and the other is the multiplicative identity. Also, there are two multiplicative isometries that interchange P and Q. One of them is the multiplicative reflection Ω_m, where m is multiplicative perpendicular of PQ. Let T_1 be the other multiplicative isometry that interchanges P and Q. Then

$$T_1 P = Q \quad \text{and} \quad T_1 Q = P.$$

Hence,

$$\begin{aligned} \Omega_m T_1 P &= \Omega_m Q \\ &= P, \end{aligned}$$

and

$$\begin{aligned} \Omega_m T_1 Q &= \Omega_m P \\ &= Q. \end{aligned}$$

Therefore,

$$\Omega_m T_1 = I_\star \quad \text{and} \quad T_1 = \Omega_m,$$

or

$$\Omega_m T_1 = \Omega_l \quad \text{and} \quad T_1 = \Omega_l \Omega_m.$$

Consequently $\mathscr{S}_\star(PQ)$ contains four elements.

3.13 Multiplicative Symmetries of a Multiplicative Angle

Suppose that $\mathscr{A}$ is a multiplicative angle, different than the multiplicative straight angle, with a vertex $P \in E^2_\star$.

Lemma 3.4. *Let T be a multiplicative affine symmetry, $T\mathscr{A} = \mathscr{A}$ and T leaves both multiplicative lines of $\mathscr{A}$ fixed. Then T leaves both multiplicative arms of $\mathscr{A}$ fixed.*

Proof. Let l_1 and l_2 be both multiplicative lines of $\mathscr{A}$, r_1 and r_2 be the associated multiplicative arms with multiplicative unit direction vectors v and w, respectively. Let also,

$$Tx = A \cdot_\star x +_\star b, \quad A \in \mathscr{M}_{\star 2\times 2}, \quad \det_\star A \neq 0_\star, \quad b \in \mathbb{R}^2_\star.$$

Since $TP = P$, we get

$$\begin{aligned} TP &= A \cdot_\star P +_\star b \\ &= P, \end{aligned}$$

whereupon

$$b = P -_\star A \cdot_\star P$$

and

$$Tx = A \cdot_\star (x -_\star P) +_\star P.$$

Since $Tl_1 = l_1$, we have

$$[A \cdot_\star v]_\star = [v]_\star.$$

Thus, there is a $\lambda \in \mathbb{R}_\star$ so that

$$A \cdot_\star v = \lambda \cdot_\star v.$$

Hence,

$$\begin{aligned} T(P +_\star v) &= TP +_\star Tv \\ &= P +_\star A \cdot_\star (v -_\star P) +_\star P \\ &= P +_\star A \cdot_\star v -_\star A \cdot_\star P +_\star P \\ &= P +_\star A \cdot_\star v -_\star P +_\star P \\ &= P +_\star A \cdot_\star v \\ &= P +_\star \lambda \cdot_\star v. \end{aligned}$$

Since $T\mathscr{A} = \mathscr{A}$ and $P +_\star v$ is in $\mathscr{A}$, $P +_\star \lambda \cdot_\star v$ must be in $\mathscr{A}$. Therefore, $\lambda \geq 0_\star$ and $Tr_1 = r_1$. As above, $Tr_2 = r_2$. This completes the proof. □

Corollary 3.4. *If T in Lemma 3.4 is a multiplicative isometry, then T is the multiplicative identity.*

Proof. We will use the notation, used in the proof of Lemma 3.4. We have

$$\begin{aligned} 1_\star &= |v|_\star \\ &= d_\star(P +_\star v, P) \\ &= d_\star(T(P +_\star v), TP) \\ &= d_\star(P +_\star \lambda \cdot_\star v, P) \\ &= \lambda \cdot_\star |v|_\star \\ &= \lambda. \end{aligned}$$

Therefore,

$$A \cdot_\star v = v.$$

As above,

$$A \cdot_\star w = w.$$

Hence, $A = I_\star$. For any $x \in E_\star^2$, we have

$$\begin{aligned} Tx &= A \cdot_\star (x -_\star P) +_\star P \\ &= x -_\star P +_\star P \\ &= x. \end{aligned}$$

Consequently, $T = I_\star$. This completes the proof. □

Theorem 3.26. *Let T be a multiplicative affine symmetry of $\mathscr{A}$. Then T permutes both multiplicative arms of $\mathscr{A}$.*

Proof. Let r_1 and r_2 be the multiplicative arms of $\mathscr{A}$ and l_1 and l_2 be the multiplicative lines containing r_1 and r_2, respectively. Let also, m be the multiplicative line that contains the multiplicative bisector of $\mathscr{A}$. By Theorem 3.21, it follows that T permutes the multiplicative lines l_1 and l_2. We will consider the following two cases.

1. Let $Tl_1 = l_1$ and $Tl_2 = l_2$. Then, by Lemma 3.4, it follows that $Tr_1 = r_1$ and $Tr_2 = r_2$.
2. Let $Tl_1 = l_2$, $Tl_2 = l_1$. Then

   $$\Omega_m T l_1 = l_1 \quad \text{and} \quad \Omega_m T l_2 = l_2.$$

   Hence,

   $$\Omega_m T r_1 = r_1 \quad \text{and} \quad \Omega_m T r_2 = r_2.$$

   Moreover,

   $$\begin{aligned} Tr_1 &= \Omega_m \Omega_m T r_1 \\ &= \Omega_m r_1 \\ &= r_2, \end{aligned}$$

   and

   $$\begin{aligned} Tr_2 &= \Omega_m \Omega_m T r_2 \\ &= \Omega_m r_2 \\ &= r_1. \end{aligned}$$

   This completes the proof.

□

Corollary 3.5. *$\mathscr{S}_\star(\mathscr{A})$ contains two elements Ω_m and $I_\star$.*

Proof. If $T \in \mathscr{S}_\star(\mathscr{A})$, then $Tr_1 = r_1$ and $Tr_2 = r_2$ and hence, $T = I_\star$. On the other hand,

$$\Omega_m T r_1 = r_1 \quad \text{and} \quad \Omega_m T r_2 = r_2.$$

Therefore, $\Omega_m T = I_\star$ and $T = \Omega_m$. This completes the proof. □

Theorem 3.27. *Let $\angle_\star PQR$ be a multiplicative angle and $d_\star(P,Q) = d_\star(Q,R)$. Let also, m be the multiplicative line containing the multiplicative bisector of $\angle_\star PQR$. Then Ω_m interchanges P and R while Q leaves fixed.*

Proof. Let u and v be the multiplicative unit directions of PQ and QR, respectively, and

$$\begin{aligned} u &= (\cos_\star \theta, \sin_\star \theta), \\ v &= (\cos_\star \phi, \sin_\star \phi). \end{aligned}$$

Then

$$\begin{aligned} P -_\star Q &= |P -_\star Q|_\star \cdot_\star u, \\ R -_\star Q &= |R -_\star Q|_\star \cdot_\star v. \end{aligned}$$

We have

$$\begin{aligned} \Omega_m P &= T_Q \left(\mathrm{Ref}_\star \left((\theta +_\star \phi)/_\star e^2\right)\right) T_{-_\star Q} P \\ &= T_Q \left(\mathrm{Ref}_\star \left((\theta +_\star \phi)/_\star e^2\right)\right) |P -_\star Q|_\star \cdot_\star u \\ &= Q +_\star |P -_\star Q|_\star \cdot_\star v \\ &= Q +_\star |R -_\star Q|_\star \cdot_\star v \\ &= Q +_\star R -_\star Q \\ &= R. \end{aligned}$$

This completes the proof. □

3.14 Multiplicative Barycentric Coordinates

Definition 3.21. *Suppose that P, Q, $R \in E_\star^2$ be three multiplicative non collinear points. Then each point $X \in E_\star^2$ can be determined by*

$$X = \lambda \cdot_\star P +_\star \mu \cdot_\star Q +_\star \nu \cdot_\star R$$

and

$$\lambda +_\star \mu +_\star \nu = 1_\star.$$

The association

$$X \to \begin{pmatrix} \lambda \\ \mu \\ \nu \end{pmatrix},$$

will be called multiplicative barycentric coordinate system, and PQR is called the multiplicative triangle of reference.

Suppose that

$$X(x_1, x_2), \quad P(a_1, a_2), \quad Q(b_1, b_2), \quad R(c_1, c_2).$$

Then

$$\begin{aligned} \lambda +_\star \mu +_\star \nu = 1_\star \quad &\Longleftrightarrow \\ \lambda \mu \nu = e. \end{aligned} \tag{3.4}$$

Also,

$$\lambda \cdot_\star P +_\star \mu \cdot_\star Q +_\star \nu \cdot_\star R = \left(e^{\log\lambda\log a_1+\log\mu\log b_1+\log\nu\log c_1}, e^{\log\lambda\log a_2+\log\mu\log b_2+\log\nu\log c_2}\right).$$

By the last equation and (3.4), we get the system

$$\begin{aligned} e &= \lambda\mu\nu \\ x_1 &= e^{\log\lambda\log a_1+\log\mu\log b_1+\log\nu\log c_1} \\ x_2 &= e^{\log\lambda\log a_2+\log\mu\log b_2+\log\nu\log c_2}, \end{aligned}$$

or

$$\begin{aligned} 1 &= \log\lambda+\log\mu+\log\nu \\ \log x_1 &= \log\lambda\log a_1+\log\mu\log b_1+\log\nu\log c_1 \\ \log x_2 &= \log\lambda\log a_2+\log\mu\log b_2+\log\nu\log c_2, \end{aligned}$$

whereupon

$$\begin{aligned} \log\nu &= 1-\log\lambda-\log\mu \\ \log x_1 &= \log\lambda\log a_1+\log\mu\log b_1+(1-\log\lambda-\log\mu)\log c_1 \\ \log x_2 &= \log\lambda\log a_2+\log\mu\log b_2+(1-\log\lambda-\log\mu)\log c_2, \end{aligned}$$

or

$$\begin{aligned} \log\nu &= 1-\log\lambda-\log\mu \\ \log x_1-\log c_1 &= (\log a_1-\log c_1)\log\lambda+(\log b_1-\log c_1)\log\mu \\ \log x_2-\log c_2 &= (\log a_2-\log c_2)\log\lambda+(\log b_2-\log c_2)\log\mu, \end{aligned}$$

or

$$\begin{aligned} \log\nu &= 1-\log\mu-\log\lambda \\ \log\frac{x_1}{c_1} &= \log\frac{a_1}{c_1}\log\lambda+\log\frac{b_1}{c_1}\log\mu \\ \log\frac{x_2}{c_2} &= \log\frac{a_2}{c_2}\log\lambda+\log\frac{b_2}{c_2}\log\mu. \end{aligned}$$

By the last system, we find

$$\begin{aligned} \log\lambda &= \frac{\log\frac{x_1}{c_1}\log\frac{b_2}{c_2}-\log\frac{x_2}{c_2}\log\frac{b_1}{c_1}}{\log\frac{a_1}{c_1}\log\frac{b_2}{c_2}-\log\frac{a_2}{c_2}\log\frac{b_1}{c_1}}, \\ \log\mu &= \frac{\log\frac{x_2}{c_2}\log\frac{a_1}{c_1}-\log\frac{x_1}{c_1}\log\frac{a_2}{c_2}}{\log\frac{a_1}{c_1}\log\frac{b_2}{c_2}-\log\frac{a_2}{c_2}\log\frac{b_1}{c_1}}, \end{aligned}$$

and

$$\begin{aligned} \log\nu &= 1-\log\lambda-\log\mu \\ &= 1-\frac{\log\frac{x_1}{c_1}\log\frac{b_2}{c_2}-\log\frac{x_2}{c_2}\log\frac{b_1}{c_1}}{\log\frac{a_1}{c_1}\log\frac{b_2}{c_2}-\log\frac{a_2}{c_2}\log\frac{b_1}{c_1}} \\ &\quad -\frac{\log\frac{x_2}{c_2}\log\frac{a_1}{c_1}-\log\frac{x_1}{c_1}\log\frac{a_2}{c_2}}{\log\frac{a_1}{c_1}\log\frac{b_2}{c_2}-\log\frac{a_2}{c_2}\log\frac{b_1}{c_1}} \\ &= \frac{\log\frac{a_1}{c_1}\log\frac{b_2}{c_2}-\log\frac{a_2}{c_2}\log\frac{b_1}{c_1}-\log\frac{x_1}{c_1}\log\frac{b_2}{a_2}+\log\frac{x_2}{c_2}\log\frac{b_1}{a_1}}{\log\frac{a_1}{c_1}\log\frac{b_2}{c_2}-\log\frac{a_2}{c_2}\log\frac{b_1}{c_1}}. \end{aligned}$$

Consequently,

$$\begin{aligned}
\lambda &= e^{\frac{\log\frac{x_1}{c_1}\log\frac{b_2}{c_2}-\log\frac{x_2}{c_2}\log\frac{b_1}{c_1}}{\log\frac{a_1}{c_1}\log\frac{b_2}{c_2}-\log\frac{a_2}{c_2}\log\frac{b_1}{c_1}}}\\
\mu &= e^{\frac{\log\frac{x_2}{c_2}\log\frac{a_1}{c_1}-\log\frac{x_1}{c_1}\log\frac{a_2}{c_2}}{\log\frac{a_1}{c_1}\log\frac{b_2}{c_2}-\log\frac{a_2}{c_2}\log\frac{b_1}{c_1}}}\\
\nu &= e^{\frac{\log\frac{a_1}{c_1}\log\frac{b_2}{c_2}-\log\frac{a_2}{c_2}\log\frac{b_1}{c_1}-\log\frac{x_1}{c_1}\log\frac{b_2}{a_2}+\log\frac{x_2}{c_2}\log\frac{b_1}{a_1}}{\log\frac{a_1}{c_1}\log\frac{b_2}{c_2}-\log\frac{a_2}{c_2}\log\frac{b_1}{c_1}}}.
\end{aligned}$$

Example 3.4. *Let*

$$X(e^4,e^6),\quad P(e^6,e^8),\quad Q(e^8,e^{10}),\quad R(e^2,e^2).$$

Then

$$\begin{aligned}
\lambda &= e^{\frac{2\cdot 5-3\cdot 4}{3\cdot 5-4\cdot 4}}\\
&= e^2,\\
\mu &= e^{\frac{3\cdot 3-2\cdot 4}{3\cdot 5-4\cdot 4}}\\
&= e^{-1},\\
\nu &= e^{\frac{3\cdot 5-4\cdot 4-2\cdot\frac{5}{4}+3\cdot\frac{4}{3}}{3\cdot 5-4\cdot 4}}\\
&= e^{-\left(-1-\frac{5}{2}+4\right)}\\
&= e^{-\frac{1}{2}}.
\end{aligned}$$

Exercise 3.8. *Find the multiplicative barycentric coordinates of $X(2,3)$ with multiplicative triangle of reference*

$$P(e,e^3),\quad Q(4,5),\quad R(e^6,1).$$

Theorem 3.28. *Let $P,Q,R \in E_\star^2$ be multiplicative points and l be the multiplicative line through the points P and Q. Suppose that PQR is the multiplicative triangle of reference and X has multiplicative barycentric coordinates (λ,μ,ν).*

1. *$\nu = 0_\star$ if and only if $X \notin l$.*
2. *$\nu > 0_\star$ if and only if $XR \cap l = \emptyset$.*

Proof. We have

$$X = \lambda \cdot_\star P +_\star \mu \cdot_\star Q +_\star \nu \cdot_\star R$$

and

$$\lambda +_\star \mu +_\star \nu = 1_\star.$$

1. (a) Suppose that $\nu = 0_\star$. Then

$$\lambda +_\star \mu = 1_\star$$

and

$$\begin{aligned}
X &= \lambda \cdot_\star P +_\star \mu \cdot_\star Q\\
&= \lambda \cdot_\star P +_\star (1_\star -_\star \lambda) \cdot_\star Q.
\end{aligned}$$

Thus, $X \in l$.

(b) Let $X \in l$. Then there is a $t \in [0_\star, 1_\star]$ such that

$$X = t \cdot_\star P +_\star (1_\star -_\star t) \cdot_\star Q.$$

Hence,

$$(t, 1_\star -_\star t, 0_\star),$$

are the multiplicative barycentric coordinates of X with the multiplicative triangle of reference PQR.

2. (a) Let $\nu > 0_\star$. Then, for $t \in [0_\star, 1_\star]$, we have

$$\begin{aligned}
&(1_\star -_\star t)X +_\star t \cdot_\star R \\
&= (1_\star -_\star t) \cdot_\star (\lambda \cdot_\star P +_\star \mu \cdot_\star Q +_\star \nu \cdot_\star R) +_\star t \cdot_\star R \\
&= (1_\star -_\star t) \cdot_\star \lambda \cdot_\star P +_\star (1_\star -_\star t) \cdot_\star \mu \cdot_\star Q \\
&\quad +_\star ((1_\star -_\star t) \cdot_\star \nu +_\star t) \cdot_\star R.
\end{aligned}$$

Note that

$$(1_\star -_\star t) \cdot_\star \nu +_\star t > 0_\star.$$

Thus, $XR \cap l = \emptyset$.

(b) Let $\nu < 0_\star$. Then

$$(1_\star -_\star t) \cdot_\star \nu +_\star t = 0_\star,$$

if and only if

$$\nu -_\star t \cdot_\star \nu +_\star t = 0_\star,$$

if and only if

$$t \cdot_\star (1_\star -_\star \nu) = -_\star \nu,$$

if and only if

$$\begin{aligned}
t &= (-_\star \nu)/_\star(1_\star -_\star \nu) \\
&< 1_\star.
\end{aligned}$$

Thus, $XR \cap l = \emptyset$. This completes the proof.

□

Definition 3.22. *A multiplicative point X is said to be in the interior of the multiplicative angle $\angle_\star PQR$ is $\lambda > 0_\star$ and $\nu > 0_\star$, where*

$$X = \lambda \cdot_\star P +_\star \mu \cdot_\star Q +_\star \nu \cdot_\star R.$$

Theorem 3.29 (The Multiplicative Crossbar Theorem). *Let X be a point in the interior of the multiplicative angle $\angle_\star PQR$. Then the multiplicative ray $\vec{QX}$ intersects the multiplicative segment PR.*

Proof. We have

$$\begin{aligned}
Q +_\star t \cdot_\star (X -_\star Q) &= (1_\star -_\star t) \cdot_\star Q +_\star t \cdot_\star X \\
&= (1_\star -_\star t) \cdot_\star Q +_\star t \cdot_\star \lambda \cdot_\star P +_\star t \cdot_\star \mu \cdot_\star Q \\
&\quad +_\star t \cdot_\star \nu \cdot_\star R \\
&= t \cdot_\star \lambda \cdot_\star P +_\star ((1_\star -_\star t) +_\star t \cdot_\star \mu) \cdot_\star Q \\
&\quad +_\star t \cdot_\star \nu \cdot_\star R.
\end{aligned}$$

Then

$$1_\star -_\star t +_\star t \cdot_\star \mu = 0_\star,$$

if and only if

$$t = 1_\star /_\star (1_\star -_\star \mu).$$

Since

$$\begin{aligned} 1_\star -_\star \mu &= \lambda +_\star \nu \\ &> 0_\star, \end{aligned}$$

we conclude that

$$t > 0_\star \quad \text{and} \quad t \cdot_\star \lambda > 0_\star, \quad t \cdot_\star \nu > 0_\star.$$

Thus,

$$Q +_\star t \cdot_\star (X -_\star Q),$$

lies on PR. This completes the proof. □

Definition 3.23. *Let l be a multiplicative line and $R \notin l$. The multiplicative half-plane determined by l and R is the set of all multiplicative points X so that $XR \cap l = \emptyset$.*

Note that any multiplicative line determines two half-planes in $E^2_\star$.

3.15 Multiplicative Addition of Multiplicative Angles

Theorem 3.30. *Let $\angle_\star PQR$ be a multiplicative angle and X be a point in its interior. Then the multiplicative radian measure of $\angle_\star PQR$ is the multiplicative sum of the multiplicative radian measures of $\angle_\star PQX$ and $\angle_\star RQX$.*

Proof. Let θ, θ_1 and θ_2 be the multiplicative radian measures of the multiplicative angles $\angle_\star PQR$, $\angle_\star PQX$ and $\angle_\star RQX$, respectively. Let also,

$$\begin{aligned} u &= P -_\star Q, \\ v &= R -_\star Q, \\ w &= X -_\star Q. \end{aligned}$$

Without loss of generality, suppose that

$$|u|_\star = |v|_\star = |w|_\star = 1_\star$$

and

$$(\mathrm{Rot}_\star \theta) \cdot_\star u = v.$$

Since X is in the interior of $\angle_\star PQR$, there are $\lambda, \mu > 0_\star$ so that

$$X -_\star Q = \lambda \cdot_\star (P -_\star Q) +_\star \mu \cdot_\star (R -_\star Q)$$

or

$$w = \lambda \cdot_\star u +_\star \mu \cdot_\star v.$$

We have the following possibilities.

1. Let

$$(\mathrm{Rot}_\star \theta_1) \cdot_\star u = w, \quad (\mathrm{Rot}_\star \theta_2) \cdot_\star w = v.$$

Then

$$\begin{aligned}(\mathrm{Rot}_\star(\theta_1 +_\star \theta_2)) \cdot_\star u &= (\mathrm{Rot}_\star \theta_2)(\mathrm{Rot}_\star \theta_1) \cdot_\star u \\ &= (\mathrm{Rot}_\star \theta_2) \cdot_\star w \\ &= v \\ &= (\mathrm{Rot}_\star \theta) \cdot_\star u.\end{aligned}$$

Thus,

$$\theta = \theta_1 +_\star \theta_2.$$

2. Let

$$(\mathrm{Rot}_\star \theta_1) \cdot_\star u = w, \quad (\mathrm{Rot}_\star \theta_2) \cdot_\star v = w.$$

Then

$$\begin{aligned}0_\star &< \sin_\star \theta_1 \\ &= \langle u^{\perp_\star}, w\rangle_\star \\ &= \langle u^{\perp_\star}, \lambda \cdot_\star u +_\star \mu \cdot_\star v\rangle_\star \\ &= \langle u^{\perp_\star}, \lambda \cdot_\star u\rangle_\star +_\star \langle u^{\perp_\star}, \mu \cdot_\star v\rangle_\star \\ &= \lambda \cdot_\star \langle u^{\perp_\star}, u\rangle_\star +_\star \mu \cdot_\star \langle u^{\perp_\star}, v\rangle_\star \\ &= \mu \cdot_\star \langle u^{\perp_\star}, v\rangle_\star.\end{aligned}$$

Since $\mu > 0_\star$, by the last inequality, we get

$$\langle u^{\perp_\star}, v\rangle_\star > 0_\star. \tag{3.5}$$

Moreover, we have

$$\begin{aligned}0_\star &< \sin_\star \theta_2 \\ &= \langle v^{\perp_\star}, w\rangle_\star \\ &= \langle v^{\perp_\star}, \lambda \cdot_\star u +_\star \mu \cdot_\star v\rangle_\star \\ &= \langle v^{\perp_\star}, \lambda \cdot_\star u\rangle_\star +_\star \langle v^{\perp_\star}, \mu \cdot_\star v\rangle_\star \\ &= \lambda \cdot_\star \langle v^{\perp_\star}, u\rangle_\star +_\star \mu \cdot_\star \langle v^{\perp_\star}, v\rangle_\star \\ &= \lambda \cdot_\star \langle v^{\perp_\star}, u\rangle_\star \\ &= \lambda \cdot_\star \langle v^{\perp_\star \perp_\star} . u^{\perp_\star}\rangle_\star \\ &= -_\star \lambda \cdot_\star \langle v, u^{\perp_\star}\rangle \\ &= -_\star \lambda \cdot_\star \langle u^{\perp_\star}, v\rangle_\star.\end{aligned}$$

By the last inequality, using that $\lambda > 0_\star$, we conclude that

$$\langle u^{\perp_\star}, v\rangle_\star < 0_\star,$$

which contradicts with (3.5).

3. Let

$$(\mathrm{Rot}_\star \theta_1) \cdot_\star w = u, \quad (\mathrm{Rot}_\star \theta_2) \cdot_\star w = v.$$

As in the second case, we have

$$\begin{aligned} 0_\star &< \sin_\star(-_\star\theta_1) \\ &= \mu \cdot_\star \langle u^{\perp_\star}, v\rangle_\star \end{aligned}$$

and

$$\begin{aligned} 0_\star &< \sin_\star(-_\star\theta_2) \\ &= -_\star\lambda \cdot_\star \langle u^{\perp_\star}, v\rangle_\star, \end{aligned}$$

which is a contradiction.

4. Let

$$(\mathrm{Rot}_\star\theta_1)\cdot_\star w = u, \quad (\mathrm{Rot}_\star\theta_2)\cdot_\star v = w.$$

As in the second case,

$$\begin{aligned} 0_\star &< \sin_\star\theta_2 \\ &= \lambda \cdot_\star \langle v^{\perp_\star}, u\rangle_\star. \end{aligned}$$

On the other hand,

$$\begin{aligned} 0_\star &< \sin_\star\theta \\ &= \langle u^{\perp_\star}, v\rangle_\star \\ &= -_\star\langle u, v^{\perp_\star}\rangle_\star, \end{aligned}$$

which is a contradiction. This completes the proof.

□

Theorem 3.31. *Let $\angle_\star PQR$ be a multiplicative straight angle and X be any point that does not lie on the multiplicative line through P and Q. Then the multiplicative sum of the multiplicative radian measures of $\angle_\star PQX$ and $\angle_\star RQX$ is e^π.*

Proof. Let θ, θ_1, θ_2, u, v, λ and μ be as in the proof of Theorem 3.30. Note that we have not any representations of w in the terms of u and v and

$$v = -_\star u, \quad v^{\perp_\star} = -_\star u^{\perp_\star}.$$

We have the following possibilities.

1. Let

$$(\mathrm{Rot}_\star\theta_1)\cdot_\star u = w, \quad (\mathrm{Rot}_\star\theta_2)\cdot_\star w = v.$$

Then

$$\begin{aligned} \theta &= \theta_1 +_\star \theta_2 \\ &= e^\pi. \end{aligned}$$

2. Let

$$(\mathrm{Rot}_\star\theta_1)\cdot_\star u = w, \quad (\mathrm{Rot}_\star\theta_2)\cdot_\star v = w.$$

Then

$$\begin{aligned} 0_\star &< \sin_\star\theta_1 \\ &= \langle u^{\perp_\star}, w\rangle_\star \\ &= -_\star\langle v^{\perp_\star}, w\rangle_\star \\ &= -\sin_\star\theta_2, \end{aligned}$$

which is a contradiction.

3. Let

$$(\mathrm{Rot}_\star \theta_1)\cdot_\star w = u, \quad (\mathrm{Rot}_\star \theta_2)\cdot_\star w = v.$$

Then

$$\begin{aligned} 0_\star &< \sin_\star(-_\star\theta_1) \\ &= \langle u^{\perp_\star}, w\rangle_\star \\ &= -_\star\langle v^{\perp_\star}, w\rangle_\star \\ &= -_\star \sin_\star(-_\star\theta_2), \end{aligned}$$

which is a contradiction.

4. Let

$$(\mathrm{Rot}_\star \theta_1)\cdot_\star w = u, \quad (\mathrm{Rot}_\star \theta_2)\cdot_\star v = w.$$

Then

$$\begin{aligned} (\mathrm{Rot}_\star(\theta_1 +_\star \theta_2))\cdot_\star v &= (\mathrm{Rot}_\star\theta_1)(\mathrm{Rot}_\star\theta_2)\cdot_\star v \\ &= (\mathrm{Rot}_\star\theta_1)\cdot_\star w \\ &= u \\ &= (\mathrm{Rot}_\star\theta)\cdot_\star v. \end{aligned}$$

Thus,

$$\begin{aligned} \theta_1 +_\star \theta_2 &= \theta \\ &= e^\pi. \end{aligned}$$

This completes the proof.

□

Definition 3.24. *The multiplicative angles $\angle_\star PQX$ and $\angle_\star RQX$ are said to be multiplicative supplements of each other.*

3.16 Multiplicative Triangles

Definition 3.25. *Let P, Q, $R \in E^2_\star$ be multiplicative nonlinear points. The multiplicative triangle $\triangle_\star PQR$ or PQR is the multiplicative rectilinear figure consisting of the multiplicative segments PQ, PR and RQ. The multiplicative segments are called the multiplicative sides of the multiplicative triangle.*

In Fig. 3.3 it is shown a multiplicative triangle. Let PQR be a multiplicative triangle. Consider PQR as a multiplicative triangle of reference for a multiplicative barycentric coordinate system. Then

1. a point X is a vertex of PQR if and only if two multiplicative barycentric coordinates are $0_\star$.
2. a point X is on PQR if it is a vertex or one barycentric coordinate is $0_\star$ and the other two are $> 0_\star$.

Out[16]=

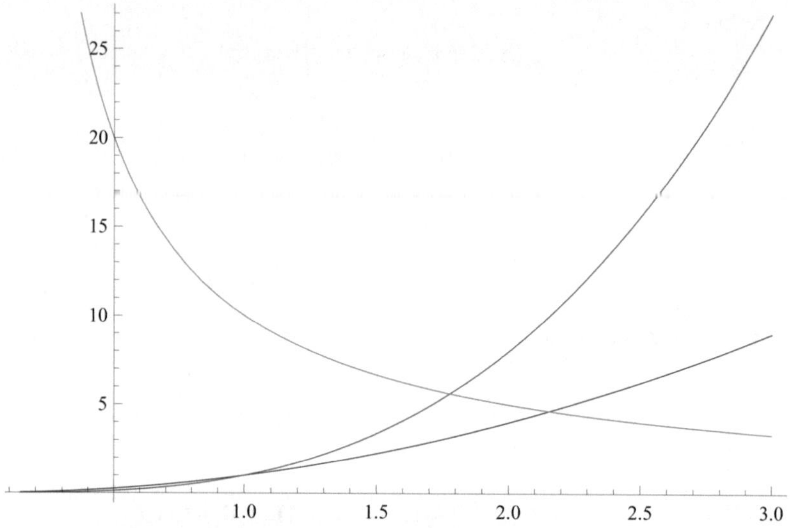

FIGURE 3.3
Multiplicative triangle.

Definition 3.26. *A point X is said to be in the interior of the multiplicative triangle PQR if it is in the interior of all three multiplicative angles determined by the points P, Q and R.*

Theorem 3.32. *Let T be a multiplicative affine transformation and PQR be a multiplicative triangle. Then* $T(\triangle_\star PQR)$ *is a multiplicative triangle.*

Proof. Let

$$P_1 = TP, \quad Q_1 = TQ, \quad R_1 = TR.$$

Since $T^{-1_\star}$ is a multiplicative affine transformation, we have that $T^{-1_\star}$ preserves the multiplicative collinearity. Because PQR is a multiplicative triangle, we have that the points P, Q are R are multiplicative non collinear points. Hence, P_1, Q_1 and R_1 are three multiplicative non collinear points and $P_1Q_1R_1$ is a multiplicative triangle. This completes the proof. □

3.17 Multiplicative Symmetries of a Multiplicative Triangle

Let PQR be a multiplicative triangle. With α we will denote the product of both multiplicative reflections

$$[R, P \longleftrightarrow Q]_\star, \quad [P, Q \longleftrightarrow R]_\star,$$

i.e., $\alpha(PQR) = RPQ$. With β we will denote the multiplicative reflection

$$[R, P \longleftrightarrow Q]_\star,$$

i.e. $\beta(PQR) = QPR$. We have

1.

$$\begin{aligned} \alpha^2(PQR) &= \alpha(\alpha(PQR)) \\ &= \alpha(RPQ) \\ &= QRP. \end{aligned}$$

2.

$$\begin{aligned}\alpha^3(PQR) &= \alpha(\alpha^2(PQR))\\ &= \alpha(QRP)\\ &= PQR,\end{aligned}$$

i.e.,

$$\alpha^3 = I_\star.$$

3.

$$\begin{aligned}\alpha\beta(PQR) &= \alpha(\beta(PQR))\\ &= \alpha(QPR)\\ &= RQP.\end{aligned}$$

4.

$$\begin{aligned}\alpha^2\beta(PQR) &= \alpha^2(\beta(PQR))\\ &= \alpha^2(QPR)\\ &= \alpha(\alpha(QPR))\\ &= \alpha(RQP)\\ &= PRQ.\end{aligned}$$

5. $\alpha^3\beta = \beta$.
6. $\alpha^4 = \alpha$.
7. $\alpha^4\beta = \alpha\beta$.
8.

$$\begin{aligned}\beta\alpha(PQR) &= \beta(\alpha(PQR))\\ &= \beta(RPQ)\\ &= PRQ.\end{aligned}$$

Thus,

$$\beta\alpha = \alpha^2\beta.$$

9.

$$\begin{aligned}\beta\alpha^2(PQR) &= \beta\alpha(\alpha(PQR))\\ &= \beta(\alpha(RPQ))\\ &= \beta(QRP)\\ &= RQP.\end{aligned}$$

Thus,

$$\beta\alpha^2 = \alpha\beta.$$

10. $\beta^2 = I_\star$.

11.

$$\begin{aligned} \beta\alpha\beta(PQR) &= \beta\alpha(\beta(PQR)) \\ &= \beta(\alpha(QPR)) \\ &= \beta(RQP) \\ &= QRP. \end{aligned}$$

Thus,

$$\beta\alpha\beta = \alpha^2.$$

12.

$$\begin{aligned} \beta\alpha^2\beta(PQR) &= \beta\alpha^2(\beta(PQR)) \\ &= \beta\alpha^2(QPR) \\ &= \beta\alpha(\alpha(QPR)) \\ &= \beta\alpha(RQP) \\ &= \beta(\alpha(RQP)) \\ &= \beta(PRQ) \\ &= RPQ. \end{aligned}$$

Thus,

$$\beta\alpha^2\beta = \alpha.$$

13.

$$\begin{aligned} \alpha\beta\alpha(PQR) &= \alpha\beta(\alpha(PQR)) \\ &= \alpha(\beta(RPQ)) \\ &= \alpha(PRQ) \\ &= QPR. \end{aligned}$$

Thus,

$$\alpha\beta\alpha = \beta.$$

14.

$$\begin{aligned} \alpha\beta\alpha^2(PQR) &= \alpha\beta\alpha(\alpha(PQR)) \\ &= \alpha\beta(\alpha(RPQ)) \\ &= \alpha(\beta(QRP)) \\ &= \alpha(RQP) \\ &= PRQ. \end{aligned}$$

Thus,

$$\alpha\beta\alpha^2 = \alpha^2\beta.$$

15. $\alpha\beta^2 = \alpha$.

16.

$$\begin{aligned} \alpha\beta\alpha\beta(PQR) &= \alpha\beta\alpha(\beta(PQR)) \\ &= \alpha\beta(\alpha(QPR)) \\ &= \alpha(\beta(RQP)) \\ &= \alpha(QRP) \\ &= PQR. \end{aligned}$$

Thus,

$$\alpha\beta\alpha\beta = I_\star.$$

17.

$$\begin{aligned}
\alpha\beta\alpha^2\beta(PQR) &= \alpha\beta\alpha^2(\beta(PQR)) \\
&= \alpha\beta\alpha^2(QPR) \\
&= \alpha\beta\alpha(\alpha(QPR)) \\
&= \alpha\beta(\alpha(RQP)) \\
&= \alpha(\beta(PRQ)) \\
&= \alpha(RPQ) \\
&= QRP.
\end{aligned}$$

Thus,

$$\alpha\beta\alpha^2\beta = \alpha^2.$$

18.

$$\begin{aligned}
\alpha^2\beta\alpha(PQR) &= \alpha^2\beta(\alpha(PQR)) \\
&= \alpha^2(\beta(RPQ)) \\
&= \alpha^2(PRQ) \\
&= \alpha(\alpha(PRQ)) \\
&= \alpha(QPR) \\
&= RQP.
\end{aligned}$$

Thus,

$$\alpha^2\beta\alpha = \alpha\beta.$$

19.

$$\begin{aligned}
\alpha^2\beta\alpha^2(PQR) &= \alpha^2\beta\alpha(\alpha(PQR)) \\
&= \alpha^2\beta(\alpha(RPQ)) \\
&= \alpha^2(\beta(QRP)) \\
&= \alpha(\alpha(RQP)) \\
&= \alpha(PRQ) \\
&= QPR.
\end{aligned}$$

Thus,

$$\alpha^2\beta\alpha^2 = \beta.$$

20. $\alpha^2\beta^2 = \alpha^2$.

21.

$$\begin{aligned}
\alpha^2\beta\alpha\beta(PQR) &= \alpha^2\beta\alpha(\beta(PQR)) \\
&= \alpha^2\beta(\alpha(QPR)) \\
&= \alpha^2(\beta(RQP)) \\
&= \alpha(\alpha(QRP)) \\
&= \alpha(PQR) \\
&= RPQ.
\end{aligned}$$

Thus,

$$\alpha^2\beta\alpha\beta = \alpha.$$

22.

$$\begin{aligned}
\alpha^2\beta\alpha^2\beta(PQR) &= \alpha^2\beta\alpha^2(\beta(PQR))\\
&= \alpha^2\beta\alpha^2(QPR)\\
&= \alpha^2\beta\alpha(\alpha(QPR))\\
&= \alpha^2\beta(\alpha(RQP))\\
&= \alpha^2(\beta(PRQ))\\
&= \alpha(\alpha(RPQ))\\
&= \alpha(QRP)\\
&= PQR.
\end{aligned}$$

Thus,

$$\alpha^2\beta\alpha^2\beta = I_\star.$$

Therefore, $\mathscr{A}\mathscr{S}_\star(\Delta_\star)$ is the group of six elements known as $S_{\star 3}$. Algebraically, $S_{\star 3}$ can be described as follows:

$$S_{\star 3} = \{I_\star,\quad \alpha,\quad \alpha^2,\quad \beta,\quad \alpha\beta,\quad \alpha^2\beta\}.$$

Definition 3.27. *A multiplicative triangle is said to be*

1. *scalene if all three multiplicative sides have different multiplicative lengths.*
2. *isosceles if exactly two multiplicative sides have equal multiplicative lengths.*
3. *equilateral if all multiplicative sides have equal multiplicative lengths.*

Theorem 3.33. $\mathscr{S}_\star(\Delta_\star)$ *consists of*

1. *the multiplicative identity if $\Delta_\star$ is scalene.*
2. $\{I_\star, \Omega\}$ *if $\Delta_\star$ is isosceles with*

$$d_\star(P,Q) = d_\star(P,R),$$

and Ω is the multiplicative reflection

$$[P, Q \longleftrightarrow R]_\star.$$

3. *all elements of $\mathscr{A}\mathscr{S}_\star(\Delta_\star)$ if $\Delta_\star$ is equilateral. In this case, two elements are nontrivial multiplicative rotations about the multiplicative centroid, three are the multiplicative reflections and the sixth is the multiplicative identity.*

Proof. If $T = [P, Q \longleftrightarrow R]_\star$ is a multiplicative isometry, then

$$\begin{aligned}
d_\star(P,Q) &= d_\star(TP,TQ)\\
&= d_\star(P,R).
\end{aligned}$$

Thus, at least two multiplicative sides of $\Delta_\star$ have the same multiplicative length. The same holds for the other two multiplicative reflections.

If T is a multiplicative isometry that permutes the vertices cyclically, then

$$\begin{aligned} d_\star(P,Q) &= d_\star(TP,TQ) \\ &= d_\star(Q,R) \\ &= d_\star(TQ,TR) \\ &= d_\star(R,P). \end{aligned}$$

So, $\Delta_\star$ must be equilateral. This completes the proof. □

Theorem 3.34. *Let $\Delta_\star$ be an equilateral multiplicative triangle with multiplicative centroid C. Then $\mathscr{S}_\star(\Delta_\star)$ consists of*

1. *the multiplicative identity.*
2. *the three multiplicative reflections in the multiplicative medians of $\Delta_\star$.*
3. *multiplicative rotations by $\pm_\star e^2/_\star e^3 \cdot_\star e^\pi$ about C.*

Proof. Note that

$$\begin{aligned} 1_\star/_\star e^3 &= e/_\star e^3 \\ &= e^{\frac{\log e}{\log e^3}} \\ &= e^{\frac{1}{3}}, \end{aligned}$$

and

$$\begin{aligned} 1_\star/_\star e^3 +_\star 1_\star/_\star e^3 +_\star 1_\star/_\star e^3 &= e^{\frac{1}{3}} +_\star e^{\frac{1}{3}} +_\star e^{\frac{1}{3}} \\ &= e^{\frac{1}{3}+\frac{1}{3}+\frac{1}{3}} \\ &= e \\ &= 1_\star, \end{aligned}$$

and

$$\begin{aligned} (e^2/_\star e^3)\cdot_\star(1_\star/_\star e^2) &= e^{\frac{2}{3}}\cdot_\star(e/_\star e^2) \\ &= e^{\frac{2}{3}}\cdot_\star e^{\frac{1}{2}} \\ &= e^{\log e^{\frac{2}{3}}\log e^{\frac{1}{2}}} \\ &= e^{\frac{2}{3}\cdot\frac{1}{2}} \\ &= e^{\frac{1}{3}}. \end{aligned}$$

Thus,

$$\begin{aligned} C &= (1_\star/_\star e^3)\cdot_\star P +_\star (e^2/_\star e^3)\cdot_\star \left((1_\star/_\star e^2)\cdot_\star Q +_\star (1_\star/_\star e^2)\cdot_\star R\right), \\ C &= (1_\star/_\star e^3)\cdot_\star Q +_\star (e^2/_\star e^3)\cdot_\star \left((1_\star/_\star e^2)\cdot_\star P +_\star (1_\star/_\star e^2)\cdot_\star R\right), \\ C &= (1_\star/_\star e^3)\cdot_\star R +_\star (e^2/_\star e^3)\cdot_\star \left((1_\star/_\star e^2)\cdot_\star P +_\star (1_\star/_\star e^2)\cdot_\star Q\right). \end{aligned}$$

Hence, C belongs to all three multiplicative medians. So, the product of two multiplicative reflections is a multiplicative rotation by θ about C. If M is a multiplicative midpoint of one of the multiplicative sides of $\Delta_\star$ and

$$T = T_M(\mathrm{Rot}_\star\theta)T_{-_\star M},$$

then T^3 is the multiplicative identity. We have

$$\begin{aligned} T^2 &= T_M(\text{Rot}_\star\theta)T_{-_\star M}T_M(\text{Rot}_\star\theta)T_{-_\star M} \\ &= T_M(\text{Rot}_\star(e^2 \cdot_\star \theta))T_{-_\star M}, \end{aligned}$$

and

$$\begin{aligned} T^3 &= T_M(\text{Rot}_\star(e^2 \cdot_\star \theta))T_{-_\star M}T_M(\text{Rot}_\star\theta)T_{-_\star M} \\ &= T_M(\text{Rot}_\star(e^3 \cdot_\star \theta))T_{-_\star M} \\ &= I_\star \\ &= T_M\left(\text{Rot}_\star\left(\pm_\star e^{2\pi}\right)\right)T_{-_\star M}. \end{aligned}$$

Hence,

$$e^3 \cdot_\star \theta = \pm_\star e^{2\pi}$$

and

$$\begin{aligned} \theta &= \left(\pm_\star e^{2\pi}\right)/_\star e^3 \\ &= e^{\pm 2\pi}/_\star e^3 \\ &= e^{\pm\frac{2}{3}\pi} \\ &= \pm_\star e^2/_\star e^3 \cdot_\star e^\pi. \end{aligned}$$

This completes the proof. □

Exercise 3.9. *Prove that for any multiplicative triangle the multiplicative centroid lies on each multiplicative median and multiplicative divides it in the multiplicative ratio $e^2/_\star e$.*

3.18 Congruence of Multiplicative Angles

Theorem 3.35. *Two multiplicative angles are congruent if and only if they have the same multiplicative radian measure.*

Proof.

1. Let $\mathscr{A}$ and $\mathscr{B}$ be congruent multiplicative angles and let T be a multiplicative isometry so that $T\mathscr{A} = \mathscr{B}$. By Theorem 3.22, it follows that T maps both multiplicative lines of $\mathscr{A}$ to the two multiplicative lines of $\mathscr{B}$. Hence, T maps the vertex of $\mathscr{A}$ to the vertex of $\mathscr{B}$. Let u and v be the multiplicative unit direction vectors of $\mathscr{A}$ and

$$Tx = A \cdot_\star x +_\star b, \quad A \in \mathscr{M}_{\star 2\times 2}, \quad \det_\star A \neq 0_\star, \quad b \in \mathbb{R}^2_\star,$$

for any $x \in \mathbb{R}^2_\star$. Then the multiplicative direction vectors of $\mathscr{B}$ must be $A \cdot_\star u$ and $A \cdot_\star v$. Since A is a multiplicative orthogonal matrix, we get

$$\langle u, v\rangle_\star = \langle A \cdot_\star u, A \cdot_\star v\rangle_\star.$$

Thus, the two multiplicative angles $\mathscr{A}$ and $\mathscr{B}$ have the same multiplicative radian measure.

2. Suppose that $\mathscr{A}$ and $\mathscr{B}$ are two multiplicative angles that have the same multiplicative radian measure θ. Assume that

$$(\mathrm{Rot}_\star\theta)\cdot_\star u = v, \quad (\mathrm{Rot}_\star\theta)\cdot_\star u_1 = v_1,$$

where u, v and u_1, v_1 are the multiplicative unit direction vectors of $\mathscr{A}$ and $\mathscr{B}$, respectively. Suppose that

$$u_1 = (\mathrm{Rot}_\star\phi)\cdot_\star u.$$

Then

$$\begin{aligned} v_1 &= (\mathrm{Rot}_\star\theta)\cdot_\star u_1 \\ &= (\mathrm{Rot}_\star\theta)(\mathrm{Rot}_\star\phi)\cdot_\star u \\ &= (\mathrm{Rot}_\star(\phi +_\star \theta))\cdot_\star u \\ &= (\mathrm{Rot}_\star\phi)(\mathrm{Rot}_\star\theta)\cdot_\star u \\ &= (\mathrm{Rot}_\star\phi)\cdot_\star v. \end{aligned}$$

Let P and Q be the vertices of both multiplicative angles $\mathscr{A}$ and $\mathscr{B}$, respectively. Then, for any $t \geq 0_\star$, we have

$$\begin{aligned} T_Q(\mathrm{Rot}_\star\phi)T_{-_\star P}\cdot_\star (P +_\star t\cdot_\star u) &= T_Q(\mathrm{Rot}_\star\phi)\cdot_\star (t\cdot_\star u) \\ &= T_Q\cdot_\star (t\cdot_\star u_1) \\ &= Q +_\star t\cdot_\star u_1, \end{aligned}$$

and as above

$$T_Q(\mathrm{Rot}_\star\phi)T_{-_\star P}\cdot_\star (P +_\star t\cdot_\star v) = Q +_\star t\cdot_\star v_1.$$

Thus, $\mathscr{A}$ and $\mathscr{B}$ are congruent. This completes the proof.

□

Definition 3.28. *A multiplicative line that intersects two multiplicative lines in distinct multiplicative points is called a multiplicative transversal to these multiplicative lines.*

Let l_1 and l_2 be two multiplicative lines that are multiplicative parallel with multiplicative unit direction v. Let m be a multiplicative transversal for l_1 and l_2 and

$$m \cap l_1 = P_1, \quad m \cap l_2 = P_2.$$

Take

$$\begin{aligned} Q &= P_1 +_\star v, \\ R &= P_2 -_\star v. \end{aligned}$$

Definition 3.29. *The multiplicative angles $\angle_\star QP_1P_2$ and $\angle_\star RP_2P_1$ are said to be alternate multiplicative angles.*

Theorem 3.36. *We have*

$$\angle_\star QP_1P_2 = \angle_\star RP_2P_1.$$

Proof. Let H_{P_2} be the multiplicative half-turn about P_2. Then

$$\begin{aligned} H_{P_2}T_{P_2 -_\star P_1}\cdot_\star P_1 &= H_{P_2}\cdot_\star P_2 \\ &= -_\star P_2 +_\star e^2\cdot_\star P_2 \\ &= P_2, \end{aligned}$$

and

$$\begin{aligned} H_{P_2}T_{P_2-_\star P_1}\cdot_\star Q &= H_{P_2}\cdot_\star(P_2-_\star P_1+_\star Q) \\ &= -_\star P_2+_\star P_1-_\star Q+_\star e^2\cdot_\star P_2 \\ &= P_2+_\star P_1-_\star Q \\ &= P_2+_\star P_1-_\star P_1-_\star v \\ &= P_2-_\star v \\ &= R, \end{aligned}$$

and

$$\begin{aligned} H_{P_2}T_{P_2-_\star P_1}\cdot_\star P_2 &= H_{P_2}\cdot_\star(P_2-_\star P_1+_\star P_2) \\ &= -_\star P_2+_\star P_1-_\star P_2+_\star e^2\cdot_\star P_2 \\ &= P_2+_\star P_1-_\star P_2 \\ &= P_1. \end{aligned}$$

Therefore,

$$H_{P_2}T_{P_2-_\star P_1}\cdot_\star \vec{P_1Q}=\vec{P_2R}$$

and

$$H_{P_2}T_{P_2-_\star P_1}\cdot_\star \vec{P_1P_2}=\vec{P_2P_1}.$$

Therefore,

$$\angle_\star QP_1P_2=\angle_\star PP_2P_1.$$

This completes the proof. □

3.19 Congruence Theorems for Multiplicative Triangles

Theorem 3.37. *Let $\Delta_\star PQR$ and $\Delta_\star P_1Q_1R_1$ be two multiplicative triangles such that*

$$\begin{aligned} d_\star(P,Q) &= d_\star(P_1,Q_1), \\ d_\star(P,R) &= d_\star(P_1,R_1), \\ d_\star(R,Q) &= d_\star(R_1,Q_1). \end{aligned}$$

Then there exists a multiplicative isometry T so that

$$TP=P_1,\quad TQ=Q_1,\quad TR=R_1.$$

Proof. We will prove the assertion in several steps.

1. Firstly, we note that for any two multiplicative lines m_1 and m_2, there is a multiplicative isometry T_1 so that $T_1m_1=m_2$.

(a) If $m_1\parallel_\star m_2$, then we take T_1 to be the multiplicative reflection in the multiplicative line being multiplicative halfway between them.

(b) If $l_1\cap l_2\neq\emptyset$, then T_1 will be the multiplicative reflection in any multiplicative bisectors of the multiplicative angles that they form.

2. Let PQ and P_1Q_1 be multiplicative collinear segments. Since

$$d_\star(P,Q) = d_\star(P_1,Q_1),$$

we have two possibilities.

(a) Let $P -_\star Q = P_1 -_\star Q_1$. Then we take

$$T_2 = T_{P_1 -_\star P}.$$

We have

$$\begin{aligned} T_2 \cdot_\star P &= T_{P_1 -_\star P} \cdot_\star P \\ &= P_1 -_\star P +_\star P \\ &= P_1, \end{aligned}$$

and

$$\begin{aligned} T_2 \cdot_\star Q &= T_{P_1 -_\star P} \cdot_\star Q \\ &= P_1 -_\star P +_\star Q \\ &= P_1 -_\star P_1 +_\star Q_1 \\ &= Q_1. \end{aligned}$$

(b) Let $P -_\star Q = Q_1 -_\star P_1$. Then we take

$$T_2 = H_{P_1} T_{P_1 -_\star P}.$$

We have

$$\begin{aligned} T_2 \cdot_\star P &= H_{P_1} T_{P_1 -_\star P} \cdot_\star P \\ &= H_{P_1} \cdot_\star (P_1 -_\star P +_\star P) \\ &= H_{P_1} \cdot_\star P_1 \\ &= -_\star P_1 +_\star e^2 \cdot_\star P_1 \\ &= P_1, \end{aligned}$$

and

$$\begin{aligned} T_2 \cdot_\star Q &= H_{P_1} T_{P_1 -_\star P} \cdot_\star Q \\ &= H_{P_1} \cdot_\star (P_1 -_\star P +_\star Q) \\ &= -_\star P_1 +_\star P -_\star Q +_\star e^2 \cdot_\star P_1 \\ &= P -_\star Q +_\star P_1 \\ &= Q_1 -_\star P_1 +_\star P_1 \\ &= Q_1. \end{aligned}$$

3. Suppose that

$$\begin{aligned} d_\star(P,R) &= d_\star(P,R_1), \\ d_\star(Q,R) &= d_\star(Q,R_1). \end{aligned}$$

Let l be the multiplicative line through the points P and Q. We have the following two possibilities.

(a) Let R and R_1 lie in the same multiplicative half-plane determined by the multiplicative line l. Then we take $T_3 = I_\star$.

(b) Let R and R_1 lie in different multiplicative half-planes determined by l. Then we take $T_3 = \Omega_l$.

4. Let now, l be the multiplicative line through the points P_1 and Q_1. We take a multiplicative isometry T_1 so that $T_1 l = l_1$. Then we choose a multiplicative isometry T_2 such that

$$T_2 T_1 \cdot_\star P = P_1 \quad \text{and} \quad T_2 T_1 \cdot_\star Q = Q_1.$$

Now, we take T_3 to be the multiplicative isometry so that

$$T_3 T_2 T_1 \cdot_\star R = R_1$$

and T_3 leaves all points of l_1 fixed. Then

$$\begin{aligned} T_3 T_2 T_1 \cdot_\star P &= T_3 \cdot_\star (T_2 T_1 \cdot_\star P) \\ &= T_3 \cdot_\star P_1 \\ &= P_1, \end{aligned}$$

and

$$\begin{aligned} T_3 T_2 T_1 \cdot_\star Q &= T_3 \cdot_\star (T_2 T_1 \cdot_\star Q) \\ &= T_3 \cdot_\star Q_1 \\ &= Q_1, \end{aligned}$$

and

$$T_3 T_2 T_1 \cdot_\star R = R_1.$$

Therefore,

$$T = T_3 T_2 T_1,$$

is the desired multiplicative isometry. This completes the proof. □

Theorem 3.38 (The Law of Multiplicative Cosines). *Let $P, Q, R \in E_\star^2$. Then*

$$\begin{aligned} d_\star(P,R)^{2_\star} &= d_\star(P,Q)^{2_\star} +_\star d_\star(Q,R)^{2_\star} \\ &\quad -_\star e^2 \cdot_\star d_\star(P,Q) \cdot_\star d_\star(Q,R) \cdot_\star \cos_\star \theta, \end{aligned}$$

where θ is the radian measure of $\angle_\star PQR$.

Proof. Let

$$x = P -_\star Q, \quad y = R -_\star Q.$$

We apply the multiplicative polarization identity to x and y and we find

$$\begin{aligned} \langle P -_\star Q, R -_\star Q \rangle_\star &= 1_\star /_\star e^2 \cdot_\star \Big(|P -_\star Q|_\star^{2_\star} +_\star |R -_\star Q|_\star^{2_\star} \\ &\quad -_\star |P -_\star R|_\star^{2_\star} \Big), \end{aligned}$$

or

$$\begin{aligned} d_\star(P,Q)^{2_\star} +_\star d_\star(R,Q)^{2_\star} - d_\star(P,R)^{2_\star} &= e^2 \cdot_\star \langle P -_\star Q, R -_\star Q \rangle_\star \\ &= e^2 \cdot_\star d_\star(P,Q) \cdot_\star d_\star(R,Q) \cdot_\star \cos_\star \theta. \end{aligned}$$

This completes the proof. □

Theorem 3.39. *Let $\Delta_\star PQR$ and $\Delta_\star P_1Q_1R_1$ be two multiplicative triangles such that*

$$\begin{aligned} d_\star(P,Q) &= d_\star(P_1,Q_1), \\ d_\star(Q,R) &= d_\star(Q_1,R_1), \end{aligned}$$

and the multiplicative radian measure of the multiplicative angles $\angle_\star PQR$ and $\angle_\star P_1Q_1R_1$ are the same. Then there exists a multiplicative affine isometry T so that

$$TP = P_1, \quad TQ = Q_1, \quad TR = R_1.$$

Proof. Let θ be the multiplicative radian measure of the multiplicative angles $\angle_\star PQR$ and $\angle_\star P_1Q_1R_1$. Then, by the law of the multiplicative cosines, we find

$$\begin{aligned} & d_\star(P,Q)^{2_\star} +_\star d_\star(R,Q)^{2_\star} -_\star d_\star(P,R)^{2_\star} \\ &= d_\star(P_1,Q_1)^{2_\star} +_\star d_\star(R_1,Q_1)^{2_\star} -_\star d_\star(P,R)^{2_\star} \\ &= e^2 \cdot_\star d_\star(P,Q) \cdot_\star d_\star(Q,R) \cdot_\star \cos_\star \theta \\ &= e^2 \cdot_\star d_\star(P_1,Q_1) \cdot_\star d_\star(Q_1,R_1) \cdot_\star \cos_\star \theta \\ &= d_\star(P_1,Q_1)^{2_\star} +_\star d_\star(R_1,Q_1)^{2_\star} -_\star d_\star(P_1,R_1)^{2_\star} \end{aligned}$$

Hence,

$$d_\star(P_1,R_1) = d_\star(P,R).$$

Now, by Theorem 3.37, it follows the desired result. This completes the proof. □

Corollary 3.6. *The base multiplicative angles of an isosceles multiplicative triangle are congruent.*

Proof. Let $\Delta_\star PQR$ be an isosceles multiplicative triangle and

$$d_\star(P,R) = d_\star(R,Q).$$

Consider $\Delta_\star PQR$ and $\Delta_\star RQP$. We have

$$\begin{aligned} d_\star(P,R) &= d_\star(R,Q), \\ d_\star(Q,R) &= d_\star(R,P) \end{aligned}$$

and the multiplicative angles $\angle_\star PRQ$ and $\angle_\star QRP$ have the same multiplicative radian measure. Hence and Theorem 3.39, it follows that the multiplicative angles $\angle_\star RPQ$ and $\angle_\star RQP$ have the same multiplicative radian measure. This completes the proof. □

3.20 Multiplicative Angle Sum of Multiplicative Triangles

Theorem 3.40. *The multiplicative sum of the multiplicative radian measure of the three multiplicative angles in any multiplicative triangle is e^π.*

Proof. Let $\Delta_\star PQR$ be a multiplicative triangle. The unique multiplicative line through the point P multiplicative parallel to the multiplicative line through Q and R consists of two multiplicative rays PA and PB, where

$$\begin{aligned} A &= P +_\star R -_\star Q, \\ B &= P +_\star Q -_\star R. \end{aligned}$$

Note that Q is in the interior of $\angle_\star BPR$. Then

$$\begin{aligned}\angle_\star BPR &= \angle_\star BPQ +_\star \angle_\star QPR\\ &= \angle_\star PQR +_\star \angle_\star QPR.\end{aligned}$$

Since $\angle_\star BPR$ and $\angle_\star APR$ are supplementary, we get

$$\begin{aligned}e^{\pi} &= \angle_\star BPR +_\star \angle_\star APR\\ &= \angle_\star PQR +_\star \angle_\star QPR +_\star \angle_\star QRP.\end{aligned}$$

This completes the proof. □

Exercise 3.10. *If two multiplicative angles of a multiplicative triangle are, respectively, congruent to two multiplicative angles of another multiplicative triangle, prove that the remaining multiplicative angles are congruent.*

3.21 Advanced Practical Problems

Problem 3.1. *Find the fixed multiplicative points and fixed multiplicative lines for the following multiplicative affine transformations.*

1.
$$A = \begin{pmatrix} 1 & e \\ e^2 & \frac{1}{2} \end{pmatrix}, \quad b = \begin{pmatrix} 1 \\ 4 \end{pmatrix}.$$

2.
$$A = \begin{pmatrix} e^3 & e^4 \\ 5 & 7 \end{pmatrix}, \quad b = \begin{pmatrix} e^3 \\ e^{11} \end{pmatrix}.$$

3.
$$A = \begin{pmatrix} e^2 & e^8 \\ e^6 & 1 \end{pmatrix}, \quad b = \begin{pmatrix} e^4 \\ 5 \end{pmatrix}.$$

4.
$$A = \begin{pmatrix} e^3 & e \\ e^4 & e^2 \end{pmatrix}, \quad b = \begin{pmatrix} e \\ e^8 \end{pmatrix}.$$

5.
$$A = \begin{pmatrix} 1 & 11 \\ 18 & 10 \end{pmatrix}, \quad b = \begin{pmatrix} 5 \\ 8 \end{pmatrix}.$$

Problem 3.2. *Let $u = (1,6)$ and $v = (10,21)$. Find the multiplicative radian measure.*

Problem 3.3. *Find the multiplicative centroid of*

$$\mathcal{F} = \{P_1(e^7,e^8),\quad P_2(3,4),\quad P_3(1,e),\quad P_4(e^{11},e),\quad P_5(e^3,e^2)\}.$$

Problem 3.4. *Find the multiplicative barycentric coordinates of $X(1,4)$ with multiplicative triangle of reference*

$$P(e^4,e^5),\quad Q(e^{-1},e^2),\quad R(e^3,e^2).$$

Problem 3.5. *Let T be a multiplicative glide reflection with axis m. If l is a multiplicative line so that $l \neq m$, prove that $Tl \parallel_\star l$ if and only if $l \parallel_\star m$ or $l \perp_\star m$.*

Problem 3.6. *Let* $P(e,1)$, $Q(3,e^2)$, $R(4,11)$. *Find the multiplicative radian measure of the following multiplicative angles*

$$\angle_\star PQR, \quad \angle_\star RPQ, \quad \angle_\star PRQ.$$

Problem 3.7. *Let* $P(3,8)$, $Q(e^3,e^2)$, $R(e^4,1)$. *Find the equations of the multiplicative lines through the multiplicative medians of* $\triangle_\star PQR$.

Problem 3.8. *Let* $P,Q,R,R_1 \in E_\star^2$ *be distinct points so that*

$$\begin{aligned} d_\star(P,R) &= d_\star(P,R_1), \\ d_\star(Q,R) &= d_\star(Q,R_1). \end{aligned}$$

Prove that $R_1 = \Omega_l R$, *where* l *is the multiplicative line through the points* P *and* Q.

Problem 3.9. *Let*

$$P\left(e^2,e^3\right), \quad Q\left(e^4,e^5\right), \quad R\left(e^6,e^7\right).$$

Determine if $\triangle_\star PQR$ *is isosceles, scalene or equilateral.*

Problem 3.10. *Let*

$$P\left(e,e^4\right), \quad Q\left(e^3,7\right), \quad R\left(e^7,e^9\right).$$

Find the multiplicative radian measure of $\angle_\star PQR$.

4

Finite Groups of Multiplicative Isometries of $E_\star^2$

In this chapter, cyclic and dihedral subgroups of $O_\star(e^2)$ are defined and some of their properties are investigated. Conjugate subgroups, orbits and stabilizers are introduced. In the chapter, regular multiplicative polygons are studied.

4.1 Cyclic and Dihedral Groups

Let $m > 0$ and $\alpha = \mathrm{Rot}_\star\left(e^{\frac{2\pi}{m}}\right)$. The smallest subgroup of $O_\star\left(e^2\right)$ containing α is denoted by $C_{\star m}$. Note that, for $m = 1$, we have

$$\begin{aligned}
\mathrm{Rot}_\star\left(e^{2\pi}\right) &= \begin{pmatrix} \cos_\star\left(e^{2\pi}\right) & -_\star \sin_\star\left(e^{2\pi}\right) \\ \sin_\star\left(e^{2\pi}\right) & \cos_\star\left(e^{2\pi}\right) \end{pmatrix} \\
&= \begin{pmatrix} e^{\cos\left(\log\left(e^{2\pi}\right)\right)} & e^{-\sin\left(\log\left(e^{2\pi}\right)\right)} \\ e^{\sin\left(\log\left(e^{2\pi}\right)\right)} & e^{\cos\left(\log\left(e^{2\pi}\right)\right)} \end{pmatrix} \\
&= \begin{pmatrix} e & 1 \\ 1 & e \end{pmatrix} \\
&= \begin{pmatrix} 1_\star & 0_\star \\ 0_\star & 1_\star \end{pmatrix} \\
&= I_\star.
\end{aligned}$$

Thus, $C_{\star 1} = \{I_\star\}$. By the properties of the multiplicative rotation, it follows that

$$\alpha^{k_\star} = \mathrm{Rot}_\star\left(e^{\frac{2\pi k}{m}}\right).$$

So,

$$C_{\star m} = \{I_\star, \quad \alpha, \quad \alpha^{2_\star}, \quad \ldots, \quad \alpha^{(m-1)_\star}\}.$$

Definition 4.1. *Any group isomorphic to $C_{\star m}$ is called cyclic group of order m.*

Let $\beta = \mathrm{Ref}_\star(0_\star)$. The smallest subgroup of $O_\star\left(e^2\right)$ containing both α and β is denoted by $D_{\star m}$.

Theorem 4.1. *In $D_{\star m}$ we have*

$$\alpha\beta\alpha = \beta \quad or \quad \beta\alpha = \alpha^{-1_\star}\beta.$$

DOI: 10.1201/9781003325284-4

Proof. By the properties of the multiplicative rotation and multiplicative reflection, we get

$$\begin{aligned}
\alpha\beta\alpha &= \mathrm{Rot}_\star\left(e^{\frac{2\pi}{m}}\right)\mathrm{Ref}_\star(0_\star)\mathrm{Rot}_\star\left(e^{\frac{2\pi}{m}}\right)\\
&= \mathrm{Rot}_\star\left(e^{\frac{2\pi}{m}}\right)\mathrm{Ref}_\star\left(0_\star -_\star (1_\star/_\star e^2)\cdot_\star e^{\frac{2\pi}{m}}\right)\\
&= \mathrm{Ref}\left(0_\star -_\star (1_\star/_\star e^2)\cdot_\star e^{\frac{2\pi}{m}} +_\star (1_\star/_\star e^2)\cdot_\star e^{\frac{2\pi}{m}}\right)\\
&= \mathrm{Ref}_\star(0_\star)\\
&= \beta.
\end{aligned}$$

This completes the proof. □

Theorem 4.2. *We have*

$$\beta\alpha^{j_\star}\beta\alpha^{k_\star} = \alpha^{-_\star j_\star}\beta^{2_\star}\alpha^{k_\star} = \alpha^{(k-j)_\star}, \quad k \geq j.$$

Proof. We have

$$\begin{aligned}
\beta\alpha^{j_\star}\beta\alpha^{k_\star} &= \beta\alpha^{j_\star}\beta\alpha^{j_\star}\alpha^{(k-j)_\star}\\
&= \beta\alpha^{(j-1)_\star}\alpha\beta\alpha\alpha^{(j-1)_\star}\alpha^{(k-j)_\star}\\
&= \beta\alpha^{(j-1)_\star}\beta\alpha^{(j-1)_\star}\alpha^{(k-j)_\star}\\
&\vdots\\
&= \beta^{2_\star}\alpha^{(k-j)_\star}\\
&= \alpha^{(k-j)_\star}.
\end{aligned}$$

This completes the proof. □

Theorem 4.3. *The index $[D_{\star m} : C_{\star m}]$ of $C_{\star m}$ in $D_{\star m}$ is equal to 2.*

Proof. Note that $C_{\star m}$ and $\beta C_{\star m}$ consist of m elements. Now, using Theorem 4.2, we get

$$D_{\star m} = C_{\star m} \cup (\beta C_{\star m}).$$

This completes the proof. □

Definition 4.2. *Any group isomorphic to $D_{\star m}$ is called a dihedral group.*

4.2 Conjugate Subgroups

Definition 4.3. *Let G be a group. Two subgroups K and H are called conjugate in G if there exists an element $g \in G$ so that*

$$K = g^{-1_\star}Hg.$$

Theorem 4.4. *Let g and T be multiplicative isometries of $E_\star^2$. Then*

1. *if T is a multiplicative reflection, so is $g^{-1_\star}Tg$.*
2. *if T is a multiplicative rotation, so is $g^{-1_\star}Tg$.*

3. *if T is a multiplicative translation, so is $g^{-1_\star}Tg$.*

Proof. 1. Let $T = \Omega_l$ be a multiplicative reflection. Then for any $x \in g^{-1_\star}l$, we have

$$\begin{aligned} g^{-1_\star}\Omega_l g \cdot_\star x &= g^{-1_\star} g \cdot_\star x \\ &= x. \end{aligned}$$

Thus, $g^{-1_\star}l$ is pointwise fixed.

2. Let $T = \Omega_l\Omega_m$ be a multiplicative rotation with center P. Then

$$\begin{aligned} g^{-1_\star}Tg &= g^{-1_\star}\Omega_l\Omega_m g \\ &= \left(g^{-1_\star}\Omega_l g\right)\left(g^{-1_\star}\Omega_m g\right). \end{aligned}$$

Since $g^{-1_\star}\Omega_l g$ and $g^{-1_\star}\Omega_m g$ are multiplicative reflections, we conclude that $g^{-1_\star}Tg$ is a product of two multiplicative reflections, and hence, it is a multiplicative rotation about $g^{-1_\star}P$.

3. Let $\Omega_l\Omega_m$ be a multiplicative translation. Then $m \parallel_\star l$. We have that $g^{-1_\star}m \parallel_\star g^{-1_\star}l$ and thus, $g^{-1_\star}Tg$ is a multiplicative translation. This completes the proof.

□

Definition 4.4. *Groups that are conjugate in $\mathscr{S}_\star(E^2_\star)$ are said to be multiplicative geometrically equivalent.*

Example 4.1. *Consider $D_{\star 1}$ and $C_{\star 2}$. Both groups are of order two and they are isomorphic. We have that $C_{\star 2}$ is generated by α and $D_{\star 1}$ is generated by β. Note that the equation*

$$g^{-1_\star}\alpha g = \beta,$$

is impossible because β has a multiplicative line of fixed points and α has exactly one fixed point. Therefore, $D_{\star 1}$ and $C_{\star 2}$ are not multiplicative geometrically equivalent.

Theorem 4.5. *Let*

$$\alpha = Rot_\star\left(e^{\frac{2\pi}{m}}\right), \quad \beta = Ref_\star 0_\star \quad and \quad \gamma = Ref_\star\theta.$$

Then the group $\{\alpha,\gamma\}$ is conjugate to the dihedral group $D_{\star m} = \{\alpha,\beta\}$.

Proof. By the properties of the multiplicative rotation and multiplication reflection, we get

$$\begin{aligned} \mathrm{Rot}_\star\theta\mathrm{Ref}_\star 0_\star\mathrm{Rot}_\star(-_\star\theta) &= \mathrm{Rot}_\star\theta\mathrm{Ref}_\star\left(0_\star -_\star (1_\star/_\star e^2)\cdot_\star(-_\star\theta)\right) \\ &= \mathrm{Rot}_\star\theta\mathrm{Ref}_\star\left((1_\star/_\star e^2)\cdot_\star\theta\right) \\ &= \mathrm{Ref}_\star\left((1_\star/_\star e^2)\cdot_\star\theta +_\star (1_\star/_\star e^2)\cdot_\star\theta\right) \\ &= \mathrm{Ref}_\star\theta. \end{aligned}$$

This completes the proof. □

Theorem 4.6. *Let $\mathscr{F}$ be a figure in $E^2_\star$ and let g be a multiplicative isometry of $E^2_\star$. Then $\mathscr{S}_\star(\mathscr{F})$ and $\mathscr{S}_\star(g\cdot_\star\mathscr{F})$ are conjugate in $\mathscr{S}_\star(E^2_\star)$.*

Proof. Let h be a multiplicative symmetry of $\mathscr{F}$. Then

$$\begin{aligned} ghg^{-1_\star}g\cdot_\star\mathscr{F} &= gh\cdot_\star\mathscr{F} \\ &= g\cdot_\star\mathscr{F}. \end{aligned}$$

Thus, $ghg^{-1_\star}$ is a multiplicative symmetry of $g\cdot_\star\mathscr{F}$. Let now, h_1 be a multiplicative symmetry of $g\cdot_\star\mathscr{F}$. Then

$$\begin{aligned} g^{-1_\star}hg\cdot_\star\mathscr{F} &= g^{-1_\star}g\cdot_\star\mathscr{F} \\ &= \mathscr{F}, \end{aligned}$$

i.e., $g^{-1_\star}hg$ is a multiplicative symmetry of $\mathscr{F}$. Therefore,

$$g^{-1_\star}\mathscr{S}_\star(g\cdot_\star\mathscr{F})g=\mathscr{F}.$$

This completes the proof. □

4.3 Orbits and Stabilizers

Definition 4.5. *Let X be a set and let G be a group of multiplicative transformations of X. Take $x\in X$ arbitrarily. Then*

$$G\cdot_\star x=\{g\cdot_\star x: g\in G\},$$

is called the orbit of x by G. When G is understood, we will write $Orbit(x)$. The set

$$G_x=\{g\in G: g\cdot_\star x=x\},$$

is called the stabilizer of x. When it is understood, we will write $Stab(x)$.

Theorem 4.7. *For any $x\in X$, we have that $Stab(x)$ is a group.*

Proof. Let $g_1,g_2,g_3\in \mathrm{Stab}(x)$. Then

$$\begin{aligned} g_1\cdot_\star x &= x, \\ g_2\cdot_\star x &= x, \end{aligned}$$

and

$$\begin{aligned} g_1g_2\cdot_\star x &= g_1(g_2\cdot_\star x) \\ &= g_1\cdot_\star x \\ &= x, \end{aligned}$$

i.e., $g_1g_2\in \mathrm{Stab}(x)$. Next,

$$g_1(g_2g_3)=(g_1g_2)g_3$$

and

$$\begin{aligned} g_1g_1^{-1_\star} &= I_\star, \\ g_1I_\star &= g_1. \end{aligned}$$

This completes the proof. □

Theorem 4.8. *For any $x\in X$, there is a bijection of the set of coset determined by $Stab(x)$ onto $Orbit(x)$.*

Proof. Consider $\tau : G \to \text{Orbit}(x)$ defined by

$$\tau g = g \cdot_\star x.$$

We have that $\tau : G \to \text{Orbit}(x)$ is surjective. Let $\pi : G \to G/G_x$ be a homomorphism. Set

$$\widetilde{\tau}\pi g = \tau g.$$

Let $\pi g = \pi \widetilde{g}$ be two representations of an elements of G/G_x. Since g, $\widetilde{g}$ belong to the same coset, we have

$$\widetilde{g}^{-1_\star} g \cdot_\star x = x.$$

Hence,

$$\begin{aligned} \tau g &= g \cdot_\star x \\ &= (\widetilde{g}\widetilde{g}^{-1_\star})(g \cdot_\star x) \\ &= \widetilde{g}(\widetilde{g}^{-1_\star} g \cdot_\star x) \\ &= \widetilde{g} \cdot_\star x \\ &= \tau \widetilde{g}. \end{aligned}$$

Thus, $\widetilde{\tau} : G/G_x \to G \cdot_\star x$ is well defined. Also, it is surjective. If

$$\widetilde{\tau}\pi g = \widetilde{\tau}\pi\widetilde{g},$$

then

$$\tau g = \tau \widetilde{g}$$

or

$$g \cdot_\star x = \widetilde{g} \cdot_\star x,$$

or

$$\widetilde{g}^{-1_\star} g \cdot_\star x = x,$$

i.e., $\widetilde{g}^{-1_\star} g \in G_x$ and $\pi g = \pi \widetilde{g}$. Therefore $\widetilde{\tau} : G/G_x \to G \cdot_\star x$ is injective. This completes the proof. □

Theorem 4.9. *Let $\mathscr{F}$ be a multiplicative rectilinear figure with only one vertex P. Then $\mathscr{S}_\star(\mathscr{F})$ is a finite group.*

Proof. Note that $\widetilde{\mathscr{F}}$ contains finite number, say m, of multiplicative lines through m. We have that $m \geq 2$. Take $Q \in \mathscr{F}$, $Q \neq P$. Then $\text{Stab}(Q)$, consisting of isometries leaving P and Q fixed, has at most two elements: the multiplicative identity and/or the multiplicative reflection $P \longleftrightarrow Q$. Next, $\text{Orbit}(Q)$ by $\mathscr{S}_\star(\mathscr{F})$ consists of points on $\mathscr{F}$ whose multiplicative distance from P is $d_\star(P,Q)$. Since $\widetilde{\mathscr{F}}$ consists m multiplicative lines through P, we can have at most $2m$ points. Thus,

$$\begin{aligned} \#\mathscr{S}_\star(\mathscr{F}) &\leq \#\text{Stab}(Q) \cdot \#\text{Orbit}(Q) \\ &= 2(2m) \\ &= 4m. \end{aligned}$$

This completes the proof. □

Definition 4.6. *Let G be a group of multiplicative transformations of X. If there is a point $x_0 \in X$ that is fixed for every multiplicative transformation in G, we call x_0 a fixed point of G.*

Let x_0 be a fixed point of G. Then

$$\text{Orbit}(x_0) = \{x_0\}, \quad \text{Stab}(x_0) = G.$$

4.4 Regular Multiplicative Polygons

Definition 4.7. *Let $m \geq 2$, $P, Q \in E_\star^2$, $P \neq Q$. For each $k \in \mathbb{N}$, set*

$$Q_k = T_P\left(Rot_\star\left(e^{\frac{2\pi k}{m}}\right)\right) T_{-_\star P} Q, \quad k \in \{1, \ldots, m\}.$$

The union of all multiplicative segments Q_kQ_{k+1}, $k \in \{1, \ldots, m-1\}$, will be called regular multiplicative polygon. Multiplicative polygon with m multiplicative sides sometimes will be abbreviated as multiplicative m-gon.

Let $\theta = e^{\frac{2\pi}{m}}$, $\delta = e^{\frac{(j-k)2\pi}{m}}$, $j, k \in \{1, \ldots, m\}$. By the definition, using that $\alpha = \mathrm{Rot}_\star\left(e^{\frac{2\pi}{m}}\right)$, we get

$$Q_k = P +_\star \alpha^{k_\star} \cdot_\star (Q -_\star P), \quad k \in \{1, \ldots, m\}.$$

Lemma 4.1. *We have*

$$\sum_{\star k=1}^{m} \alpha^{k_\star} = O_\star.$$

Proof. We have

$$\begin{aligned}
\alpha^{k_\star} &= \mathrm{Rot}_\star\left(e^{\frac{2k\pi}{m}}\right) \\
&= \begin{pmatrix} \cos_\star\left(e^{\frac{2k\pi}{m}}\right) & -_\star \sin_\star\left(e^{\frac{2k\pi}{m}}\right) \\ \sin_\star\left(e^{\frac{2k\pi}{m}}\right) & \cos_\star\left(e^{\frac{2k\pi}{m}}\right) \end{pmatrix} \\
&= \begin{pmatrix} e^{\cos\left(\frac{2k\pi}{m}\right)} & e^{-\sin\left(\frac{2k\pi}{m}\right)} \\ e^{\sin\left(\frac{2k\pi}{m}\right)} & e^{\cos\left(\frac{2k\pi}{m}\right)} \end{pmatrix}, \quad k \in \{1, \ldots, m\}.
\end{aligned}$$

Hence,

$$\begin{aligned}
\sum_{\star k=1}^{m} \alpha^{k_\star} &= \sum_{\star k=1}^{m} \begin{pmatrix} e^{\cos\left(\frac{2k\pi}{m}\right)} & e^{-\sin\left(\frac{2k\pi}{m}\right)} \\ e^{\sin\left(\frac{2k\pi}{m}\right)} & e^{\cos\left(\frac{2k\pi}{m}\right)} \end{pmatrix} \\
&= \begin{pmatrix} e^{\sum\limits_{k=1}^{m} \cos\left(\frac{2k\pi}{m}\right)} & e^{-\sum\limits_{k=1}^{m} \sin\left(\frac{2k\pi}{m}\right)} \\ e^{\sum\limits_{k=1}^{m} \sin\left(\frac{2k\pi}{m}\right)} & e^{\sum\limits_{k=1}^{m} \cos\left(\frac{2k\pi}{m}\right)} \end{pmatrix} \\
&= \begin{pmatrix} e^0 & e^0 \\ e^0 & e^0 \end{pmatrix} \\
&= \begin{pmatrix} 0_\star & 0_\star \\ 0_\star & 0_\star \end{pmatrix} \\
&= O_\star.
\end{aligned}$$

This completes the proof. □

Theorem 4.10. *Let $\{Q_k\}_{k=1}^{m}$ be the vertices of multiplicative m-gon. Then*

$$(1_\star /_\star m) \cdot_\star \sum_{\star k=1}^{m} Q_k = P.$$

Proof. We have, using Lemma 4.1,

$$\begin{aligned}(1_\star/_\star m)\cdot_\star \sum_{\star k=1}^{m} Q_k &= (1_\star/_\star m)\cdot_\star \sum_{\star k=1}^{m}(P+_\star \alpha^{k_\star}\cdot_\star(Q-_\star P))\\ &= (1_\star/_\star m)\cdot_\star m\cdot_\star P\\ &\quad +_\star(1_\star/_\star m)\cdot_\star(Q-_\star P)\cdot_\star \sum_{\star k=1}^{m}\alpha^{k_\star}\\ &= P.\end{aligned}$$

This completes the proof. □

Set

$$J=\mathrm{Rot}_\star\left(e^{\frac{\pi}{2}}\right).$$

Then

$$\begin{aligned}J &= \begin{pmatrix}\cos_\star\left(e^{\frac{\pi}{2}}\right) & -_\star\sin_\star\left(e^{\frac{\pi}{2}}\right)\\ \sin_\star\left(e^{\frac{\pi}{2}}\right) & \cos_\star\left(e^{\frac{\pi}{2}}\right)\end{pmatrix}\\ &= \begin{pmatrix} e^{\cos\left(\log\left(e^{\frac{\pi}{2}}\right)\right)} & e^{-\sin\left(\log\left(e^{\frac{\pi}{2}}\right)\right)}\\ e^{\sin\left(\log\left(e^{\frac{\pi}{2}}\right)\right)} & e^{\cos\left(\log\left(e^{\frac{\pi}{2}}\right)\right)}\end{pmatrix}\\ &= \begin{pmatrix} e^{\cos\left(\frac{\pi}{2}\right)} & e^{-\sin\left(\frac{\pi}{2}\right)}\\ e^{\sin\left(\frac{\pi}{2}\right)} & e^{\cos\left(\frac{\pi}{2}\right)}\end{pmatrix}\\ &= \begin{pmatrix} 1 & e^{-1}\\ e & 1\end{pmatrix}\\ &= \begin{pmatrix} 0_\star & -_\star 1_\star\\ 1_\star & 0_\star\end{pmatrix},\end{aligned}$$

and

$$\begin{aligned}J^{2_\star} &= \mathrm{Rot}_\star\left(e^{\frac{\pi}{2}}\right)\mathrm{Rot}_\star\left(e^{\frac{\pi}{2}}\right)\\ &= \mathrm{Rot}_\star\left(e^{\pi}\right)\\ &= \begin{pmatrix}\cos_\star(e^\pi) & -_\star\sin_\star(e^\pi)\\ \sin_\star(e^\pi) & \cos_\star(e^\pi)\end{pmatrix}\\ &= \begin{pmatrix} e^{\cos(\log(e^\pi))} & e^{-\sin(\log(e^\pi))}\\ e^{\sin(\log(e^\pi))} & e^{\cos(\log(e^\pi))}\end{pmatrix}\\ &= \begin{pmatrix} e^{-1} & 1\\ 1 & e^{-1}\end{pmatrix}\\ &= \begin{pmatrix} -_\star 1_\star & 0_\star\\ 0_\star & -_\star 1_\star\end{pmatrix}\\ &= -_\star I_\star,\end{aligned}$$

i.e.,

$$J^{2_\star} = -_\star I_\star. \tag{4.1}$$

Next, if $v=(v_1,v_2)$, then

$$\begin{aligned} J\cdot_\star v &= \begin{pmatrix} 1 & e^{-1} \\ e & 1 \end{pmatrix} \cdot_\star \begin{pmatrix} v_1 \\ v_2 \end{pmatrix} \\ &= \begin{pmatrix} e^{\log 1 \log v_1 + \log e^{-1} \log v_2} \\ e^{\log e \log v_1 + \log 1 \log v_2} \end{pmatrix} \\ &= \begin{pmatrix} e^{-\log v_2} \\ e^{\log v_1} \end{pmatrix}. \end{aligned}$$

Hence,

$$\begin{aligned} \langle v, J\cdot_\star v\rangle_\star &= e^{\log v_1 \log e^{-\log v_2} + \log v_2 \log e^{\log v_1}} \\ &= e^{-\log v_1 \log v_2 + \log v_1 \log v_2} \\ &= e^0 \\ &= 0_\star, \end{aligned}$$

i.e.,

$$\langle v, J\cdot_\star v\rangle_\star = 0_\star. \tag{4.2}$$

Moreover,

$$\begin{aligned} &\cos_\star\left((\delta -_\star \theta)/_\star e^2\right)\cdot_\star I_\star +_\star \sin_\star\left((\delta -_\star \theta)/_\star e^2\right)\cdot_\star J \\ &= \begin{pmatrix} \cos_\star\left((\delta -_\star \theta)/_\star e^2\right) & 0_\star \\ 0_\star & \cos_\star\left((\delta -_\star \theta)/_\star e^2\right) \end{pmatrix} \\ &\quad +_\star \sin_\star\left((\delta -_\star \theta)/_\star e^2\right)\cdot_\star \begin{pmatrix} \cos_\star\left(e^{\frac{\pi}{2}}\right) & -_\star \sin_\star\left(e^{\frac{\pi}{2}}\right) \\ \sin_\star\left(e^{\frac{\pi}{2}}\right) & \cos_\star\left(e^{\frac{\pi}{2}}\right) \end{pmatrix} \\ &= \begin{pmatrix} \cos_\star\left((\delta -_\star \theta)/_\star e^2\right) & 0_\star \\ 0_\star & \cos_\star\left((\delta -_\star \theta)/_\star e^2\right) \end{pmatrix} \\ &\quad +_\star \begin{pmatrix} 0_\star & -_\star \sin_\star\left((\delta -_\star \theta)/_\star e^2\right) \\ \sin_\star\left((\delta -_\star \theta)/_\star e^2\right) & 0_\star \end{pmatrix} \\ &= \begin{pmatrix} \cos_\star\left((\delta -_\star \theta)/_\star e^2\right) & -_\star \sin_\star\left((\delta -_\star \theta)/_\star e^2\right) \\ \sin_\star\left((\delta -_\star \theta)/_\star e^2\right) & \cos_\star\left((\delta -_\star \theta)/_\star e^2\right) \end{pmatrix} \\ &= \begin{pmatrix} \cos_\star\left(e^{\frac{\pi}{2}}\right) & -_\star \sin_\star\left(e^{\frac{\pi}{2}}\right) \\ \sin_\star\left(e^{\frac{\pi}{2}}\right) & \cos_\star\left(e^{\frac{\pi}{2}}\right) \end{pmatrix} \\ &= \mathrm{Rot}_\star\left((\delta -_\star \theta)/_\star e^2\right), \end{aligned}$$

i.e.,

$$\cos_\star\left((\delta -_\star \theta)/_\star e^2\right)\cdot_\star I_\star +_\star \sin_\star\left((\delta -_\star \theta)/_\star e^2\right)\cdot_\star J = \mathrm{Rot}_\star\left((\delta -_\star \theta)/_\star e^2\right). \tag{4.3}$$

Also, we have

$$
\begin{aligned}
& e^2 \cdot_\star \left(\sin_\star(\theta /_\star e^2)\right) \cdot_\star J \cdot_\star \left(\mathrm{Rot}_\star(\theta /_\star e^2)\right) \\
&= e^2 \cdot_\star \left(\sin_\star(\theta /_\star e^2)\right) \cdot_\star \mathrm{Rot}_\star\left(e^{\frac{\pi}{2}}\right) \cdot_\star \mathrm{Rot}_\star(\theta /_\star e^2) \\
&= e^2 \cdot_\star \left(\sin_\star(\theta /_\star e^2)\right) \cdot_\star \mathrm{Rot}_\star\left(e^{\frac{\pi}{2}}\right) \cdot_\star \mathrm{Rot}_\star\left(e^{\frac{\pi}{m}}\right) \\
&= e^2 \cdot_\star \left(\sin_\star(\theta /_\star e^2)\right) \cdot_\star \mathrm{Rot}_\star\left(e^{\frac{\pi}{2}} +_\star e^{\frac{\pi}{m}}\right) \\
&= e^2 \cdot_\star \left(\sin_\star(\theta /_\star e^2)\right) \cdot_\star \mathrm{Rot}_\star\left(e^{\frac{\pi(m+2)}{2m}}\right) \\
&= e^2 \cdot_\star \left(\sin_\star\left(e^{\frac{\pi}{m}}\right)\right) \cdot_\star \begin{pmatrix} \cos_\star\left(e^{\frac{\pi}{m}+\frac{\pi}{2}}\right) & -_\star \sin_\star\left(e^{\frac{\pi}{m}+\frac{\pi}{2}}\right) \\ \sin_\star\left(e^{\frac{\pi}{m}+\frac{\pi}{2}}\right) & \cos_\star\left(e^{\frac{\pi}{m}+\frac{\pi}{2}}\right) \end{pmatrix} \\
&= e^2 \cdot_\star \left(\sin_\star\left(e^{\frac{\pi}{m}}\right)\right) \cdot_\star \begin{pmatrix} e^{\cos\left(\frac{\pi}{m}+\frac{\pi}{2}\right)} & e^{-\sin\left(\frac{\pi}{m}+\frac{\pi}{2}\right)} \\ e^{\sin\left(\frac{\pi}{m}+\frac{\pi}{2}\right)} & e^{\cos\left(\frac{\pi}{m}+\frac{\pi}{2}\right)} \end{pmatrix} \\
&= e^2 \cdot_\star e^{\sin\frac{\pi}{m}} \cdot_\star \begin{pmatrix} e^{-\sin\frac{\pi}{m}} & e^{-\cos\frac{\pi}{m}} \\ e^{\cos\frac{\pi}{m}} & e^{-\sin\frac{\pi}{m}} \end{pmatrix} \\
&= \begin{pmatrix} e^{-2\left(\sin\frac{\pi}{m}\right)^2} & e^{-2\sin\frac{\pi}{m}\cos\frac{\pi}{m}} \\ e^{2\sin\frac{\pi}{m}\cos\frac{\pi}{m}} & e^{-2\left(\sin\frac{\pi}{m}\right)^2} \end{pmatrix} \\
&= \begin{pmatrix} e^{\cos\frac{2\pi}{m}-1} & e^{-\sin\frac{2\pi}{m}} \\ e^{\sin\frac{2\pi}{m}} & e^{\cos\frac{2\pi}{m}-1} \end{pmatrix} \\
&= \begin{pmatrix} \cos_\star\left(e^{\frac{2\pi}{m}}\right) -_\star 1_\star & -_\star \sin_\star\left(e^{\frac{2\pi}{m}}\right) \\ \sin_\star\left(e^{\frac{2\pi}{m}}\right) & \cos_\star\left(e^{\frac{2\pi}{m}}\right) -_\star 1_\star \end{pmatrix} \\
&= \mathrm{Rot}_\star\theta -_\star I_\star,
\end{aligned}
$$

i.e.,

$$e^2 \cdot_\star \left(\sin_\star(\theta -_\star e^2)\right) \cdot_\star J \cdot_\star \left(\mathrm{Rot}_\star(\theta /_\star e^2)\right) = \mathrm{Rot}_\star\theta -_\star I_\star. \tag{4.4}$$

As above,

$$e^2 \cdot_\star \left(\sin_\star(\delta /_\star e^2)\right) \cdot_\star J \cdot_\star \left(\mathrm{Rot}_\star(\delta /_\star e^2)\right) = \mathrm{Rot}_\star\delta -_\star I_\star. \tag{4.5}$$

Theorem 4.11. *Let $\{Q_k\}_{k=1}^m$ be the vertices of a multiplicative m-gon and l_k, $k \in \{1,\dots,m\}$, be the multiplicative lines through the points Q_k and Q_{k+1}. Then the whole multiplicative polygon lies on the same multiplicative side of l_k, $k \in \{1,\dots,m\}$.*

Proof. Take $k \in \{1,\dots,m\}$ arbitrarily. Let

$$v = Q_k -_\star P.$$

We write

$$
\begin{aligned}
Q_{k+1} &= P +_\star (\mathrm{Rot}_\star\theta) \cdot_\star v, \\
Q_j &= P +_\star (\mathrm{Rot}_\star\delta) \cdot_\star v.
\end{aligned}
$$

Then, using (4.4), we find

$$
\begin{aligned}
Q_{k+1} -_\star Q_k &= P +_\star (\mathrm{Rot}_\star\theta) \cdot_\star v -_\star P -_\star v \\
&= (\mathrm{Rot}_\star\theta -_\star I_\star) \cdot_\star v \\
&= e^2 \cdot_\star \left(\sin_\star(\theta /_\star e^2)\right) \cdot_\star J \cdot_\star \left(\mathrm{Rot}_\star(\theta /_\star e^2)\right) \cdot_\star v
\end{aligned}
$$

and using (4.5), we obtain

$$\begin{aligned} Q_j -_\star Q_k &= P +_\star (\mathrm{Rot}_\star \delta) \cdot_\star v -_\star Q_k \\ &= (\mathrm{Rot}_\star \delta) \cdot_\star v -_\star I_\star \cdot_\star v \\ &= (\mathrm{Rot}_\star \delta -_\star I_\star) \cdot_\star v \\ &= e^2 \cdot_\star \left(\sin_\star(\delta /_\star e^2)\right) \cdot_\star J \cdot_\star \left(\mathrm{Rot}_\star(\delta /_\star e^2)\right) \cdot_\star v. \end{aligned}$$

Note that

$$N = J \cdot_\star (Q_{k+1} -_\star Q_k),$$

is the multiplicative normal vector to l_k. Therefore, the equation of the multiplicative line l_k is as follows:

$$\langle x -_\star Q_k, N \rangle_\star = 0_\star.$$

Now, using (4.1), (4.2) and (4.3), we obtain

$$\begin{aligned} \langle Q_j -_\star Q_k, N \rangle_\star &= \langle e^2 \cdot_\star \left(\sin_\star(\delta /_\star e^2)\right) \cdot_\star J \cdot_\star \left(\mathrm{Rot}_\star(\delta /_\star e^2)\right) \cdot_\star v, \\ &\quad e^2 \cdot_\star \left(\sin_\star(\theta /_\star e^2)\right) \cdot_\star J \cdot_\star \left(\mathrm{Rot}_\star(\theta /_\star e^2)\right) \cdot_\star J \cdot_\star v \rangle_\star \\ &= e^4 \cdot_\star \sin_\star(\delta /_\star e^2) \cdot_\star \sin_\star(\theta /_\star e^2) \\ &\quad \cdot_\star \langle \mathrm{Rot}_\star(\delta /_\star e^2) \cdot_\star J \cdot_\star v, J \cdot_\star (\mathrm{Rot}_\star(\theta /_\star e^2) \cdot_\star J \cdot_\star v \rangle_\star \\ &= e^4 \cdot_\star \sin_\star(\delta /_\star e^2) \cdot_\star \sin_\star(\theta /_\star e^2) \\ &\quad \cdot_\star \langle \mathrm{Rot}_\star(\delta /_\star e^2) \cdot_\star J \cdot_\star v, -_\star I_\star \cdot_\star (\mathrm{Rot}_\star(\theta /_\star e^2) \cdot_\star v \rangle_\star \\ &= e^4 \cdot_\star \sin_\star(\delta /_\star e^2) \cdot_\star \sin_\star(\theta /_\star e^2) \\ &\quad \cdot_\star \langle \mathrm{Rot}_\star(\delta /_\star e^2) \cdot_\star J \cdot_\star v, (\mathrm{Rot}_\star(-_\star(\theta /_\star e^2)) \cdot_\star v \rangle_\star \\ &= e^4 \cdot_\star \sin_\star(\delta /_\star e^2) \cdot_\star \sin_\star(\theta /_\star e^2) \\ &\quad \cdot_\star \langle \mathrm{Rot}_\star((\delta -_\star \theta) /_\star e^2) \cdot_\star v, J \cdot_\star v \rangle_\star \\ &= e^4 \cdot_\star \sin_\star(\delta /_\star e^2) \cdot_\star \sin_\star(\theta /_\star e^2) \\ &\quad \cdot_\star \langle \left(\cos_\star((\delta -_\star \theta) /_\star e^2) \cdot_\star I_\star +_\star \sin_\star((\delta -_\star \theta) /_\star e^2) \cdot_\star J\right) \cdot_\star v, J \cdot_\star v \rangle_\star \\ &= e^4 \cdot_\star \sin_\star(\delta /_\star e^2) \cdot_\star \sin_\star(\theta /_\star e^2) \cdot_\star \sin_\star((\delta -_\star \theta) /_\star e^2) \\ &\quad \cdot_\star \langle J \cdot_\star v, J \cdot_\star v \rangle_\star \\ &= e^4 \cdot_\star \sin_\star(\delta /_\star e^2) \cdot_\star \sin_\star(\theta /_\star e^2) \cdot_\star \sin_\star((\delta -_\star \theta) /_\star e^2) \\ &\quad \cdot_\star |J \cdot_\star v|_\star^{2_\star} \\ &\geq 0_\star, \end{aligned}$$

because

$$\sin_\star(\delta /_\star e^2) \cdot_\star \sin_\star((\delta -_\star \theta) /_\star e^2) \cdot_\star \sin_\star((\delta -_\star \theta) /_\star e^2) \geq 0_\star.$$

This completes the proof. □

By Theorem 4.11, it follows that the multiplicative line l_k does not intersect any multiplicative segment of the form $Q_j Q_{j+1}$ except the following cases.

1. $Q_j = Q_k$, $Q_{j+1} = Q_{k+1}$.
2. $Q_j = Q_{k+1}$.
3. $Q_{j+1} = Q_k$.

Also, by Theorem 4.11, we get that the vertices of the multiplicative m-gon are precisely m points, Q_j, $j \in \{1, \ldots, m\}$.

4.5 Similar Regular Multiplicative Polygons

Theorem 4.12. *Any two regular multiplicative m-gons are similar.*

Proof. Let P, Q and $\widetilde{P}$, $\widetilde{Q}$ be the center and a vertex of the multiplicative m-gons $\mathscr{P}$ and $\widetilde{\mathscr{P}}$. Then

$$T_{-_\star P}\mathscr{P} \quad \text{and} \quad T_{-_\star \widetilde{P}}\widetilde{\mathscr{P}},$$

are regular multiplicative m-gons. Let ϕ and ϕ_1 be chosen so that

$$\mathscr{P}_0 = \mathrm{Rot}_\star \phi T_{-_\star P}\mathscr{P} \quad \text{and} \quad \widetilde{\mathscr{P}_0} = \mathrm{Rot}_\star \phi_1 T_{-_\star \widetilde{P}}\widetilde{\mathscr{P}}$$

be regular multiplicative m-gons with vertex

$$A(q, 0_\star) \quad \text{and} \quad \widetilde{A}(q_1, 0_\star),$$

respectively. Define

$$Sx = (q_1 /_\star q)x, \quad x \in E^2_\star.$$

Let A_j and $\widetilde{A}_j$ be vertices of $\mathscr{P}_0$ and $\widetilde{\mathscr{P}_0}$. Then

$$\begin{aligned} SA_j &= S\mathrm{Rot}_\star(j \cdot_\star \theta)A \\ &= \mathrm{Rot}_\star(j \cdot_\star \theta)SA \\ &= \mathrm{Rot}_\star(j \cdot_\star \theta)\widetilde{A} \\ &= \widetilde{A}_j. \end{aligned}$$

If a_j and $\widetilde{a}_j$ are edges of $\mathscr{P}_0$ and $\widetilde{\mathscr{P}_0}$, respectively, then

$$Sa_j = \widetilde{a}_j.$$

Thus,

$$S\mathscr{P}_0 = \widetilde{\mathscr{P}_0}.$$

Hence,

$$S\mathrm{Rot}_\star \phi T_{-_\star P}\mathscr{P}_0 = \mathrm{Rot}_\star \phi_1 T_{-_\star \widetilde{P}}\widetilde{\mathscr{P}_0},$$

whereupon

$$T_{\widetilde{P}}\mathrm{Rot}_\star(-_\star \phi_1)S\mathrm{Rot}_\star \phi T_{-_\star P}\mathscr{P}_0 = \widetilde{\mathscr{P}_0}.$$

This completes the proof. □

Theorem 4.13. *The symmetry groups of any two regular multiplicative m-gons are conjugate in* $\mathscr{S}_\star(E^2_\star)$.

Proof. Let T be the similarity constructed in Theorem 4.12. Take $g \in \mathscr{S}_\star(\mathscr{P})$. Then

$$g = T_P g_0 T_{-_\star P},$$

where $g_0 \in O_\star(e^2)$. Now, using that S commutes with any element of $O_\star(e^2)$, we obtain

$$\begin{aligned} TgT^{-1} &= T_{\widetilde{P}}\mathrm{Rot}_\star(-_\star \phi_1)S\mathrm{Rot}_\star \phi T_P g_0 T_{-_\star P} \\ &\quad T_P\mathrm{Rot}_\star(-_\star \phi)S^{-1_\star}\mathrm{Rot}_\star \phi_1 T_{-_\star \widetilde{P}} \\ &= T_{\widetilde{P}}\mathrm{Rot}_\star(-_\star \phi_1)S\mathrm{Rot}_\star \phi g_0 \mathrm{Rot}_\star(-_\star \phi)S^{-1_\star}\mathrm{Rot}_\star(\phi_1)T_{-_\star \widetilde{P}} \\ &= T_{\widetilde{P}}\mathrm{Rot}_\star(\phi -_\star \phi_1)Sg_0S^{-1_\star}\mathrm{Rot}_\star(\phi_1 -_\star \phi)T_{-_\star \widetilde{P}} \\ &= T_{\widetilde{P}}\mathrm{Rot}_\star(\phi -_\star \phi_1)T_{-_\star P}g_0T_P\mathrm{Rot}\star(\phi_1 -_\star \phi)T_{-_\star \widetilde{P}} \\ &= \widetilde{T}g_0\widetilde{T}^{-1_\star}, \end{aligned}$$

where $\widetilde{T}$ is a multiplicative isometry. Thus, $\mathscr{S}_\star(\mathscr{P})$ and $\mathscr{S}_\star(\widetilde{\mathscr{P}})$ are conjugate in $\mathscr{S}_\star(E_\star^2)$. This completes the proof. □

4.6 Advanced Practical Problems

Problem 4.1. *A multiplicative parallelogram is a multiplicative rectilinear figure containing of four multiplicative segments AB, BC, CD and DA so that*

$$AB \parallel_\star CD, \quad BC \parallel_\star DA.$$

Prove that there is a multiplicative affine transformation relating any two multiplicative parallelograms.

Problem 4.2. *Find the multiplicative affine symmetry group of the multiplicative parallelogram.*

Problem 4.3. *A multiplicative rhombus is a multiplicative parallelogram in which all four multiplicative sides have equal multiplicative lengths. Find the multiplicative affine symmetry group of the multiplicative rhombus.*

Problem 4.4. *A multiplicative rectangle is a multiplicative parallelogram in which adjacent multiplicative sides are multiplicative perpendicular. Find the multiplicative affine symmetry group of the multiplicative rectangle.*

Problem 4.5. *A multiplicative square is a multiplicative parallelogram that it is both a multiplicative rhombus and a multiplicative rectangle. Find the multiplicative affine symmetry group of the multiplicative square.*

5

Multiplicative Geometry on the Multiplicative Sphere

In this chapter, multiplicative spheres, multiplicative lines on multiplicative spheres, multiplicative reflections on multiplicative spheres and multiplicative rectilinear figures on multiplicative spheres are defined and some of their properties are obtained.

5.1 The Space $E^3_\star$

Set

$$\mathbb{R}^3_\star = \{x = (x_1, x_2, x_3) : x_1, x_2, x_3 \in \mathbb{R}_\star\}.$$

Definition 5.1. *For $x = (x_1, x_2, x_3)$, $y = (y_1, y_2, y_3) \in \mathbb{R}^3_\star$, define*

$$\begin{aligned}
x +_\star y &= (x_1 +_\star y_1, x_2 +_\star y_2, x_3 +_\star y_3),\\
x -_\star y &= (x_1 -_\star y_1, x_2 -_\star y_2, x_3 -_\star y_3),\\
c \cdot_\star x &= (c \cdot_\star x_1, c \cdot_\star x_2, c \cdot_\star x_3),\\
\langle x, y\rangle_\star &= e^{\log x_1 \log y_1 + \log x_2 \log y_2 + \log x_3 \log y_3},\\
|x|_\star &= e^{\left((\log x_1)^2 + (\log x_2)^2 + (\log x_3)^2\right)^{\frac{1}{2}}}
\end{aligned}$$

and multiplicative distance

$$d_\star(x, y) = |x -_\star y|_\star.$$

The symbol $E^3_\star$ will be used to be denoted the set $\mathbb{R}^3_\star$ equipped with the multiplicative distance $d_\star$.

Example 5.1. *Let*

$$\begin{aligned}
x &= (1, e, 3),\\
y &= (2, 3, 4).
\end{aligned}$$

Then

$$\begin{aligned}
x +_\star y &= (1 +_\star 2, e +_\star 3, 3 +_\star 4)\\
&= (2, 3e, 12),
\end{aligned}$$

and

$$\begin{aligned}
x -_\star y &= (1 -_\star 2, e -_\star 3, 3 -_\star 4)\\
&= \left(\frac{1}{2}, \frac{e}{3}, \frac{3}{4}\right),
\end{aligned}$$

DOI: 10.1201/9781003325284-5

and

$$\begin{aligned} 3\cdot_\star x &= (3\cdot_\star 1, 3\cdot_\star e, 3\cdot_\star 3) \\ &= \left(e^{\log 3 \log 1}, e^{\log 3 \log e}, e^{(\log 3)^2}\right) \\ &= \left(1, 3, e^{(\log 3)^2}\right), \end{aligned}$$

and

$$\begin{aligned} \langle x, y\rangle_\star &= e^{\log 1 \log 2 + \log e \log 3 + \log 3 \log 4} \\ &= e^{\log 3(1+2\log 2)}, \end{aligned}$$

and

$$\begin{aligned} |x|_\star &= e^{\left((\log 1)^2 + (\log e)^2 + (\log 3)^2\right)^{\frac{1}{2}}} \\ &= e^{\left(1+(\log 3)^2\right)^{\frac{1}{2}}}. \end{aligned}$$

Exercise 5.1. *Let*

$$\begin{aligned} x &= (3, e^2, 11), \\ y &= (4, e, e^5). \end{aligned}$$

Find

1. $x +_\star y$.
2. $x -_\star y$.
3. $4 \cdot_\star x$.
4. $5 \cdot_\star y$.
5. $\langle x, y\rangle_\star$.
6. $|x|_\star$.
7. $|y|_\star$.

5.2 The Multiplicative Cross Product

Definition 5.2. *Let $u = (u_1, u_2, u_3)$, $v = (v_1, v_2, v_3) \in E^3_\star$. Then we define the multiplicative cross product of u and v, denoted by $u \times_\star v$, to be the vector*

$$u \times_\star v = \left(e^{\log u_2 \log v_3 - \log v_2 \log u_3}, e^{\log v_1 \log u_3 - \log u_1 \log v_3}, e^{\log u_1 \log v_2 - \log v_1 \log u_2}\right).$$

Example 5.2. *Let*

$$\begin{aligned} u &= (e, e^2, e^3), \\ v &= (e^2, e^4, e^6). \end{aligned}$$

Then

$$\begin{aligned} u \times_\star v &= \left(e^{\log e^2 \log e^6 - \log e^4 \log e^3}, e^{\log e^2 \log e^3 - \log e \log e^6}, e^{\log e \log e^4 - \log e^2 \log e^2}\right) \\ &= \left(e^{12-12}, e^{6-6}, e^{4-4}\right) \\ &= (1,1,1). \end{aligned}$$

Exercise 5.2. *Let*

$$\begin{aligned} u &= \left(e^3, e, e^4\right), \\ v &= \left(e^8, e, e^{15}\right). \end{aligned}$$

Find $u \times_\star v$.

Below, we deduce some of the properties of the multiplicative cross product. Suppose that

$$u = (u_1, u_2, u_3), \quad v = (v_1, v_2, v_3), \quad w = (w_1, w_2, w_3), \quad z = (z_1, z_2, z_3) \in E_\star^2.$$

Let also, $a \in \mathbb{R}_\star$. Then we have the following.

1. $\langle u \times_\star v, u \rangle_\star = \langle u \times_\star v, v \rangle_\star = 0_\star$.

   Proof. We have

$$\begin{aligned} &\log u_1 \log u_2 \log v_3 - \log u_1 \log v_2 \log u_3 + \log u_2 \log v_1 \log u_3 \\ &\qquad - \log u_2 \log u_1 \log v_3 + \log u_3 \log u_1 \log v_2 - \log u_3 \log v_1 \log u_2 \\ &= 0, \end{aligned}$$

   and

$$\begin{aligned} \langle u \times_\star v, u \rangle_\star &= e^0 \\ &= 1 \\ &= 0_\star. \end{aligned}$$

   Next,

$$\begin{aligned} &\log v_1 \log u_2 \log v_3 - \log v_1 \log v_2 \log u_3 + \log v_2 \log v_1 \log u_3 \\ &\qquad - \log v_2 \log u_1 \log v_3 + \log v_3 \log u_1 \log v_2 - \log v_3 \log v_1 \log u_2 \\ &= 0, \end{aligned}$$

   and

$$\begin{aligned} \langle u \times_\star v, v \rangle_\star &= e^0 \\ &= 1 \\ &= 0_\star. \end{aligned}$$

   This completes the proof. □

2. $u \times_\star v = -_\star v \times_\star u$.

Proof. We have

$$\begin{aligned}
v \times_\star u &= \left(e^{\log v_2 \log u_3 - \log u_2 \log v_3}, e^{\log u_1 \log v_3 - \log v_1 \log u_3}, e^{\log v_1 \log u_2 - \log u_1 \log v_2}\right) \\
&= \Big(e^{-(\log u_2 \log v_3 - \log v_2 \log u_3)}, e^{-(\log v_1 \log u_3 - \log u_1 \log v_3)}, \\
&\quad e^{-(\log u_1 \log v_2 - \log v_1 \log u_2)}\Big) \\
&= \Big(-_\star e^{\log u_2 \log v_3 - \log v_2 \log u_3}, -_\star e^{\log v_1 \log u_3 - \log u_1 \log v_3}, \\
&\quad -_\star e^{\log u_1 \log v_2 - \log v_1 \log u_2}\Big) \\
&= -_\star \left(e^{\log u_2 \log v_3 - \log v_2 \log u_3}, e^{\log v_1 \log u_3 - \log u_1 \log v_3}, e^{\log u_1 \log v_2 - \log v_1 \log u_2}\right) \\
&= -_\star u \times_\star v.
\end{aligned}$$

This completes the proof. □

3. $\langle u \times_\star v, w\rangle_\star = \langle u, v \times_\star w\rangle_\star$.

Proof. We have

$$u \times_\star v = \left(e^{\log u_2 \log v_3 - \log v_2 \log u_3}, e^{\log v_1 \log u_3 - \log u_1 \log v_3}, e^{\log u_1 \log v_2 - \log v_1 \log u_2}\right)$$

and

$$\begin{aligned}
\langle u \times_\star v, w\rangle_\star &= \exp\Big(\log w_1 \log u_2 \log v_3 - \log w_1 \log v_2 \log u_3 \\
&\quad + \log w_2 \log v_1 \log u_3 - \log w_2 \log u_1 \log v_3 \\
&\quad + \log w_3 \log u_1 \log v_3 - \log w_3 \log v_1 \log u_2\Big).
\end{aligned}$$

Next,

$$v \times_\star w = \left(e^{\log v_2 \log w_3 - \log w_2 \log v_3}, e^{\log w_1 \log v_3 - \log v_1 \log w_3}, e^{\log v_1 \log w_2 - \log w_1 \log v_2}\right)$$

and

$$\begin{aligned}
\langle u, v \times_\star w\rangle_\star &= \exp\Big(\log u_1 \log v_2 \log w_3 - \log u_1 \log w_2 \log v_3 \\
&\quad + \log u_2 \log w_1 \log v_3 - \log u_2 \log v_1 \log w_3 \\
&\quad + \log u_3 \log v_1 \log w_2 - \log u_3 \log w_1 \log v_2\Big).
\end{aligned}$$

Thus,

$$\langle u \times_\star v, w\rangle_\star = \langle u, v \times_\star w\rangle_\star.$$

This completes the proof. □

4. $(u \times_\star v) \times_\star w = \langle u, w\rangle_\star \cdot_\star v -_\star \langle v, w\rangle_\star \cdot_\star u$.

Proof. We have

$$u \times_\star v = \left(e^{\log u_2 \log v_3 - \log v_2 \log u_3}, e^{\log v_1 \log u_3 - \log u_1 \log v_3}, e^{\log u_1 \log v_2 - \log v_1 \log u_2}\right)$$

and

$$\begin{aligned}(u \times_\star v) \times_\star w = \Big(& e^{(\log v_1 \log u_3 - \log u_1 \log v_3)\log w_3 - \log w_2(\log u_1 \log v_2 - \log v_1 \log u_2)} \\ & e^{(\log u_1 \log v_2 - \log v_1 \log u_2)\log w_1 - (\log u_2 \log v_3 - \log v_2 \log u_3)\log w_3}, \\ & e^{(\log u_2 \log v_3 - \log v_2 \log u_3)\log w_2 - \log w_1(\log v_1 \log u_3 - \log u_1 \log v_3)}\Big).\end{aligned}$$

Moreover,

$$\begin{aligned}\langle u, w\rangle_\star \cdot_\star v &= e^{\log u_1 \log w_1 + \log u_2 \log w_2 + \log u_3 \log w_3} \cdot_\star v \\ &= \Big(e^{\log v_1(\log u_1 \log w_1 + \log u_2 \log w_2 + \log u_3 \log w_3)}, \\ &\quad e^{\log v_2(\log u_1 \log w_1 + \log u_2 \log w_2 + \log u_3 \log w_3)}, \\ &\quad e^{\log v_3(\log u_1 \log w_1 + \log u_2 \log w_2 + \log u_3 \log w_3)}\Big),\end{aligned}$$

and

$$\begin{aligned}\langle v, w\rangle_\star \cdot_\star u &= e^{\log v_1 \log w_1 + \log v_2 \log w_2 + \log v_3 \log w_3} \cdot_\star u \\ &= \Big(e^{\log u_1(\log v_1 \log w_1 + \log v_2 \log w_2 + \log v_3 \log w_3)}, \\ &\quad e^{\log u_2(\log v_1 \log w_1 + \log v_2 \log w_2 + \log v_3 \log w_3)}, \\ &\quad e^{\log u_3(\log v_1 \log w_1 + \log v_2 \log w_2 + \log v_3 \log w_3)}\Big).\end{aligned}$$

Note that

$$\begin{aligned}& \log v_1(\log u_1 \log w_1 + \log u_2 \log w_2 + \log u_3 \log w_3) \\ & \quad - \log u_1(\log v_1 \log w_1 + \log v_2 \log w_2 + \log v_3 \log w_3) \\ &= (\log v_1 \log u_3 - \log u_1 \log v_3)\log w_3 + (\log u_2 \log v_1 - \log u_1 \log v_2)\log w_2\end{aligned}$$

and

$$\begin{aligned}& \log v_2(\log u_1 \log w_1 + \log u_2 \log w_2 + \log u_3 \log w_3) \\ & \quad - \log u_2(\log v_1 \log w_1 + \log v_2 \log w_2 + \log v_3 \log w_3) \\ &= \log w_1(\log u_1 \log v_2 - \log u_2 \log v_1) - \log w_3(\log u_2 \log v_3 - \log u_3 \log v_2),\end{aligned}$$

and

$$\begin{aligned}& \log v_3(\log u_1 \log w_1 + \log u_2 \log w_2 + \log u_3 \log w_3) \\ & \quad - \log u_3(\log v_1 \log w_1 + \log v_2 \log w_2 + \log v_3 \log w_3) \\ &= \log w_2(\log u_2 \log v_3 - \log u_3 \log v_2) - \log w_1(\log u_3 \log v_1 - \log v_3 \log u_1).\end{aligned}$$

Therefore,

$$(u \times_\star v) \times_\star w = \langle u, w\rangle_\star \cdot_\star v -_\star \langle v, w\rangle_\star \cdot_\star u.$$

This completes the proof. □

5. $u \times_\star (a \cdot_\star u) = 0_\star$.

Proof. We have

$$a \cdot_\star u = \left(e^{\log a \log u_1}, e^{\log a \log u_2}, e^{\log a \log u_3}\right)$$

and

$$\begin{aligned} u \times_\star (a \cdot_\star u) &= \Big(e^{\log a \log u_2 \log u_3 - \log a \log u_2 \log u_3}, e^{\log a \log u_1 \log u_3 - \log a \log u_1 \log u_3}, \\ &\quad e^{\log a \log u_1 \log u_2 - \log a \log u_1 \log u_2}\Big) \\ &= (e^0, e^0, e^0) \\ &= (1, 1, 1) \\ &= (0_\star, 0_\star, 0_\star). \end{aligned}$$

This completes the proof. □

6.

$$\begin{aligned} \langle u \times_\star v, w \times_\star z \rangle_\star &= \langle u, w \rangle_\star \cdot_\star \langle v, z \rangle_\star \\ &\quad -_\star \langle v, w \rangle_\star \cdot_\star \langle u, z \rangle_\star. \end{aligned}$$

Proof. We have

$$\begin{aligned} \langle u \times_\star v, w \times_\star z \rangle_\star &= \langle u, v \times_\star (w \times_\star z) \rangle_\star \\ &= -_\star \langle u, (w \times_\star z) \times_\star v \rangle_\star \\ &= -_\star \langle u, \langle w, v \rangle_\star \cdot_\star z -_\star \langle z, v \rangle_\star \cdot_\star w \rangle_\star \\ &= -_\star \langle u, \langle w, v \rangle_\star \cdot_\star z \rangle_\star +_\star \langle u, \langle z, v \rangle_\star \cdot_\star w \rangle_\star \\ &= \langle u, w \rangle_\star \cdot_\star \langle z, v \rangle_\star -_\star \langle u, z \rangle_\star \cdot_\star \langle w, v \rangle_\star. \end{aligned}$$

This completes the proof. □

7. $|u \times_\star v|_\star^{2_\star} = |u|_\star^{2_\star} \cdot_\star |v|_\star^{2_\star} -_\star \langle u, v \rangle_\star^{2_\star}$.

Proof. Take $w \times_\star z = u \times v$ in the equality in the previous point and we get the desired result. This completes the proof. □

8. Let

$$\begin{aligned} e_1 &= (1_\star, 0_\star, 0_\star), \\ e_2 &= (0_\star, 1_\star, 0_\star), \\ e_3 &= (0_\star, 0_\star, 1_\star). \end{aligned}$$

Then

$$\begin{aligned} e_1 \times_\star e_2 &= e_3, \\ e_2 \times e_3 &= e_1, \\ e_3 \times_\star e_1 &= e_2. \end{aligned}$$

Proof. We have

$$\begin{aligned} e_1 \times_\star e_2 &= \left(e^{\log 1 \log 1 - \log e \log 1}, e^{\log 1 \log 1 - \log e \log 1}, e^{\log e \log e - \log 1 \log 1}\right) \\ &= (1, 1, e) \\ &= (0_\star, 0_\star, 1_\star). \end{aligned}$$

The other two equalities we leave to the reader as an exercise. This completes the proof. □

Let $u, v \in E^2_\star$ and $u \times_\star v \neq 0_\star$. Assume that there are $\lambda, \mu, \nu \in \mathbb{R}_\star$ so that

$$\lambda \cdot_\star u +_\star \mu \cdot_\star v +_\star \nu \cdot_\star (u \times_\star v) = 0_\star. \tag{5.1}$$

We take multiplicative inner product with $u \times_\star v$ of both sides of (5.1) and we get

$$\nu \cdot_\star |u \times_\star v|_\star^{2_\star} = 0_\star.$$

Then $\nu = 0_\star$. Hence,

$$\lambda \cdot_\star (u \times_\star v) = 0_\star,$$

and

$$\mu \cdot_\star (u \times_\star v) = 0_\star.$$

Consequently, $\lambda = \mu = 0_\star$.

Exercise 5.3. *Prove that*

$$\begin{aligned} (e_1 \times_\star e_2) \times_\star e_3 &= 0_\star = \langle e_1, e_3 \rangle_\star \cdot_\star e_2 -_\star \langle e_2, e_3 \rangle_\star \cdot_\star e_1, \\ (e_2 \times_\star e_3) \times_\star e_3 &= -_\star e_2 = \langle e_2, e_3 \rangle_\star \cdot_\star e_3 -_\star \langle e_3, e_3 \rangle_\star \cdot_\star e_2, \\ (e_3 \times_\star e_1) \times_\star e_3 &= e_1 = \langle e_3, e_3 \rangle_\star \cdot_\star e_1 -_\star \langle e_1, e_3 \rangle_\star \cdot_\star e_3. \end{aligned}$$

5.3 Multiplicative Orthonormal Bases

Definition 5.3. *Let $u, v \in E^3_\star$. We say that u and v are multiplicative orthogonal and we write $u \perp_\star v$ if $\langle u, v \rangle_\star = 0_\star$.*

Definition 5.4. *A triple $\{u, v, w\}$ of mutually multiplicative orthogonal multiplicative unit vectors is called a multiplicative orthonormal basis.*

If $\{u, v, w\}$ is a multiplicative orthonormal basis, then any $x \in E^3_\star$ can be represented in the form

$$x = \langle x, u \rangle_\star \cdot_\star u +_\star \langle x, v \rangle_\star \cdot_\star v +_\star \langle x, w \rangle_\star \cdot_\star w.$$

Theorem 5.1. *If u is any multiplicative unit vector, then there are vectors v and w so that $\{u, v, w\}$ is a multiplicative orthonormal basis.*

Proof. Take $\xi \in E^3_\star$ such that $\xi \neq \pm_\star u$. Let

$$v = \frac{u \times_\star \xi}{|u \times_\star \xi|_\star} \quad \text{and} \quad w = u \times_\star v.$$

We have $u \perp_\star v$, $u \perp_\star w$, $v \perp_\star w$ and

$$\begin{aligned} |w|_\star^{2_\star} &= |u \times_\star v|_\star^{2_\star} \\ &= |u|_\star^{2_\star} \cdot_\star |v|_\star^{2_\star} -_\star \langle u, v \rangle_\star^{2_\star} \\ &= 1_\star \cdot_\star 1_\star -_\star 0_\star \\ &= 1_\star. \end{aligned}$$

This completes the proof. □

5.4 Multiplicative Planes

Definition 5.5. *A multiplicative plane is a set $\Pi_\star$ of points of $E_\star^3$ with the following properties.*

1. *$\Pi_\star$ is not contained in one multiplicative line.*
2. *The multiplicative line joining any two points of $\Pi_\star$ lies in $\Pi_\star$.*
3. *Not every point of $E_\star^3$ is in $\Pi_\star$.*

In Fig. 5.1 it is shown a multiplicative plane.

Theorem 5.2. *Let v and w be not multiplicative proportional and P be any point of $E_\star^3$. Then*

$$P +_\star [v, w],$$

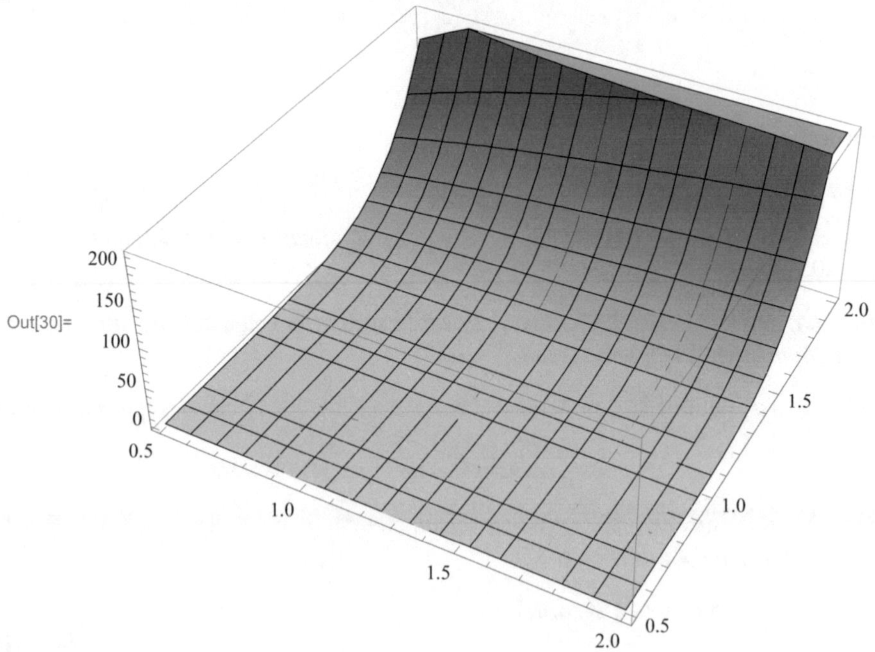

FIGURE 5.1
A multiplicative plane.

is a multiplicative plane. We speak of the multiplicative plane through P spanned by $\{v,w\}$. *Here*

$$[v,w] = \{t \cdot_\star v +_\star s \cdot_\star w : t,s \in \mathbb{R}_\star\},$$

is called the multiplicative span of $\{v,w\}$.

Proof. Let

$$\alpha = P +_\star [v,w].$$

Set

$$Q = P +_\star v, \quad R = P +_\star w.$$

Then $Q -_\star P$ and $R -_\star P$ are not multiplicative proportional. Thus, P, Q and R are not multiplicative collinear and they do not lie on one multiplicative line. Set

$$X = P +_\star v \times_\star w.$$

Since $\{v, w, v \times_\star w\}$ is a multiplicative linearly independent system, we conclude that $X \notin \alpha$. Thus, α does not contain any point of $E^3_\star$. Now, suppose that

$$\begin{aligned} Y &= P +_\star y_1 \cdot_\star v +_\star y_2 \cdot_\star w, \\ Z &= P +_\star z_1 \cdot_\star v +_\star z_2 \cdot_\star w, \end{aligned}$$

are two points of α. Take $t \in \mathbb{R}_\star$ arbitrarily. Then

$$\begin{aligned} &(1_\star -_\star t) \cdot_\star Y +_\star t \cdot_\star Z \\ &= (1_\star -_\star t) \cdot_\star P +_\star y_1 \cdot_\star (1_\star -_\star t) \cdot_\star v +_\star y_2 \cdot_\star (1_\star -_\star t) \cdot_\star w \\ &\quad +_\star t \cdot_\star P +_\star t \cdot_\star z_1 \cdot_\star v +_\star t \cdot_\star z_2 \cdot_\star w \\ &= P +_\star ((1_\star -_\star t) \cdot_\star y_1 +_\star t \cdot_\star z_1) \cdot_\star v \\ &\quad +_\star ((1_\star -_\star t) \cdot_\star y_2 +_\star t \cdot_\star z_2) \cdot_\star w. \end{aligned}$$

Hence,

$$(1_\star -_\star t) \cdot_\star Y +_\star t \cdot_\star Z \in \alpha.$$

Therefore, α is a multiplicative plane. This completes the proof. □

Theorem 5.3. *Let P, Q and R be multiplicative non collinear points. Then there exists a unique multiplicative plane through P, Q and R. In this case, we speak for the multiplicative plane PQR.*

Proof. Let

$$\begin{aligned} v &= Q -_\star P, \\ w &= R -_\star P. \end{aligned}$$

Then $P +_\star [v,w]$ is a multiplicative plane containing the points P, Q and R. Assume that there is other multiplicative plane β containing the points P, Q and R. Then

$$\begin{aligned} P +_\star \lambda \cdot_\star v +_\star \mu \cdot_\star w &= P +_\star \lambda \cdot_\star (Q -_\star P) +_\star \mu \cdot_\star (R -_\star P) \\ &= (1_\star -_\star \lambda -_\star \mu) \cdot_\star P +_\star \lambda \cdot_\star Q +_\star \mu \cdot_\star R \\ &= (1_\star -_\star \lambda -_\star \mu) \cdot_\star P \\ &\quad +_\star (\lambda +_\star \mu) \cdot_\star ((\lambda /_\star (\lambda +_\star \mu)) \cdot_\star Q +_\star (\mu /_\star (\lambda +_\star \mu)) \cdot_\star R). \end{aligned}$$

Let

$$X = (\lambda /_\star (\lambda +_\star \mu)) \cdot_\star Q +_\star (\mu /_\star (\lambda +_\star \mu)) \cdot_\star R.$$

Thus, X belongs to the multiplicative line through Q and R. Therefore, any plane containing P, Q and R contains $P +_\star [v,w]$. Suppose that β contains a point S that does not belong in $P +_\star [v,w]$. Then $\{S -_\star P, v, w\}$ is a multiplicative linearly independent system. Let $Z \in E^3_\star$ be arbitrarily chosen. Then, for some $\lambda, \mu, \nu \in E^3_\star$, we have

$$Z -_\star P = \lambda \cdot_\star v +_\star \mu \cdot_\star w +_\star \nu \cdot_\star (S -_\star P),$$

and

$$\begin{aligned} Z &= (1_\star -_\star \nu) \cdot_\star P \\ &\quad +_\star (1_\star -_\star \nu) \cdot_\star (P +_\star (\lambda /_\star (1_\star -_\star \nu)) \cdot_\star v +_\star (\mu /_\star (1_\star -_\star \nu)) \cdot_\star w). \end{aligned}$$

Therefore, any point of $E^3_\star$ is on a multiplicative line joining S and a point of β, i.e., β contains any point of $E^3_\star$. This is a contradiction. This completes the proof. □

Theorem 5.4. *If N is a multiplicative unit vector and $P \in E^3_\star$, then*

$$\{X : \langle X -_\star P, N \rangle_\star = 0_\star\},$$

is a multiplicative plane. We speak of the multiplicative plane through P and multiplicative normal N.

Proof. By Theorem 5.1, it follows that there are vectors v and w so that $\{N, v, w\}$ is a multiplicative orthonormal basis. Then any point $X \in E^3_\star$ can be represented in the form

$$\begin{aligned} X -_\star P &= \langle X -_\star P, N \rangle_\star \cdot_\star N +_\star \langle X -_\star P, v \rangle_\star \cdot_\star v \\ &\quad +_\star \langle X -_\star P, w \rangle_\star \cdot_\star w. \end{aligned}$$

Hence, $X -_\star P$ lies in $\{v, w\}$ if and only if

$$\langle X -_\star P, N \rangle_\star = 0_\star.$$

This completes the proof. □

5.5 Incidence Multiplicative Geometry of the Multiplicative Sphere

Definition 5.6. *The set*

$$S^2_\star = \{X \in E^3_\star : |X|^{2_\star}_\star = 1_\star\},$$

is said to be the multiplicative sphere.

Fig. 5.2 shows the multiplicative sphere.

Definition 5.7. *Let ξ be a multiplicative unit vector. Then*

$$l = \{x \in S^2_\star : \langle \xi, x \rangle_\star = 0_\star\}$$

is said to be the multiplicative line with pole ξ. We also say that l is the polar multiplicative line of ξ.

Definition 5.8. *Two points $P, Q \in S^2_\star$ are said to be multiplicative antipodal if $P = -_\star Q$.*

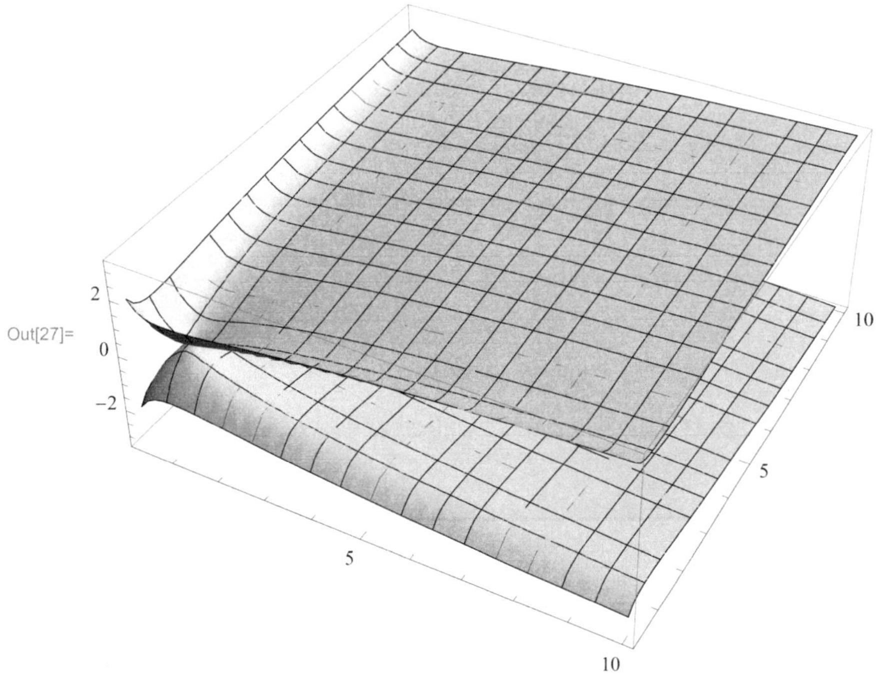

FIGURE 5.2
The multiplicative sphere.

Theorem 5.5.

1. *If ξ is a pole of l, then $-_\star\xi$ is a pole of l.*
2. *If $P \in l$, then $-_\star P \in l$.*

Proof.

1. We have
$$\langle \xi, x\rangle_\star = 0_\star, \quad x \subset S^2_\star \cap l,$$
and hence,
$$\begin{aligned}\langle -_\star\xi, x\rangle_\star &= -_\star\langle \xi, x\rangle_\star \\ &= 0_\star, \quad x \in S^2_\star \cap l.\end{aligned}$$
2. It is a direct consequence of 1. This completes the proof.

□

Theorem 5.6. *Let $P, Q \in S^2_\star$, $P \neq \pm_\star Q$. Then there exists a unique multiplicative line containing P and Q.*

Proof. Since $P \neq \pm_\star Q$, we have that $P \times_\star Q \neq 0_\star$. Take
$$\xi = \frac{P \times_\star Q}{|P \times_\star Q|}.$$
Then the multiplicative line l with pole ξ passes through P and Q. Let m be another multiplicative line through P and Q with a pole η. Then
$$\langle \eta, P\rangle_\star = \langle \eta, Q\rangle_\star = 0_\star.$$

Hence,

$$\begin{aligned}(P\times_\star Q)\times_\star\eta &= \langle P,\eta\rangle_\star\cdot_\star Q-_\star\langle Q,\eta\rangle_\star\cdot_\star P\\ &= 0_\star.\end{aligned}$$

Then there is a constant $\lambda\in\mathbb{R}_\star$ so that

$$\eta=\lambda\cdot_\star(P\times_\star Q).$$

Since $|\eta|_\star=1_\star$, we get

$$\lambda=\pm_\star(1_\star/_\star|P\times_\star Q|_\star).$$

Then $\eta=\pm_\star\xi$ and $l=m$. This completes the proof. □

Theorem 5.7. *Let l and m be two distinct multiplicative lines of $S^2_\star$. Then l and m have exactly two points of intersection that are multiplicative antipodal.*

Proof. Let ξ and η be poles of l and m, respectively. Since $l\neq m$, we have that $\xi\times_\star\eta\neq 0_\star$. Thus, both points of intersection of l and m are

$$\pm_\star\frac{\xi\times_\star\eta}{|\xi\times_\star\eta|_\star}.$$

If there is a third point on the intersection of l and m, by the uniqueness theorem, Theorem 5.6, we get $l=m$. This is a contradiction. This completes the proof. □

Corollary 5.1. *No two lines of $S^2_\star$ can be multiplicative parallel.*

5.6 The Multiplicative Distance

Let $P(p_1,p_2,p_3)$, $Q(q_1,q_2,q_3)\in S^2_\star$. Then

$$|P|_\star=1_\star.$$

Hence,

$$e^{(\log p_1)^2+(\log p_2)^2+(\log p_3)^2}=e,$$

and

$$(\log p_1)^2+(\log p_2)^2+(\log p_3)^2=1.$$

Therefore,

$$-1\leq\log p_1,\log p_2,\log p_3\leq 1. \tag{5.2}$$

Next, by the Cauchy–Schwartz inequality, we find

$$\begin{aligned}|\langle P,Q\rangle_\star|_\star &\leq |P|_\star\cdot_\star|Q|_\star\\ &= 1_\star.\end{aligned}$$

From here,

$$\left|e^{\log p_1\log q_1+\log p_2\log q_2+\log p_3\log q_3}\right|_\star\leq e.$$

Consequently,

$$-1\leq\log p_1\log q_1+\log p_2\log q_2+\log p_3\log q_3\leq 1. \tag{5.3}$$

Definition 5.9. *We define the multiplicative distance as follows:*

$$d_\star(P,Q) = \cos_\star^{-1_\star} \langle P,Q\rangle_\star.$$

By the definition, it follows that

$$\begin{aligned} d_\star(P,Q) &= e^{\arccos(\log(\langle P,Q\rangle_\star))} \\ &= e^{\arccos\left(\log\left(e^{\log p_1 \log q_1 + \log p_2 \log q_2 + \log p_3 \log q_3}\right)\right)} \\ &= e^{\arccos(\log p_1 \log q_1 + \log p_2 \log q_2 + \log p_3 \log q_3)}. \end{aligned}$$

Example 5.3. *Let* $P\left(e^{\frac{1}{\sqrt{2}}}, e^{\frac{1}{\sqrt{2}}}, e\right)$, $Q\left(e^{\frac{1}{\sqrt{3}}}, e^{\frac{1}{\sqrt{3}}}, e^{\frac{1}{\sqrt{3}}}\right)$. *We have*

$$\begin{aligned} |P|_\star^{2_\star} &= e^{\left(\log e^{\frac{1}{\sqrt{2}}}\right)^2 + \left(\log e^{\frac{1}{\sqrt{2}}}\right)^2 + (\log 1)^2} \\ &= e^{\frac{1}{2}+\frac{1}{2}} \\ &= e \\ &= 1_\star, \end{aligned}$$

and

$$\begin{aligned} |Q|_\star^{2_\star} &= e^{\left(\log e^{\frac{1}{\sqrt{3}}}\right)^2 + \left(\log e^{\frac{1}{\sqrt{3}}}\right)^2 + \left(\log e^{\frac{1}{\sqrt{3}}}\right)^2} \\ &= e^{\frac{1}{3}+\frac{1}{3}+\frac{1}{3}} \\ &= e \\ &= 1_\star. \end{aligned}$$

Therefore, $P,Q \in S_\star^2$. *Next,*

$$\begin{aligned} d_\star(P,Q) &= e^{\arccos\left(\log e^{\frac{1}{\sqrt{2}}} \log e^{\frac{1}{\sqrt{3}}} + \log e^{\frac{1}{\sqrt{2}}} \log e^{\frac{1}{\sqrt{3}}} + \log e \log 1\right)} \\ &= e^{\arccos\left(\frac{1}{\sqrt{6}} + \frac{1}{\sqrt{6}}\right)} \\ &= e^{\arccos \frac{2}{\sqrt{6}}}. \end{aligned}$$

Exercise 5.4. *Let* $P\left(e^{\frac{1}{2}}, e^{\frac{1}{2}}, e^{\frac{1}{\sqrt{2}}}\right)$, $Q\left(e^{\frac{1}{\sqrt{6}}}, e^{\frac{1}{\sqrt{6}}}, e^{\frac{\sqrt{2}}{\sqrt{3}}}\right)$.

1. *Prove that* $P,Q \in S_\star^2$.
2. *Find* $d_\star(P,Q)$.

Theorem 5.8. *Let* $P,Q,R \in S_\star^2$. *Then*

1. $d_\star(P,Q) \geq 0_\star$.
2. $d_\star(P,Q) = 0_\star$ *if and only if* $P = Q$.
3. $d_\star(P,Q) = d_\star(Q,P)$.
4. $d_\star(P,Q) \leq d_\star(P,R) +_\star d_\star(R,Q)$.

Proof.

1. By (5.3), it follows that

$$\arccos(\log p_1 \log q_1 + \log p_2 \log q_2 + \log p_3 \log q_3) \geq 0$$

and then

$$\begin{aligned} d_\star(P,Q) &= e^{\arccos(\log p_1 \log q_1 + \log p_2 \log q_2 + \log p_3 \log q_3)} \\ &\geq e^0 \\ &= 1 \\ &= 0_\star. \end{aligned}$$

2. We have that

$$d_\star(P,Q) = 0_\star,$$

if and only if

$$\arccos(\log p_1 \log q_1 + \log p_2 \log q_2 + \log p_3 \log q_3) = 0,$$

if and only if

$$\log p_1 \log q_1 + \log p_2 \log q_2 + \log p_3 \log q_3 = 0,$$

if and only if

$$e^{\log p_1 \log q_1 + \log p_2 \log q_2 + \log p_3 \log q_3} = e^0,$$

if and only if

$$\langle P,Q \rangle_\star = 0_\star,$$

if and only if $P = Q$.

3. We have

$$\begin{aligned} d_\star(P,Q) &= e^{\arccos(\log p_1 \log q_1 + \log p_2 \log q_2 + \log p_3 \log q_3)} \\ &= e^{\arccos(\log q_1 \log p_1 + \log q_2 \log p_2 + \log q_3 \log p_3)} \\ &= d_\star(Q,P). \end{aligned}$$

4. Let

$$\begin{aligned} r &= d_\star(P,Q), \\ p &= d_\star(Q,R), \\ q &= d_\star(P,R). \end{aligned}$$

By the Cauchy–Schwartz inequality, we get

$$\langle P \times_\star Q, Q \times_\star R \rangle_\star^{2_\star} \leq |P \times_\star R|_\star^{2_\star} \cdot_\star |Q \times_\star R|_\star^{2_\star}.$$

Note that

$$\begin{aligned} \langle P \times_\star Q, Q \times_\star R \rangle_\star^{2_\star} &= (\langle P,Q \rangle_\star \cdot_\star \langle R,R \rangle_\star -_\star \langle R,Q \rangle_\star \cdot_\star \langle P,Q \rangle_\star)^{2_\star} \\ &= (\cos_\star r -_\star \cos_\star p \cdot_\star \cos_\star q)^{2_\star}, \end{aligned}$$

and

$$\begin{aligned}
&|P \times_\star R|_\star^{2_\star} \cdot_\star |Q \times_\star R|_\star^{2_\star} \\
&= \left(|P|_\star^{2_\star} \cdot_\star |R|_\star^{2_\star} -_\star \langle P,R\rangle_\star^{2_\star}\right)^{2_\star} \cdot_\star \left(|Q|_\star^{2_\star} \cdot_\star |R|_\star^{2_\star} -_\star \langle Q,R\rangle_\star^{2_\star}\right)^{2_\star} \\
&= \left(1_\star -_\star (\cos_\star q)^{2_\star}\right) \cdot_\star \left(1_\star -_\star (\cos_\star p)^{2_\star}\right) \\
&= (\sin_\star q)^{2_\star} \cdot_\star (\sin_\star p)^{2_\star}.
\end{aligned}$$

Therefore,

$$(\cos_\star r -_\star \cos_\star p \cdot_\star \cos_\star q)^{2_\star} \le (\sin_\star q)^{2_\star} \cdot_\star (\sin_\star p)^{2_\star},$$

whereupon

$$\cos_\star r -_\star \cos_\star p \cdot_\star \cos_\star q \le \sin_\star p \cdot_\star \sin_\star q$$

or

$$\begin{aligned}
\cos_\star r &\le \cos_\star p \cdot_\star \cos_\star q +_\star \sin_\star p \cdot_\star \sin_\star q \\
&= \cos_\star (q -_\star p).
\end{aligned}$$

Hence,

$$r \ge q -_\star p,$$

or

$$d_\star(P,Q) \ge d_\star(P,R) -_\star d_\star(Q,R),$$

or

$$d_\star(P,R) \le d_\star(P,Q) +_\star d_\star(R,Q).$$

We have equality, by the Cauchy–Schwartz inequality, if and only if $P \times_\star R$ and $Q \times_\star R$ are multiplicative proportional if and only if P, Q and R are multiplicative collinear. This completes the proof.

□

Let l be a multiplicative line with a pole ξ. Then, by Theorem 5.1, it follows that there are $P, Q \in S_\star^2$, $P \neq Q$, so that $\{\xi, P, Q\}$ is a multiplicative orthonormal basis. Set

$$f(t) = \cos_\star t \cdot_\star P +_\star \sin_\star t \cdot_\star Q.$$

Then

$$l = \{f(t) : t \in \mathbb{R}_\star\}.$$

Exercise 5.5. *Prove that*

$$d_\star(f(t_1), f(t_2)) = d_\star(t_1, t_2) \quad \textit{if} \quad 0_\star \le |t_1 -_\star t_2|_\star \le e^\pi.$$

Definition 5.10. *Two multiplicative lines on $S_\star^2$ are said to be multiplicative perpendicular if their poles are multiplicative orthogonal.*

If P is a pole of a multiplicative line l on $S_\star^2$, then any multiplicative line through P on $S_\star^2$ will be multiplicative perpendicular on l.

5.7 Multiplicative Motions on $S^2_\star$

Definition 5.11. *For any multiplicative line l with a pole ξ, define the multiplicative reflection in l as follows.*

$$\Omega_l X = X -_\star e^2 \cdot_\star \langle X, \xi \rangle_\star \cdot_\star \xi, \quad X \in E^3_\star.$$

Theorem 5.9. *Let $\xi \in E^3_\star$, $\langle \xi, \xi \rangle_\star = 1_\star$ and consider the map*

$$Tx = x -_\star e^2 \cdot_\star \langle x, \xi \rangle_\star \cdot_\star \xi, \quad x \in E^3_\star.$$

Then

1. *T is linear.*
2. *$\langle Tx, Ty \rangle_\star = \langle x, y \rangle_\star$, $x, y \in E^3_\star$.*
3. *$TTx = x$, $x \in E^3_\star$.*

Proof.

1. Let $x, y \in E^3_\star$, $a, b \in \mathbb{R}_\star$. Then

$$\begin{aligned} T(a \cdot_\star x +_\star b \cdot_\star y) &= a \cdot_\star x +_\star b \cdot_\star y \\ &\quad -_\star e^2 \cdot_\star \langle a \cdot_\star x +_\star b \cdot_\star y, \xi \rangle_\star \cdot_\star \xi \\ &= a \cdot_\star x +_\star b \cdot_\star y -_\star e^2 \cdot_\star \langle a \cdot_\star x, \xi \rangle_\star \cdot_\star \xi \\ &\quad -_\star e^2 \cdot_\star \langle b \cdot_\star y, \xi \rangle_\star \cdot_\star \xi \\ &= a \cdot_\star x +_\star b \cdot_\star y -_\star a \cdot_\star e^2 \cdot_\star \langle x, \xi \rangle_\star \cdot_\star \xi \\ &\quad -_\star b \cdot_\star e^2 \cdot_\star \langle y, \xi \rangle_\star \cdot_\star \xi \\ &= a \cdot_\star (x -_\star e^2 \cdot_\star \langle x, \xi \rangle_\star \cdot_\star \xi) \\ &\quad +_\star b \cdot_\star (y -_\star e^2 \cdot_\star \langle y, \xi \rangle_\star \cdot_\star \xi) \\ &= a \cdot_\star Tx +_\star b \cdot_\star Ty. \end{aligned}$$

2. Let $x, y \in E^2_\star$. Then

$$\begin{aligned} \langle Tx, Ty \rangle_\star &= \langle x -_\star e^2 \cdot_\star \langle x, \xi \rangle_\star \cdot_\star \xi, y -_\star e^2 \cdot_\star \langle y, \xi \rangle_\star \cdot_\star \xi \rangle_\star \\ &= \langle x, y \rangle_\star -_\star \langle x, e^2 \cdot_\star \langle y, \xi \rangle_\star \cdot_\star \xi \rangle_\star \\ &\quad -_\star \langle e^2 \cdot_\star \langle x, \xi \rangle_\star \cdot_\star \xi, y \rangle_\star \\ &\quad +_\star \langle e^2 \cdot_\star \langle x, \xi \rangle_\star \cdot_\star \xi, e^2 \cdot_\star \langle y, \xi \rangle_\star \cdot_\star \xi \rangle_\star \\ &= \langle x, y \rangle_\star -_\star e^2 \cdot_\star \langle y, \xi \rangle_\star \cdot_\star \langle x, \xi \rangle_\star \\ &\quad -_\star e^2 \cdot_\star \langle x, \xi \rangle_\star \cdot_\star \langle y, \xi \rangle_\star \\ &\quad +_\star e^4 \cdot_\star \langle x, \xi \rangle_\star \cdot_\star \langle y, \xi \rangle_\star \cdot_\star \langle \xi, \xi \rangle_\star \\ &= \langle x, y \rangle_\star -_\star e^4 \cdot_\star \langle y, \xi \rangle_\star \cdot_\star \langle x, \xi \rangle_\star \\ &\quad +_\star e^4 \cdot_\star \langle x, \xi \rangle_\star \cdot_\star \langle y, \xi \rangle_\star \\ &= \langle x, y \rangle_\star. \end{aligned}$$

3. Let $x \in E^3_\star$. Then

$$\begin{aligned} TTx &= Tx -_\star e^2 \cdot_\star \langle Tx, \xi \rangle_\star \cdot_\star \xi \\ &= x -_\star e^2 \cdot_\star \langle x, \xi \rangle_\star \cdot_\star \xi \\ &\quad -_\star e^2 \cdot_\star \langle x -_\star e^2 \cdot_\star \langle x, \xi \rangle_\star \cdot_\star \xi, \xi \rangle_\star \cdot_\star \xi \\ &= x -_\star e^2 \cdot_\star \langle x, \xi \rangle_\star \cdot_\star \xi \\ &\quad -_\star e^2 \cdot_\star \langle x, \xi \rangle_\star \cdot_\star \xi \\ &\quad +_\star e^4 \cdot_\star \langle x, \xi \rangle_\star \cdot_\star \langle \xi, \xi \rangle_\star \cdot_\star \xi \\ &= x -_\star e^4 \cdot_\star \langle x, \xi \rangle_\star \cdot_\star \xi \\ &\quad +_\star e^4 \cdot_\star \langle x, \xi \rangle_\star \cdot_\star \xi \\ &= x. \end{aligned}$$

This completes the proof.

□

Theorem 5.10. *We have $\Omega_l X = X$ if and only if $X \in l$.*

Proof.

1. Let $\Omega_l X = X$. Then

$$X = X -_\star e^2 \cdot_\star \langle X, \xi \rangle_\star \cdot_\star \xi,$$

whereupon

$$e^2 \cdot_\star \langle X, \xi \rangle_\star \cdot_\star \xi = 0_\star,$$

or

$$\langle X, \xi \rangle_\star = 0_\star.$$

Thus, $X \in l$.

2. Let $X \in l$. Then $\langle X, \xi \rangle_\star = 0_\star$ and $\Omega_l X = X$. This completes the proof.

□

Let l and m be two multiplicative lines with poles ξ and η, respectively. Suppose that $l \cap m \neq \emptyset$. Take P to be an intersection point of l and m. Set $e_3 = P$. By Theorem 5.1, it follows that there are two vectors e_1 and e_2 so that $\{e_1, e_2, e_3\}$ is a multiplicative orthonormal basis. Next, there are θ and ϕ so that

$$\begin{aligned} \xi &= (-_\star \sin_\star \theta) \cdot_\star e_1 +_\star (\cos_\star \theta) \cdot_\star e_2, \\ \eta &= (-_\star \sin_\star \phi) \cdot_\star e_1 +_\star (\cos_\star \phi) \cdot_\star e_2. \end{aligned}$$

We have

$$\begin{aligned} \Omega_l e_1 &= e_1 -_\star e^2 \cdot_\star \langle e_1, \xi \rangle_\star \cdot_\star \xi \\ &= e_1 +_\star e^2 \cdot_\star \sin_\star \theta \cdot_\star (-_\star \sin_\star \theta \cdot_\star e_1 +_\star \cos_\star \theta \cdot_\star e_2) \\ &= \left(1_\star -_\star e^2 \cdot_\star (\sin_\star \theta)^{2_\star}\right) \cdot_\star e_1 +_\star e^2 \cdot_\star \sin_\star \theta \cdot_\star \cos_\star \theta \cdot_\star e_2 \\ &= \cos_\star (e^2 \cdot_\star \theta) \cdot_\star e_1 +_\star \sin_\star (e^2 \cdot_\star \theta) \cdot_\star e_2 \end{aligned}$$

and

$$\begin{aligned} \Omega_l e_2 &= e_2 -_\star e^2 \cdot_\star \langle e_2, \xi \rangle_\star \cdot_\star \xi \\ &= e_2 -_\star e^2 \cdot_\star \cos_\star \theta \cdot_\star (-_\star \sin_\star \theta \cdot_\star e_1 +_\star \cos_\star \theta \cdot_\star e_2) \\ &= \left(e^2 \cdot_\star \sin_\star \theta \cdot_\star \cos_\star \theta\right) \cdot_\star e_1 +_\star \left(1_\star -_\star e^2 \cdot_\star (\cos_\star \theta)^{2_\star}\right) \cdot_\star e_2 \\ &= \sin_\star (e^2 \cdot_\star \theta) \cdot_\star e_1 -_\star \cos_\star (e^2 \cdot_\star \theta) \cdot_\star e_2, \end{aligned}$$

and

$$\Omega_l e_3 = e_3.$$

Thus, Ω_l can be represented in the terms of the basis $\{e_1, e_2, e_3\}$ as follows:

$$\begin{pmatrix} \cos_\star(e^2 \cdot_\star \theta) & \sin_\star(e^2 \cdot_\star \theta) & 0_\star \\ \sin_\star(e^2 \cdot_\star \theta) & -_\star \cos_\star(e^2 \cdot_\star \theta) & 0_\star \\ 0_\star & 0_\star & 1_\star \end{pmatrix} = \begin{pmatrix} \mathrm{Ref}_\star(\theta) & 0_\star \\ 0_\star & 1_\star \end{pmatrix}.$$

As above, the reflection Ω_m can be represented in the terms of the basis $\{e_1, e_2, e_3\}$ as follows

$$\begin{pmatrix} \mathrm{Ref}_\star(\phi) & 0_\star \\ 0_\star & 1_\star \end{pmatrix}.$$

Then $\Omega_l \Omega_m$ can be represented in the following way.

$$\begin{pmatrix} \mathrm{Ref}_\star(e^2 \cdot_\star (\theta -_\star \phi)) & 0_\star \\ 0_\star & 1_\star \end{pmatrix}.$$

Definition 5.12. *If α and β be two multiplicative lines passing through a point P, then the multiplicative isometry $\Omega_\alpha \Omega_\beta$ is called a multiplicative rotation about P. The set of all multiplicative rotations about P will be denoted by $ROT_\star(P)$.*

Definition 5.13. *Let l be any multiplicative line and let m and n be multiplicative lines that are multiplicative perpendicular to l. Then the multiplicative isometry $\Omega_m \Omega_n$ will be called a multiplicative translation along l. The set of all multiplicative translations along l will be denoted by $TRANS(l)$.*

Unlike of the case of $E^2_\star$, $S^2_\star$ does not have multiplicative parallel multiplicative lines. If two multiplicative lines are multiplicative perpendicular to l, then they intersect in the multiplicative pole of l. Thus, every multiplicative translation of $S^2_\star$ is a multiplicative rotation and any multiplicative rotation is a multiplicative translation of $S^2_\star$.

5.8 Multiplicative Orthogonal Transformations

Definition 5.14. *A map $T : E^3_\star \to E^3_\star$ is said to be multiplicative orthogonal if*

1. *T is multiplicative linear.*
2. *$\langle Tx, Ty_\star = \langle x, y\rangle_\star$ for any $x, y \in E^3_\star$.*

Theorem 5.11. *A map $T : E^3_\star \to E^3_\star$ is multiplicative orthogonal if and only if its matrix is multiplicative orthogonal.*

Proof.

1. Let T be multiplicative orthogonal with respect to the multiplicative orthonormal basis $\{e_1, e_2, e_3\}$. Then

$$\begin{aligned} \langle Te_i, Te_j\rangle_\star &= \langle \sum_{\star k=1}^{3} a_{ki} \cdot_\star e_k, \sum_{\star l=1}^{3} a_{lj} \cdot_\star e_l\rangle_\star \\ &= \sum_{\star k,l=1}^{3} a_{ki} \cdot_\star a_{lj} \cdot_\star \langle e_k, e_l\rangle_\star \end{aligned}$$

$$\begin{aligned} &= \sum_{\star k=1}^{3} a_{ki} \cdot_\star a_{kj} \\ &= (A^T \cdot_\star A)_{ij}, \quad i,j \in \{1,2,3\}. \end{aligned}$$

Since

$$\begin{aligned} \langle Te_i, Te_j \rangle_\star &= \langle e_i, e_j \rangle_\star \\ &= (I_\star)_{ij}, \quad i,j \in \{1,2,3\}, \end{aligned}$$

we get that

$$A^T \cdot_\star A = I_\star.$$

Let A be a multiplicative orthogonal matrix. Then, as in point 1, we have

$$\langle Te_i, Te_j \rangle_\star = \langle e_i, e_j \rangle_\star, \quad i,j \in \{1,2,3\}.$$

Take

$$\begin{aligned} x &= \sum_{\star l=1}^{3} x_l \cdot_\star e_l, \\ y &= \sum_{\star j=1}^{3} y_j \cdot_\star e_j. \end{aligned}$$

Then

$$\begin{aligned} Tx &= T\left(\sum_{\star l=1}^{3} x_l \cdot_\star e_l \right) \\ &= \sum_{\star l=1}^{3} x_l \cdot_\star Te_l, \end{aligned}$$

and

$$\begin{aligned} Ty &= T\left(\sum_{\star j=1}^{3} y_j \cdot_\star e_j \right) \\ &= \sum_{\star j=1}^{3} y_j \cdot_\star Te_j. \end{aligned}$$

Hence,

$$\begin{aligned} \langle Tx, Ty \rangle_\star &= \langle \sum_{\star l=1}^{3} x_l \cdot_\star Te_l, \sum_{\star j=1}^{3} y_j \cdot_\star Te_j \rangle_\star \\ &= \sum_{\star l,j=1}^{3} x_l \cdot_\star y_j \cdot_\star \langle Te_l, Te_j \rangle_\star \\ &= \sum_{\star l,j=1}^{3} x_l \cdot_\star y_j \cdot_\star \langle e_l, e_j \rangle_\star \\ &= \langle \sum_{\star l=1}^{3} x_l \cdot_\star e_l, \sum_{\star j=1}^{3} y_j \cdot_\star e_j \rangle_\star \\ &= \langle x, y \rangle_\star. \end{aligned}$$

This completes the proof. □

Theorem 5.12.

1. *The set $O_\star(e^3)$ of all multiplicative orthogonal transformations of $E_\star^3$ is a group.*
2. *The set $SO_\star(e^3)$ of all multiplicative orthogonal transformations with multiplicative determinant $1_\star$ is a subgroup of $O_\star(e^3)$.*

Proof.

1. Let A and B be multiplicative orthogonal matrices. Then

$$\begin{aligned}(A\cdot_\star B)^T\cdot_\star(A\cdot B) &= B^T\cdot_\star A^T\cdot_\star A\cdot_\star B\\ &= B^T\cdot_\star B\\ &= I_\star,\end{aligned}$$

and

$$\begin{aligned}(A^{-1_\star})^T\cdot_\star A &= (A^T)^T\cdot_\star A^T\\ &= (A\cdot_\star A^T)^T\\ &= I_\star.\end{aligned}$$

2. Let A and B be multiplicative orthogonal matrices with multiplicative determinants $1_\star$. Then

$$\begin{aligned}\det_\star(A\cdot_\star B) &= \det_\star A\cdot_\star \det_\star B\\ &= 1_\star\cdot_\star 1_\star\\ &= 1_\star,\end{aligned}$$

and

$$\begin{aligned}\det_\star(A^{-1_\star}) &= 1_\star/_\star(\det_\star A)\\ &= 1_\star/_\star 1_\star\\ &= 1_\star.\end{aligned}$$

This completes the proof.

□

5.9 The Euler Theorem

Theorem 5.13 (The Euler Theorem)**.** *For each $T\in SO_\star(e^3)$, there is a $x\in S_\star^2$ so that $Tx = x$.*

Proof. Consider the equation

$$\det_\star(T-_\star\lambda\cdot_\star I_\star)\cdot_\star x = 0_\star.$$

This equation has at least one solution $\lambda\in\mathbb{R}_\star$. We have

$$Tx = \lambda\cdot_\star x,$$

and

$$\begin{aligned} 1_\star &= (\det_\star(Tx))^{2_\star} \\ &= \lambda^{2_\star} \cdot_\star |x|_\star^{2_\star} \\ &= \lambda^{2_\star}. \end{aligned}$$

Thus, $\lambda = \pm_\star 1_\star$. But

$$\lambda_1 \cdot_\star \lambda_2 \cdot_\star \lambda_3 = 1_\star.$$

Therefore $\lambda = 1_\star$ and $Tx = x$. This completes the proof. □

Corollary 5.2. *For any $T \in SO_\star(e^3)$, the restriction of T to $S_\star^2$ is a multiplicative rotation.*

Proof. By the Euler theorem, it follows that there is $e_3 \in S_\star^2$ so that $Te_3 = e_3$. Let $\{e_1, e_2, e_3\}$ be a multiplicative orthonormal basis. Then there is a $\theta \in \mathbb{R}_\star$ so that

$$\begin{aligned} Te_1 &= (\cos_\star \theta) \cdot_\star e_1 + \sin_\star \theta \cdot_\star e_2, \\ Te_2 &= \pm_\star((-_\star \sin_\star \theta) \cdot_\star e_1 +_\star \cos_\star \theta \cdot_\star e_2), \\ Te_3 &= e_3. \end{aligned}$$

Since $\det_\star T = 1_\star$, we have to use the multiplicative positive sign in Te_2. The matrix of T is

$$\begin{pmatrix} \cos_\star \theta & -_\star \sin_\star \theta & 0_\star \\ \sin_\star \theta & \cos_\star \theta & 0_\star \\ 0_\star & 0_\star & 1_\star \end{pmatrix}.$$

Since T can be factored into the multiplicative product of two multiplicative reflections, we have that it is a multiplicative rotation. This completes the proof. □

Definition 5.15. *The multiplicative antipodal map is a multiplicative transformation E defined by*

$$Ex = -_\star x.$$

With respect to any multiplicative orthonormal basis, E has the matrix

$$\begin{pmatrix} -_\star 1_\star & 0_\star & 0_\star \\ 0_\star & -_\star 1_\star & 0_\star \\ 0_\star & 0_\star & -_\star 1_\star \end{pmatrix}.$$

Any multiplicative antipodal map is a multiplicative glide reflection since it can be factored as follows:

$$\begin{pmatrix} -_\star 1_\star & 0_\star & 0_\star \\ 0_\star & 1_\star & 0_\star \\ 0_\star & 0_\star & 1_\star \end{pmatrix} \cdot_\star \begin{pmatrix} 1_\star & 0_\star & 0_\star \\ 0_\star & -_\star 1_\star & 0_\star \\ 0_\star & 0_\star & 1_\star \end{pmatrix} \cdot_\star \begin{pmatrix} 1_\star & 0_\star & 0_\star \\ 0_\star & 1_\star & 0_\star \\ 0_\star & 0_\star & -_\star 1_\star \end{pmatrix}.$$

Theorem 5.14. *Every multiplicative orthogonal transformation restricts to a multiplicative motion of $S_\star^2$.*

Proof. Let B be any multiplicative orthogonal transformation. If $\det_\star T = 1_\star$, then T is a multiplicative rotation and hence, it is a multiplicative motion. If $\det_\star T = -_\star 1_\star$, then T is a multiplicative rotation and from here, T is a multiplicative rotation. This completes the proof. □

5.10 Multiplicative Isometries

Definition 5.16. *A function $T : S^2_\star \to S^2_\star$ is said to be a multiplicative isometry if*

$$d_\star(Tx, Ty) = d_\star(x, y),$$

for any $x, y \in S^2_\star$.

Theorem 5.15. *For any multiplicative isometry T_0 of $S^2_\star$ there is a multiplicative orthogonal transformation T coinciding with T_0 on $S^2_\star$.*

Proof. Let $\{e_1, e_2, e_3\}$ be a multiplicative orthonormal basis of $E^3_\star$. Since T_0 is a multiplicative isometry, we have

$$\langle T_0 e_i, T_0 e_j\rangle_\star = \langle e_i, e_j\rangle_\star, \quad j = 1, 2, 3.$$

Note that any point of $E^3_\star$ is of the form $\lambda \cdot_\star x$, where $x \in S^2_\star$ and $\lambda \geq 0_\star$. Define $T : E^3_\star \to E^3_\star$ as follows:

$$\begin{aligned} T(\lambda \cdot_\star x) &= \lambda \cdot_\star T_0 x \quad \text{if} \quad \lambda \cdot_\star x \neq 0_\star, \\ T(0_\star) &= 0_\star. \end{aligned}$$

We will check that T is a multiplicative orthogonal transformation. For any $x \in S^2_\star$, we have

$$Tx = T_0 x,$$

and

$$\begin{aligned} Tx &= \sum_{i=1_\star}^{3} \langle Tx, Te_i\rangle_\star \cdot_\star Te_i \\ &= \sum_{i=1_\star}^{3} \langle x, e_i\rangle_\star \cdot_\star Te_i, \end{aligned}$$

because $\{Te_1, Te_2, Te_3\}$ is a multiplicative orthonormal basis. Hence,

$$\begin{aligned} T(\lambda \cdot_\star x) &= \lambda \cdot_\star T_0 x \\ &= \lambda \cdot_\star \sum_{i=1_\star}^{3} \langle x, e_i\rangle_\star \cdot_\star Te_i \\ &= \sum_{i=1_\star}^{3} \langle \lambda \cdot_\star x, e_i\rangle_\star \cdot_\star Te_i. \end{aligned}$$

Thus, for any $u \in E^3_\star$, $u \neq 0_\star$, we have

$$Tu = \sum_{i=1_\star}^{3} \langle u, e_i\rangle_\star \cdot_\star Te_i.$$

Hence, $T : E^3_\star \to E^3_\star$ is linear in u. Let now, $u, v \in E^3_\star$. Then

$$\begin{aligned} \langle Tu, Tv\rangle_\star &= \sum_{i=1_\star}^{3} \langle u, e_i\rangle_\star \cdot_\star \langle v, e_j\rangle_\star \cdot_\star \langle Te_i, Te_j\rangle_\star \\ &= \sum_{i=1_\star}^{3} \langle u, e_i\rangle_\star \cdot_\star \langle v, e_j\rangle_\star \cdot_\star \langle e_i, e_j\rangle_\star \\ &= \langle u, v\rangle_\star. \end{aligned}$$

This completes the proof. □

5.11 Multiplicative Segments

Definition 5.17. *A subset σ of $S^2_\star$ is called a multiplicative segment if there exist points P and Q such that $\langle P,Q\rangle_\star = 0_\star$ and $t_1, t_2 \in \mathbb{R}_\star$ with $t_1 < t_2$, $t_2 -_\star t_1 < e^{2\pi}$ and*

$$\sigma = \{(\cos_\star t)\cdot_\star P +_\star (\sin_\star t)\cdot_\star Q : t_1 \le t \le t_2\}.$$

Theorem 5.16. *Let σ be a multiplicative segment determined by P, Q, t_1, t_2 and P_1, Q_1, s_1, s_2. Then*

1. *$t_2 -_\star t_1 = s_2 -_\star s_1$. This number will be called multiplicative length of the multiplicative segment .*
2. *If*

$$\begin{aligned}\alpha(t) &= \{(\cos_\star t)\cdot_\star P +_\star (\sin_\star t)\cdot_\star Q : t_1 \le t \le t_2\},\\ \alpha_1(s) &= \{(\cos_\star s)\cdot_\star P_1 +_\star (\sin_\star s)\cdot_\star Q_1 : s_1 \le s \le s_2\},\end{aligned}$$

then

$$\{\alpha(t_1), \alpha(t_2)\} = \{\alpha_1(s_1), \alpha_1(s_2)\}.$$

These points will be called the end points of σ. All other points will be called the interior points of σ.

3. *$P \times_\star Q = P_1 \times_\star Q_1$. These points will be called multiplicative poles of the multiplicative line on which σ lies.*

Proof. Note that there is a $\phi \in \mathbb{R}_\star$ so that

$$P_1 = \alpha(\phi), \quad Q_1 = \alpha\left(\phi +_\star e^{\frac{\pi}{2}}\right).$$

Hence,

$$\begin{aligned}\alpha_1(u) &= \{(\cos_\star u)\cdot_\star \alpha(\phi) +_\star (\sin_\star u)\cdot_\star \alpha\left(\phi +_\star e^{\frac{\pi}{2}}\right)\}\\ &= \{(\cos_\star u)\cdot_\star ((\cos\star\phi)\cdot_\star P +_\star (\sin_\star \phi)\cdot_\star Q)\\ &\quad +_\star (\sin_\star u)\cdot_\star ((\cos_\star \left(\phi + e^{\frac{\pi}{2}}\right)\cdot_\star P +_\star \left(\sin_\star \left(\phi +_\star e^{\frac{\pi}{2}}\right)\right)\cdot_\star Q)\}\\ &= \{\cos_\star u \cdot_\star \cos_\star \phi \cdot_\star P +_\star \cos_\star u \cdot_\star \sin_\star \phi \cdot_\star Q\\ &\quad -_\star \sin_\star u \cdot_\star \sin_\star \phi \cdot_\star P +_\star \sin_\star u \cdot_\star \cos_\star \phi \cdot_\star Q\}\\ &= \{\cos_\star(u +_\star \phi)\cdot_\star P +_\star (\sin_\star(u +_\star \phi))\cdot_\star Q\}\\ &= \alpha(u +_\star \phi).\end{aligned}$$

Moreover,

$$\begin{aligned}P_1 \times_\star Q_1 &= (\cos_\star \phi \cdot_\star P +_\star \sin_\star \phi \cdot_\star Q) \times_\star (-_\star \sin_\star \phi \cdot_\star P +_\star \cos_\star \phi \cdot_\star Q)\\ &= (\cos_\star \phi)^{2_\star} \cdot_\star (P \times_\star Q) -_\star (\sin_\star \phi)^{2_\star}(Q \times_\star P)\\ &= (\cos_\star \phi)^{2_\star} \cdot_\star (P \times_\star Q) +_\star (\sin_\star \phi)^{2_\star}(P \times_\star Q)\\ &= P \times_\star Q.\end{aligned}$$

This completes the proof. □

Corollary 5.3. *Let A, B be two points so that $\langle A,B\rangle_\star = 0_\star$. Let σ be a multiplicative segment that lies on AB. Then there exist $a, b \in \mathbb{R}_\star$, $a < b$, so that*

$$\sigma = \{(\cos_\star t)\cdot_\star A +_\star (\sin_\star t)\cdot_\star B : a \le t \le b\}.$$

Proof. Let

$$\sigma = \{(\cos_\star t)\cdot_\star P +_\star (\sin_\star t)\cdot_\star Q : t_1 \le t \le t_2\},$$

where

$$\begin{aligned} P &= (\cos_\star \phi)\cdot_\star A +_\star (\sin_\star \phi)\cdot_\star B,\\ Q &= (-_\star \sin_\star \phi)\cdot_\star A +_\star (\cos_\star \phi)\cdot_\star B, \end{aligned}$$

for some $\phi \in \mathbb{R}_\star$. Hence,

$$\begin{aligned} &(\cos_\star t)\cdot_\star P +_\star (\sin_\star t)\cdot_\star B\\ &\quad= (\cos_\star t)\cdot_\star (\cos_\star \phi \cdot_\star A +_\star \sin_\star \phi \cdot_\star B)\\ &\qquad +_\star (\sin_\star t)\cdot_\star (-_\star \sin_\star \phi \cdot_\star A +_\star \cos_\star \phi \cdot_\star B)\\ &\quad= \cos_\star (t +_\star \phi)\cdot_\star A +_\star \sin_\star (t +_\star \phi)\cdot_\star B. \end{aligned}$$

Then we take $a = t_1 +_\star \phi \bmod e^{2\pi}$ in $\left[1, e^{2\pi}\right)$ and

$$b = a +_\star (t_2 -_\star t_1).$$

This completes the proof. □

Theorem 5.17. *Let A and B be not multiplicative antipodal points. Then there exist exactly two multiplicative segments having A and B as end points. The union of both multiplicative segments is the multiplicative line AB. The intersection of both multiplicative segments is* $\{A, B\}$.

Proof. Let ξ be a multiplicative unit vector in the direction of $A \times_\star B$. Take $Q = \xi \times_\star A$. Then there is a unique $\phi \in \left(1, e^{2\pi}\right)$ so that

$$B = (\cos_\star \phi)\cdot_\star A +_\star (\sin_\star \phi)\cdot_\star Q.$$

Then the multiplicative segments

$$\{(\cos_\star t)\cdot_\star A +_\star (\sin_\star t)\cdot_\star Q : 0_\star \le t \le \phi\},$$

and

$$\{(\cos_\star t)\cdot_\star A -_\star (\sin_\star t)\cdot_\star Q : 0_\star \le t \le e^{2\pi} -_\star \phi\}, \tag{5.4}$$

have A and B as end points. The multiplicative segment (5.4) can be written as follows:

$$\{\cos_\star t \cdot_\star A +_\star \sin_\star t \cdot_\star Q : 0_\star \le t \le \phi -_\star e^{2\pi}\}.$$

Their union is the multiplicative line AB and their intersection is $\{A, B\}$. This completes the proof. □

5.12 Multiplicative Rays, Multiplicative Angles and Multiplicative Triangles

Definition 5.18. *We define a multiplicative ray to be a multiplicative half-line with one end point removed.*

We can define multiplicative angle just as in the case of $E^2_\star$. Then a multiplicative straight angle is just a multiplicative line with one end point removed.

Definition 5.19. *The multiplicative radian measure of a multiplicative angle* $\angle_\star PQR$ *is defined as*

$$\cos_\star^{-1_\star}\langle (Q\times_\star P)/_\star|Q\times_\star P|_\star, (Q\times_\star R)/_\star|Q\times_\star R|_\star\rangle_\star.$$

Definition 5.20. *If P, Q and R are three multiplicative non collinear points, the multiplicative triangle* $\triangle_\star PQR$ *is defined to be the union of the three minor multiplicative segments PQ, PR and QR. The multiplicative segments are called multiplicative sides of the multiplicative triangle, and the multiplicative length of each multiplicative side is equal to the multiplicative distance between its end points.*

The interior of a multiplicative triangle and multiplicative rectilinear figures we define just as in the case of $E_\star^2$. All definitions and proofs are either exactly the same or very similar.

5.13 Multiplicative Spherical Trigonometry

Theorem 5.18 (Multiplicative Spherical Version of Law of Cosines)**.** *Let ABC be a multiplicative triangle on* $S_\star^2$. *Let also, a be the multiplicative length of BC, b be the multiplicative length of AC and c be the multiplicative length of AB. Then*

$$\cos_\star A = (\cos_\star a -_\star \cos_\star b \cdot_\star \cos_\star c)/_\star(\sin_\star b \cdot_\star \sin_\star c),$$

where A is the multiplicative radian measure of $\angle_\star BAC$.

Proof. We have

$$\begin{aligned} |B|_\star &= 1_\star, \\ |C|_\star &= 1_\star \end{aligned}$$

and

$$\begin{aligned} \langle B,C\rangle_\star &= \cos_\star a, \\ \langle A,C\rangle_\star &= \cos_\star b, \\ \langle A,B\rangle_\star &= \cos_\star c. \end{aligned}$$

Hence,

$$\begin{aligned} |B\times_\star C|_\star^{2_\star} &= |B|_\star^{2_\star}\cdot_\star|C|_\star^{2_\star} -_\star \langle B,C\rangle_\star^{2_\star} \\ &= 1_\star -_\star (\cos_\star a)^{2_\star} \\ &= (\sin_\star a)^{2_\star} \end{aligned}$$

and

$$\begin{aligned} |A\times_\star B|_\star^{2_\star} &= |A|_\star^{2_\star}\cdot_\star|B|_\star^{2_\star} -_\star \langle A,B\rangle_\star^{2_\star} \\ &= 1_\star -_\star (\cos_\star c)^{2_\star} \\ &= (\sin_\star c)^{2_\star}, \end{aligned}$$

and

$$\begin{aligned} |A\times_\star C|_\star^{2_\star} &= |A|_\star^{2_\star}\cdot_\star|C|_\star^{2_\star}-_\star\langle A,C\rangle_\star^{2_\star}\\ &= 1_\star-_\star(\cos_\star b)^{2_\star}\\ &= (\sin_\star b)^{2_\star}. \end{aligned}$$

Therefore,

$$\begin{aligned} |B\times_\star C|_\star &= \sin_\star a,\\ |A\times_\star B|_\star &= \sin_\star c,\\ |A\times_\star C|_\star &= \sin_\star b. \end{aligned}$$

Next,

$$\begin{aligned} \cos_\star A &= \langle (A\times_\star B)/_\star|A\times_\star B|_\star,(A\times_\star C)/_\star|A\times_\star C|_\star\rangle_\star\\ &= (1_\star/_\star(|A\times_\star B|_\star\cdot_\star|A\times_\star C|_\star))\\ &\quad \cdot_\star\langle A\times_\star B,A\times_\star C\rangle_\star\\ &= (1_\star/_\star(\sin_\star b\cdot_\star\sin_\star c))\cdot_\star(\langle B,C\rangle_\star-_\star\langle A,C\rangle_\star\cdot_\star\langle A,B\rangle_\star)\\ &= (1_\star/_\star(\sin_\star b\cdot_\star\sin_\star c))\cdot_\star(\cos_\star a-_\star\cos_\star b\cdot_\star\cos_\star c). \end{aligned}$$

This completes the proof. □

Theorem 5.19 (Multiplicative Spherical Version of Law of Sines). *Let ABC be a multiplicative triangle on $S^2_\star$. Let also, a be the multiplicative length of BC, b be the multiplicative length of AC and c be the multiplicative length of AB. Then*

$$\sin_\star A/_\star\sin_\star a=\sin_\star B/_\star\sin_\star b=\sin_\star C/_\star\sin_\star c,$$

where A is the multiplicative radian measure of $\angle_\star BAC$, B is the multiplicative radian measure of $\angle_\star CBA$ and C is the multiplicative radian measure of $\angle_\star ACB$.

Proof. By the proof of Theorem 5.18, we get

$$\begin{aligned} 1_\star-_\star\cos_\star A &= 1_\star-_\star(1_\star/_\star(\sin_\star b\cdot_\star\sin_\star c))\cdot_\star(\cos_\star a-_\star\cos_\star b\cdot_\star\cos_\star c)\\ &= (-_\star\cos_\star a+_\star\cos_\star b\cdot_\star\cos_\star c+_\star\sin_\star b\cdot_\star\sin_\star c)\\ &\quad /_\star(\sin_\star b\cdot_\star\sin_\star c)\\ &= (-_\star\cos_\star a+_\star\cos_\star(b-_\star c))/_\star(\sin_\star b\cdot_\star\sin_\star c). \end{aligned}$$

Hence,

$$e^2\cdot_\star\left(\sin_\star(A/_\star e^2)\right)^{2_\star}=(-_\star\cos_\star a+_\star\cos_\star(b-_\star c))/_\star(\sin_\star b\cdot_\star\sin_\star c).$$

Next,

$$\begin{aligned} 1_\star+_\star\cos_\star A &= 1_\star+_\star(1_\star/_\star(\sin_\star b\cdot_\star\sin_\star c))\cdot_\star(\cos_\star a-_\star\cos_\star b\cdot_\star\cos_\star c)\\ &= (\cos_\star a-_\star(\cos_\star b\cdot_\star\cos_\star c-_\star\sin_\star b\cdot_\star\sin_\star c))\\ &\quad /_\star(\sin_\star b\cdot_\star\sin_\star c)\\ &= (\cos_\star a-_\star\cos_\star(b+_\star c))/_\star(\sin_\star b\cdot_\star\sin_\star c), \end{aligned}$$

whereupon

$$e^2\cdot_\star\left(\cos_\star(A/_\star e^2)\right)^{2_\star}=(\cos_\star a-_\star\cos_\star(b+_\star c))/_\star(\sin_\star b\cdot_\star\sin_\star c).$$

Therefore,

$$\begin{aligned}
(\sin_\star A)^{2_\star} &= e^4 \cdot_\star \left(\sin_\star (A/_\star e^2)\right)^{2_\star} \cdot_\star \left(\cos_\star (A/_\star e^2)\right)^{2_\star} \\
&= (1_\star/_\star(\sin_\star b \cdot_\star \sin_\star c))^{2_\star} \\
&\quad \cdot_\star (\cos_\star a -_\star \cos_\star (b +_\star c)) \cdot_\star (-_\star \cos_\star a +_\star \cos_\star (b -_\star c)) \\
&= \left(1_\star/_\star(\sin_\star b \cdot_\star \sin_\star c)^{2_\star}\right) \\
&\quad \cdot_\star e^4 \cdot_\star \sin_\star \left((a +_\star b +_\star c)/_\star e^2\right) \cdot_\star \sin_\star \left((a -_\star b -_\star c)/_\star e^2\right) \\
&\quad \cdot_\star \sin_\star \left((b -_\star c +_\star a)/_\star e^2\right) \cdot_\star \sin_\star \left((b -_\star c -_\star a)/_\star e^2\right) \\
&= \left(1_\star/_\star(\sin_\star b \cdot_\star \sin_\star c)^{2_\star}\right) \\
&\quad \cdot_\star e^4 \cdot_\star \sin_\star s \cdot_\star \sin_\star (s -_\star a) \cdot_\star \sin_\star (s -_\star b) \cdot_\star \sin_\star (s -_\star c),
\end{aligned}$$

where

$$a +_\star b +_\star c = e^2 \cdot_\star s.$$

Hence,

$$\begin{aligned}
\sin_\star A/_\star \sin_\star a &= (1_\star/_\star(\sin_\star a \cdot_\star \sin_\star b \cdot_\star \sin_\star c)) \\
&\quad \cdot_\star e^2 \cdot_\star (\sin_\star s \cdot_\star \sin_\star (s -_\star a) \cdot_\star \sin_\star (s -_\star b) \cdot_\star \sin_\star (s -_\star c))^{\frac{1}{2}_\star}.
\end{aligned}$$

As above,

$$\begin{aligned}
\sin_\star B/_\star \sin_\star b &= (1_\star/_\star(\sin_\star a \cdot_\star \sin_\star b \cdot_\star \sin_\star c)) \\
&\quad \cdot_\star e^2 \cdot_\star (\sin_\star s \cdot_\star \sin_\star (s -_\star a) \cdot_\star \sin_\star (s -_\star b) \cdot_\star \sin_\star (s -_\star c))^{\frac{1}{2}_\star},
\end{aligned}$$

and

$$\begin{aligned}
\sin_\star C/_\star \sin_\star c &= (1_\star/_\star(\sin_\star a \cdot_\star \sin_\star b \cdot_\star \sin_\star c)) \\
&\quad \cdot_\star e^2 \cdot_\star (\sin_\star s \cdot_\star \sin_\star (s -_\star a) \cdot_\star \sin_\star (s -_\star b) \cdot_\star \sin_\star (s -_\star c))^{\frac{1}{2}_\star}.
\end{aligned}$$

This completes the proof. □

5.14 A Multiplicative Congruence Theorem

Theorem 5.20. *Let $P, Q \in S^2_\star$. Then there is a unique multiplicative reflection interchanging them.*

Proof. Let

$$\xi = (P -_\star Q)/_\star |P -_\star Q|_\star.$$

Let also, l be the multiplicative line with pole ξ. We have

$$\begin{aligned}
|P -_\star Q|_\star^{2_\star} &= \langle P -_\star Q, P -_\star Q \rangle_\star \\
&= \langle P, P \rangle_\star -_\star e^2 \langle P, Q \rangle_\star +_\star \langle Q, Q \rangle_\star \\
&= e^2 \cdot_\star (1_\star -_\star \langle P, Q \rangle_\star).
\end{aligned}$$

Then, using the definition for Ω_l, we get

$$\begin{aligned}
\Omega_l P &= P -_\star e^2 \cdot_\star \langle P, (P -_\star Q)/_\star |P -_\star Q|_\star\rangle_\star \cdot_\star ((P -_\star Q)/_\star |P -_\star Q|_\star) \\
&= P -_\star (e^2 /_\star |P -_\star Q|_\star^{2_\star}) \cdot_\star \langle P, P -_\star Q\rangle_\star \cdot_\star (P -_\star Q) \\
&= P -_\star e^2 /_\star (e^2 \cdot_\star (1_\star -_\star \langle P, Q\rangle_\star)) \cdot_\star (\langle P, P\rangle_\star -_\star \langle P, Q\rangle_\star) \cdot_\star (P -_\star Q) \\
&= P -_\star (e^2 /_\star (e^2 \cdot_\star (1_\star -_\star \langle P, Q\rangle_\star))) \cdot_\star (1_\star -_\star \langle P, Q\rangle_\star) \cdot_\star (P -_\star Q) \\
&= P -_\star (P -_\star Q) \\
&= P -_\star P +_\star Q \\
&= Q.
\end{aligned}$$

Suppose that there are two multiplicative reflections Ω_l and Ω_m so that

$$\begin{aligned}
\Omega_l P &= Q, \\
\Omega_l Q &= P, \\
\Omega_m P &= Q, \\
\Omega_m Q &= P.
\end{aligned}$$

Then

$$\begin{aligned}
\Omega_l \Omega_m P &= P, \\
\Omega_l \Omega_m Q &= Q.
\end{aligned}$$

If P and Q are not multiplicative antipodal, then

$$\Omega_l \Omega_m l = l$$

and $l = m$. Let P and Q be multiplicative antipodal. Then the pole ξ of l satisfies

$$P -_\star e^2 \cdot_\star \langle P, \xi\rangle_\star \cdot_\star \xi = -_\star P.$$

Hence,

$$P = \langle P, \xi\rangle_\star \cdot_\star \xi.$$

Therefore, l is a polar multiplicative line of P. The same arguments we apply to m. Thus, $l = m$. This completes the proof. □

Definition 5.21. *Let α be a multiplicative segment. The multiplicative perpendicular bisector of α is the unique multiplicative line l such that Ω_l interchanges the end points of α.*

Definition 5.22. *The multiplicative midpoint of a multiplicative segment is the unique point of intersection with its multiplicative perpendicular bisector.*

Definition 5.23. *Consider $\angle_\star PQR$. The multiplicative ray QX, where*

$$X = (P +_\star R)/_\star |P +_\star R|_\star,$$

is called the multiplicative bisector of $\angle_\star PQR$.

5.15 Multiplicative Right Triangles

Theorem 5.21 (The Multiplicative Pythagoras Theorem). *Let $\triangle_\star ABC$ be a multiplicative triangle on $S_\star^2$ with*

$$\begin{aligned} a &= d_\star(B,C), \\ b &= d_\star(A,C), \\ c &= d_\star(A,B). \end{aligned}$$

If $AC \perp_\star BC$, then

$$\cos_\star a = \cos_\star b \cdot_\star \cos_\star c.$$

Proof. Let ξ be a pole of the multiplicative line through A and B. Then $\{\xi, A, \xi \times_\star A\}$ is a multiplicative orthonormal basis. Then

$$\begin{aligned} C &= (\cos_\star b) \cdot_\star C +_\star (\sin_\star b) \cdot_\star \xi, \\ B &= (\cos_\star c) \cdot_\star A +_\star (\sin_\star c) \cdot_\star (\xi \times_\star A). \end{aligned}$$

Then

$$\begin{aligned} \cos_\star a &= \cos_\star d_\star(B,C) \\ &= \cos_\star \cos_\star^{-1_\star} \langle B, C \rangle_\star \\ &= \langle B, C \rangle_\star \\ &= \cos_\star b \cdot_\star \cos_\star c. \end{aligned}$$

This completes the proof. □

5.16 Advanced Practical Problems

Problem 5.1. *Let $P\left(e^{\frac{1}{\sqrt{3}}}, e^{\frac{1}{\sqrt{3}}}, e^{\frac{1}{\sqrt{3}}}\right)$.*

1. *Prove that $P \in S_\star^2$.*
2. *Find e_1, e_2 so that $\{e_1, e_2, P\}$ is a multiplicative orthonormal basis.*

Problem 5.2. *Let*

$$\begin{aligned} x &= \left(\frac{1}{3}, e^{\frac{1}{4}}, 18\right), \\ y &= (e^3, e^2, e^4). \end{aligned}$$

Find

1. $x +_\star y$.
2. $x -_\star y$.
3. $4 \cdot_\star x$.

4. $5 \cdot_\star y$.
5. $\langle x, y \rangle_\star$.
6. $|x|_\star$.
7. $|y|_\star$.

Problem 5.3. *Let*

$$\begin{aligned} u &= \left(4, e^2, e^8\right), \\ v &= \left(5, e^3, e^7\right). \end{aligned}$$

Find $u \times_\star v$.

Problem 5.4. *Let* $P\left(e^{\frac{1}{\sqrt{7}}}, e^{\frac{1}{\sqrt{7}}}, e^{\sqrt{\frac{5}{7}}}\right)$, $Q\left(e^{\frac{1}{\sqrt{5}}}, e^{\frac{1}{\sqrt{5}}}, e^{\sqrt{\frac{3}{5}}}\right)$.

1. *Prove that* $P, Q \in S^2_\star$.
2. *Find* $d_\star(P, Q)$.

Problem 5.5. *Let* T *be a multiplicative isometry so that* $TP = P$ *and* $TQ = -_\star Q$. *Prove that* $\langle P, Q \rangle_\star = 0_\star$.

6

The Projective Multiplicative Plane $P^2_\star$

In this chapter, the projective multiplicative plane $P^2_\star$, multiplicative perpendicular points, multiplicative lines, multiplicative perpendicular multiplicative lines, multiplicative poles, multiplicative polarities, multiplicative conics, multiplicative tangents, multiplicative secants and a multiplicative cross product are defined and some of their properties are obtained. Multiplicative analogues of the Desargues and Pappus theorems, as well as, the fundamental theorem of the projective multiplicative geometry are proved.

6.1 Definition: Incidence Properties of $P^2_\star$

Definition 6.1. *The projective multiplicative plane $P^2_\star$ is the set of all pairs $\{x, -_\star x\}$, where $x \in S^2_\star$.*

Let $\pi : S^2_\star \to P^2_\star$ be defined as follows:

$$\pi x = \{x, -_\star x\}.$$

Definition 6.2. *A multiplicative line of $P^2_\star$ is a set of the form πl, where l is a multiplicative line of $S^2_\star$.*

If ξ is a pole of the multiplicative line l of $S^2_\star$, then $\pi\xi$ is the pole of πl. Note that πx lies on πl if and only if $\langle x, \xi \rangle_\star = 0_\star$. Two points of $P^2_\star$ are multiplicative perpendicular if their representatives on $S^2_\star$ are multiplicative perpendicular. The multiplicative lines of $P^2_\star$ are multiplicative perpendicular if their poles are multiplicative perpendicular.

Theorem 6.1.

1. *Two multiplicative lines of $P^2_\star$ have exactly one point of intersection.*
2. *Two points of $P^2_\star$ lie on exactly one multiplicative line.*

Proof.

1. Let $\pi\xi$ and $\pi\eta$ be poles of the multiplicative lines of $P^2_\star$. Since $\pi\xi \neq \pi\eta$, ξ and η are not antipodal. Next, $\xi \times_\star \eta$ and $-_\star\xi \times_\star \eta$ determine the two points of intersection of the corresponding multiplicative lines of $S^2_\star$. But $\pi(\xi \times_\star \eta)$ and $\pi(-_\star\xi \times \eta)$ are the same point of $P^2_\star$.
2. Let πX and πY be two points of $P^2_\star$. Then X and Y are not antipodal and they lie on a unique multiplicative line of $S^2_\star$. Thus, πX and πY lie only on πl. This completes the proof. □

6.2 Multiplicative Homogeneous Coordinates

Definition 6.3. *Let $\{e_1, e_2, e_3\}$ be a basis in $E^3_\star$. Then any $x \in E^3_\star$ can be represented in the form*

$$x = x_1 \cdot_\star e_1 +_\star x_2 \cdot_\star e_2 +_\star x_3 \cdot_\star e_3.$$

DOI: 10.1201/9781003325284-6

If $\lambda \in \mathbb{R}_\star$, $\lambda \neq 0_\star$ *and*

$$\lambda \cdot_\star x = u_1 \cdot_\star e_1 +_\star u_2 \cdot_\star e_2 +_\star u_3 \cdot_\star e_3,$$

then (u_1, u_2, u_3) *is said to be multiplicative homogeneous vector of* πx. *We say that* u_1, u_2 *and* u_3 *are multiplicative homogeneous coordinates of* πx.

Theorem 6.2. *Let* $P, Q, R, S \in P_\star^2$, *no three of which are multiplicative collinear. Then there is a basis of* $E_\star^3$ *so that the four points have coordinates* $(1_\star, 0_\star, 0_\star)$, $(0_\star, 1_\star, 0_\star)$, $(0_\star, 0_\star, 1_\star)$ *and* $(1_\star, 1_\star, 1_\star)$.

Proof. Let $v_1, v_2, v_3 \in E_\star^3$ be any representatives of the points P, Q and R, respectively. Since P, Q and R are not multiplicative collinear, we have that v_1, v_2 and v_3 are multiplicative linearly independent. Let now, v_4 be any representative of the point S. Then there are $k_1, k_2, k_3 \in \mathbb{R}_\star$ so that

$$v_4 = k_1 \cdot_\star v_1 +_\star k_2 \cdot_\star v_2 +_\star k_3 \cdot_\star v_3. \tag{6.1}$$

Set

$$\begin{aligned} e_1 &= k_1 \cdot_\star v_1, \\ e_2 &= k_2 \cdot_\star v_2, \\ e_3 &= k_3 \cdot_\star v_3. \end{aligned}$$

Then $\{e_1, e_2, e_3\}$ is a basis and

$$\begin{aligned} e_1 &= (1_\star, 0_\star, 0_\star), \\ e_2 &= (0_\star, 1_\star, 0_\star), \\ e_3 &= (0_\star, 0_\star, 1_\star). \end{aligned}$$

By (6.1), we find

$$\begin{aligned} v_4 &= e_1 +_\star e_2 +_\star e_3 \\ &= (1_\star, 0_\star, 0_\star) +_\star (0_\star, 1_\star, 0_\star) +_\star (0_\star, 0_\star, 1_\star) \\ &= (1_\star, 1_\star, 1_\star). \end{aligned}$$

This completes the proof. □

6.3 The Desargues Theorem and the Pappus Theorem

Theorem 6.3 (The Desargues Theorem)**.** *Let* PQR *and* $P_1Q_1R_1$ *be multiplicative triangles so that* PP_1, QQ_1 *and* RR_1 *are multiplicative concurrent. Then* $PQ \cap P_1Q_1$, $PR \cap P_1R_1$ *and* $QR \cap Q_1R_1$ *are multiplicative collinear.*

Proof. Let X be the given point of multiplicative concurrence. By Theorem 6.2, it follows that we may choose a basis in $E_\star^3$ so that in the associated multiplicative homogeneous coordinate system we have

$$\begin{aligned} P &= (1_\star, 0_\star, 0_\star), \\ Q &= (0_\star, 1_\star, 0_\star), \\ R &= (0_\star, 0_\star, 1_\star), \\ X &= (1_\star, 1_\star, 1_\star). \end{aligned}$$

If X is a multiplicative collinear with any of these points, then the multiplicative sides (such as PQ and P_1Q_1) coincide and the assertion follows. Suppose that there are no three points of P, Q, R and X that are multiplicative collinear. We can take

$$\begin{aligned} P_1 &= (p, 1_\star, 1_\star), \\ Q_1 &= (1_\star, q, 1_\star), \\ R_1 &= (1_\star, 1_\star, r). \end{aligned}$$

Then the equation of PQ is $x_3 = 0_\star$ and the equation of P_1Q_1 is

$$(1_\star -_\star q) \cdot_\star x_1 +_\star (1_\star -_\star p) \cdot_\star x_2 +_\star (p \cdot_\star q -_\star 1_\star) \cdot_\star x_3 = 0_\star,$$

and

$$PQ \cap P_1Q_1 = L = (p -_\star 1_\star, 1_\star -_\star q, 0_\star).$$

The equation of PR is $x_2 = 0_\star$ and the equation of P_1R_1 is

$$(1_\star -_\star r_\star) \cdot_\star x_1 +_\star (p \cdot_\star r -_\star 1_\star) \cdot_\star x_2 +_\star (1_\star -_\star p) \cdot_\star x_3 = 0_\star,$$

and

$$PR \cap P_1R_2 = M = (1_\star -_\star p, 0_\star, r -_\star 1_\star).$$

Next, the equation of QR is $x_1 = 0_\star$ and the equation of Q_1R_1 is

$$(r \cdot_\star q -_\star 1_\star) \cdot_\star x_1 +_\star (1_\star -_\star r) \cdot_\star x_2 +_\star (1_\star -_\star q) \cdot_\star x_3 = 0_\star,$$

and

$$QR \cap Q_1R_1 = N = (0_\star, q -_\star 1_\star, 1_\star -_\star r).$$

We have that

$$L +_\star M +_\star N = 0_\star,$$

and thus, L, M and N are multiplicative collinear. This completes the proof. □

Theorem 6.4 (The Pappus Theorem). *Let $A_1B_1C_1$ and $A_2B_2C_2$ be multiplicative collinear triples and $A_1B_1 \cap A_2B_2 = C_3$, $A_1C_1 \cap A_2C_2 = B_3$, $B_1C_1 \cap B_2C_2 = A_3$. Then A_3, B_3 and C_3 are multiplicative collinear.*

Proof. Let in the multiplicative homogeneous coordinate system, we have

$$\begin{aligned} A_1 &= (1_\star, 0_\star, 0_\star), \\ A_2 &= (0_\star, 1_\star, 0_\star), \\ A_3 &= (0_\star, 0_\star, 1_\star), \\ C_1 &= (1_\star, 1_\star, 1_\star), \\ B_1 &= (p, 1_\star, 1_\star), \\ B_3 &= (1_\star, q, 1_\star), \\ B_2 &= (1_\star, 1_\star, r). \end{aligned}$$

Then

$$\begin{aligned} C_2 &= B_1A_3 \cap B_3A_1 = (p \cdot_\star q, q, 1_\star), \\ C_3 &= A_1B_2 \cap A_2B_1 = (p \cdot_\star r, 1_\star, r). \end{aligned}$$

Since A_2, B_2 and C_2 are multiplicative collinear, we have

$$\begin{aligned} C_2 &= (0_\star, \lambda, 0_\star) +_\star (1_\star, 1_\star, r) \\ &= (1_\star, \lambda +_\star 1_\star, r). \end{aligned}$$

Therefore,

$$\begin{aligned} p \cdot_\star q &= 1_\star, \\ q &= 1_\star, \\ r &= 1_\star. \end{aligned}$$

Thus,

$$p \cdot_\star q \cdot_\star r = 1_\star.$$

Note that A_3B_3 consists of points of the form

$$(0_\star, 0_\star, \lambda) +_\star (1_\star, q, 1_\star) = (1_\star, q, 1_\star +_\star \lambda).$$

Hence, C_3 must be on A_3B_3. This completes the proof. □

6.4 The Projective Multiplicative Group

With $\mathrm{PGL}_\star(e^2)$, we will denote the group of each multiplicative collineations of $P^2_\star$. Note that each multiplicative invertible linear map $A : E^3_\star \to E^3_\star$ determines a unique multiplicative collineation $\widetilde{A}$ in $\mathrm{PGL}_\star(e^2)$ as follows:

$$\widetilde{A}\pi x = \pi A x.$$

The mapping $A \to \widetilde{A}$ is a homomorphism of $\mathrm{GL}_\star(e^2) \to \mathrm{PGL}_\star(e^2)$ with a kernel

$$K = \{k \cdot_\star I_\star : k \neq 0_\star \in \mathbb{R}_\star\}.$$

Also, this map is surjective.

Definition 6.4. *Any element of $PGL_\star(e^2)$ is called projective multiplicative collineation or projective multiplicative transformation.*

6.5 The Fundamental Theorem of the Projective Multiplicative Geometry

Theorem 6.5. *Let $PQRS$ and $P_1Q_1R_1S_1$ be two multiplicative quadrangles. Then there is a unique multiplicative transformation $T \in PGL_\star(e^2)$ so that*

$$TP = P_1, \quad TQ = Q_1, \quad TR = R_1, \quad TS = S_1.$$

Proof. Choose multiplicative homogeneous coordinates for P, Q, R and S, i.e.,

$$\begin{aligned} P &= (1_\star, 0_\star, 0_\star), \\ Q &= (0_\star, 1_\star, 0_\star), \\ R &= (0_\star, 0_\star, 1_\star), \\ S &= (1_\star, 1_\star, 1_\star). \end{aligned}$$

Let

$$\begin{aligned} P_1 &= (x_1, y_1, z_1),\\ Q_1 &= (x_2, y_2, z_2),\\ R_1 &= (x_3, y_3, z_3) \end{aligned}$$

and

$$A = \begin{pmatrix} x_1 & y_1 & z_1 \\ x_2 & y_2 & z_2 \\ x_3 & y_3 & z_3 \end{pmatrix}.$$

Then

$$\begin{pmatrix} x_1 & y_1 & z_1 \\ x_2 & y_2 & z_2 \\ x_3 & y_3 & z_3 \end{pmatrix} \cdot_\star \begin{pmatrix} 1_\star & 0_\star & 0_\star \\ 0_\star & 1_\star & 0_\star \\ 0_\star & 0_\star & 1_\star \end{pmatrix} = \begin{pmatrix} x_1 & y_1 & z_1 \\ x_2 & y_2 & z_2 \\ x_3 & y_3 & z_3 \end{pmatrix}$$

and

$$\begin{pmatrix} x_1 & y_1 & z_1 \\ x_2 & y_2 & z_2 \\ x_3 & y_3 & z_3 \end{pmatrix} \cdot_\star \begin{pmatrix} 1_\star \\ 1_\star \\ 1_\star \end{pmatrix} = \begin{pmatrix} x_1 +_\star y_1 +_\star z_1 \\ x_2 +_\star y_2 +_\star z_2 \\ x_3 +_\star y_3 +_\star z_3 \end{pmatrix}.$$

Moreover, for any $\lambda, \mu, \nu \in \mathbb{R}_\star$, we have

$$\begin{pmatrix} \lambda \cdot_\star x_1 & \mu \cdot_\star y_1 & \nu \cdot_\star z_1 \\ \lambda \cdot_\star x_2 & \mu \cdot_\star y_2 & \nu \cdot_\star z_2 \\ \lambda \cdot_\star x_3 & \mu \cdot_\star y_3 & \nu \cdot_\star z_3 \end{pmatrix} \cdot_\star \begin{pmatrix} 1_\star \\ 1_\star \\ 1_\star \end{pmatrix} = \begin{pmatrix} \lambda \cdot_\star x_1 +_\star \mu \cdot_\star y_1 +_\star \nu \cdot_\star z_1 \\ \lambda \cdot_\star x_2 +_\star \mu \cdot_\star y_2 +_\star \nu \cdot_\star z_2 \\ \lambda \cdot_\star x_3 +_\star \mu \cdot_\star y_3 +_\star \nu \cdot_\star z_3 \end{pmatrix}.$$

We choose $\lambda, \mu, \nu \in \mathbb{R}_\star$ so that

$$S_1 = (\lambda \cdot_\star x, \mu \cdot_\star y, \nu \cdot_\star z).$$

Then the projective multiplicative collineation with matrix

$$\begin{pmatrix} \lambda \cdot_\star x_1 & \mu \cdot_\star y_1 & \nu \cdot_\star z_1 \\ \lambda \cdot_\star x_2 & \mu \cdot_\star y_2 & \nu \cdot_\star z_2 \\ \lambda \cdot_\star x_3 & \mu \cdot_\star y_3 & \nu \cdot_\star z_3 \end{pmatrix}$$

is the required multiplicative transformation. The uniqueness we leave to the reader as an exercise. This completes the proof. □

Corollary 6.1. *Let $\{P, Q, R\}$ and $\{P_1, Q_1, R_1\}$ be two multiplicative noncollinear triples of points. Let l be a multiplicative line not containing any of these points. Then there is a unique multiplicative transformation T so that*

$$TP = P_1, \quad TQ = Q_1, TR = R_1 \quad \text{and} \quad Tl = l.$$

Proof. Let

$$l \cap PQ = A, \quad l \cap P_1Q_1 = A_1, \quad l \cap PR = B, \quad l \cap P_1R_1 = B_1.$$

By Theorem 6.5, it follows that there is a unique multiplicative transformation T so that

$$TR = R_1, \quad TQ = Q_1, \quad TA = A_1, \quad TB = B_1.$$

Since $A, B, A_1, B_1 \in l$, we have that $Tl = l$. Next,

$$AQ \cap BR = P, \quad A_1Q_1 \cap B_1R_1 = P_1.$$

Then $TP = P_1$. This completes the proof. □

6.6 Multiplicative Polarities

One of the reason, for invention of $P_\star^2$ is to simplify the incidence multiplicative geometry of $E_\star^2$. We regard the plane $x_3 = 1_\star$ consisting of all points in $E_\star^3$ of the form $(x_1, x_2, 1_\star)$ as a model of $E_\star^2$. Every multiplicative line through the multiplicative origin $(0_\star, 0_\star, 0_\star)$ of $E_\star^3$ that is not multiplicative parallel to $E_\star^2$ meets $E_\star^2$ in a unique point. If (x_1, x_2, x_3) are the multiplicative homogeneous coordinates for such a point of $P_\star^2$, then

$$(x_1 /_\star x_3, x_2 /_\star x_3, 1_\star),$$

is the corresponding point of $E_\star^2$. Conversely, each point of $E_\star^2$ determines a unique point of $P_\star^2$. Every multiplicative line of $E_\star^2$ determines a unique multiplicative plane through the multiplicative origin in $E_\star^3$ and hence, a unique multiplicative line of $P_\star^2$. Every multiplicative line of $P_\star^2$ determines a unique multiplicative plane through the multiplicative origin in $E_\star^3$ and hence, a unique multiplicative line of $E_\star^2$. The exception is the multiplicative plane through the multiplicative origin parallel to $E_\star^2$.

Suppose that b is a symmetric, multiplicative nonnegative, bilinear function on $E_\star^3$. Set

$$b_{ij} = b(e_i, e_j), \quad i, j \in \{1, 2, 3\}.$$

Then

$$\begin{aligned} b(x,y) &= \sum_{i,j=1_\star}^{3} x_i \cdot_\star y_j \cdot_\star b(e_i, e_j) \\ &= \sum_{i,j=1_\star}^{3} b_{ij} \cdot_\star x_i \cdot_\star y_j \\ &= x^T \cdot B \cdot_\star y \\ &= \langle x, B \cdot_\star y \rangle_\star, \end{aligned}$$

where $B = (b_{ij})$. Each such b define a relation

$$\widetilde{b} \subset P_\star^2 \times P_\star^2,$$

containing of those pairs $\{\pi x, \pi y\}$ such that $b(x,y) = 0_\star$.

Definition 6.5. *The relation $\widetilde{b}$ is called a multiplicative polarity. If $b(x,y) = 0_\star$, we say that πx and πy are multiplicative conjugate.*

Definition 6.6. *For a given y, the set*

$$\{\pi x : b(x,y) = 0_\star\},$$

is called multiplicative polar line of πy. We call πy to be the pole of the multiplicative line with respect to b.

Definition 6.7. *The set of self-conjugate points is called multiplicative conic determined by $\widetilde{b}$.*

Example 6.1. *Let*

$$B = \begin{pmatrix} e^2 & 0_\star & 0_\star \\ 0_\star & e^2 & 0_\star \\ 0_\star & 0_\star & e^{-1} \end{pmatrix}.$$

Then

$$\begin{aligned} x^T \cdot_\star B \cdot_\star x &= (x_1, x_2, x_3) \cdot_\star \begin{pmatrix} e^2 & 0_\star & 0_\star \\ 0_\star & e^2 & 0_\star \\ 0_\star & 0_\star & e^{-1} \end{pmatrix} \cdot_\star \begin{pmatrix} x_1 \\ x_2 \\ x_3 \end{pmatrix} \\ &= (x_1, x_2, x_3) \cdot_\star \begin{pmatrix} e^2 \cdot_\star x_1 \\ e^2 \cdot_\star x_2 \\ e^{-1} \cdot_\star x_3 \end{pmatrix} \\ &= e^2 \cdot_\star x_1^{2_\star} +_\star e^2 \cdot_\star x_2^{2_\star} +_\star e^{-1} \cdot_\star x_3^{2_\star} \\ &= e^{2(\log x_1)^2} +_\star e^{2(\log x_2)^2} +_\star e^{-(\log x_3)^2} \\ &= e^{2(\log x_1)^2 + 2(\log x_2)^2 - (\log x_3)^2} \\ &= 0_\star \\ &= e^0, \end{aligned}$$

if and only if

$$2(\log x_1)^2 + 2(\log x_2)^2 - (\log x_3)^2 = 0.$$

Then the multiplicative conic determined by $\widetilde{b}$ *is given by*

$$\{\pi x : 2(\log x_1)^2 + 2(\log x_2)^2 - (\log x_3)^2 = 0\}.$$

The multiplicative conic in $P_\star^2$ *corresponds to the multiplicative line*

$$2(\log x_1)^2 + 2(\log x_2)^2 = 1.$$

Example 6.2. *Let*

$$B = \begin{pmatrix} 0_\star & 0_\star & e^{-2} \\ o_\star & 1_\star & 0_\star \\ e^{-2} & 0_\star & 0_\star \end{pmatrix}.$$

Then

$$\begin{aligned} x^T \cdot_\star B \cdot_\star x &= (x_1, x_2, x_3) \cdot_\star \begin{pmatrix} 0_\star & 0_\star & e^{-2} \\ o_\star & 1_\star & 0_\star \\ e^{-2} & 0_\star & 0_\star \end{pmatrix} \cdot_\star \begin{pmatrix} x_1 \\ x_2 \\ x_3 \end{pmatrix} \\ &= (x_1, x_2, x_3) \cdot_\star \begin{pmatrix} e^{-2} \cdot_\star x_3 \\ x_2 \\ e^{-2} \cdot_\star x_1 \end{pmatrix} \\ &= e^{-2} \cdot_\star x_1 \cdot_\star x_3 +_\star x_2^{2_\star} +_\star e^{-2} \cdot_\star x_1 \cdot_\star x_3 \\ &= e^{-2\log x_1 \log x_3} +_\star e^{(\log x_2)^2} +_\star e^{-2\log x_1 \log x_3} \\ &= e^{-2\log x_1 \log x_3 + (\log x_2)^2 - 2\log x_1 \log x_3} \\ &= e^{-4\log x_1 \log x_3 + (\log x_2)^2} \\ &= 0_\star \\ &= e^0, \end{aligned}$$

if and only if

$$-4\log x_1 \log x_3 + (\log x_2)^2 = 0.$$

Then the multiplicative conic is the set

$$\{\pi x : -4\log x_1 \log x_3 + (\log x_2)^2 = 0\}.$$

The multiplicative conic in $P_\star^2$ corresponds to the multiplicative line

$$4\log x_1 = (\log x_2)^2.$$

Exercise 6.1. *Find the multiplicative conic determined by $\widetilde{b}$, where*

$$B = \begin{pmatrix} e^{-3} & 1_\star & 0_\star \\ 1_\star & e^{-2} & 0_\star \\ 0_\star & e^{-4} & 1_\star \end{pmatrix}.$$

Definition 6.8. *Two multiplicative lines are said to be conjugate if the multiplicative pole of one lies on the other.*

Definition 6.9. *A multiplicative line that passes through its own multiplicative pole is said to be self-conjugate.*

Theorem 6.6. *Let P and Q be points of $P_\star^2$ with multiplicative polar lines l and m, respectively. Then P lies on m if and only if Q lies on l.*

Proof. Let

$$P = \pi x, \quad Q = \pi y.$$

Then P lies on m if and only if

$$b(x,y) = 0_\star.$$

Next, Q lies on l if and only if

$$b(y,x) = 0_\star.$$

Now, using that b is symmetric, we complete the proof. □

Theorem 6.7. *Let P and Q be self-conjugate points of $P_\star^2$, $P \neq Q$. Then the multiplicative line through P and Q cannot be a self-conjugate line.*

Proof. Let l and m be multiplicative polar lines of P and Q, respectively. Since $P \neq Q$, the multiplicative lines l and m are different. Let R be an intersection point of l and m. Note that R does not lie on the multiplicative line through P and Q. Since R lies on l and m, we have that R and P are conjugate and R and Q are conjugate. Therefore, the multiplicative polar line n of R passes through the points P and Q. Thus, n is the multiplicative line through P and Q. From here, we conclude that n is not self-conjugate multiplicative line. This completes the proof. □

Theorem 6.8. *A multiplicative line l contains exactly one self-conjugate point if and only if l is a self-conjugate line.*

Proof. 1. Let l contains exactly one self-conjugate point $P = \pi x$. Take $Q = \pi y$ to be another point that lies on l. Then for any $\lambda \in \mathbb{R}_\star$, we have

$$\begin{aligned} b(x +_\star \lambda \cdot_\star y, x +_\star \lambda \cdot_\star y) &= b(x,x) +_\star e^2 \cdot_\star \lambda \cdot_\star b(x,y) +_\star \lambda^{2_\star} \cdot_\star b(y,y) \\ &= \lambda \cdot_\star \left(e^2 \cdot_\star b(x,y) +_\star \lambda \cdot_\star b(y,y)\right). \end{aligned} \tag{6.2}$$

Since l contains exactly one self-conjugate point, we must have

$$b(x,y) = 0_\star.$$

Otherwise, the equation

$$b(x +_\star \lambda \cdot_\star y, x +_\star \lambda \cdot_\star y) = 0_\star, \tag{6.3}$$

can be solved for $\lambda \neq 0_\star$. Hence, $Q = \pi y$ is the multiplicative pole of l. Because Q lies on l, we conclude that l is self-conjugate.

2. Let l be a self-conjugate multiplicative line. Then l passes through its own multiplicative pole $Q = \pi y$. Then Q is self-conjugate. If we assume that there is other self-conjugate point P that lies on l, by Theorem 6.7, it follows that l can not be self-conjugate. This is a contradiction. Consequently, l has exactly one self-conjugate point. This completes the proof.

□

Definition 6.10. *Let $\widetilde{b}$ be a multiplicative polarity defining a multiplicative conic K. A multiplicative line that is self-conjugate with respect to K is said to be a multiplicative tangent of K. The multiplicative pole of this multiplicative line is called the point of multiplicative contact.*

Theorem 6.9. *A multiplicative line meets a multiplicative conic in at most two points.*

Proof. We consider the quadratic function of the form (6.2). Since the quadratic equation (6.3) has at most two solutions, the proof is completed. □

Definition 6.11. *A multiplicative line that meets a multiplicative conic twice is called a multiplicative secant.*

6.7 Multiplicative Cross Product

Let $u, v \in \mathbb{E}^3_\star$. Then there exists a unique vector w so that

$$b(w,z) = (|\det{}_\star B|_\star)^{\frac{1}{2}\star} \cdot_\star \det{}_\star(z,u,v).$$

Definition 6.12. *The vector w is said to be the multiplicative cross product of u and v and we write $w = u \times_\star v$.*

We have

$$\begin{aligned} b(u \times_\star v, w) &= (|\det{}_\star B|_\star)^{\frac{1}{2}\star} \cdot_\star \det{}_\star(w,u,v) \\ &= (|\det{}_\star B|_\star)^{\frac{1}{2}\star} \cdot_\star \det{}_\star(u,v,w) \\ &= b(v \times_\star w, u) \\ &= b(u, v \times_\star v), \end{aligned}$$

and

$$\begin{aligned} b(u, u \times_\star v) &= b(u \times_\star v, u) \\ &= (|\det{}_\star B|_\star)^{\frac{1}{2}\star} \cdot_\star \det{}_\star(u,u,v) \\ &= 0_\star, \end{aligned}$$

and

$$\begin{aligned} b(v, u \times_\star v) &= b(u \times_\star v, v) \\ &= (|\det{}_\star B|_\star)^{\frac{1}{2}\star} \cdot_\star \det{}_\star(v,u,v) \\ &= 0_\star. \end{aligned}$$

Theorem 6.10. *Let πu and πv be points of $P^2_\star$. Then the multiplicative line joining πu and πv has a multiplicative pole $\pi(u \times_\star v)$.*

Proof. We have

$$b(u, u \times_\star v) = 0_\star,$$

and

$$b(v, u \times_\star v) = 0_\star.$$

This completes the proof. □

Definition 6.13. *A multiplicative triangle $\triangle_\star PQR$ of $P^2_\star$ is said to be self-polar, if each vertex is a multiplicative pole of the multiplicative side opposite of it. Any self-polar multiplicative triangle gives rise to a multiplicative basis $\{e_1, e_2, e_3\}$ of $E^3_\star$ such that*

$$\begin{aligned} b(e_i, e_j) &= 0_\star, \\ b(e_i, e_i) &= \pm_\star 1_\star, \quad i, j \in \{1,2,3\}, \quad i \neq j. \end{aligned}$$

Such a multiplicative basis is said to be multiplicative orthonormal basis with respect to b.

Let $\{e_1, e_2, e_3\}$ be a multiplicative orthonormal basis with respect to b. Then

$$\begin{aligned} e_1 \times_\star e_2 &= b(e_3, e_3) \cdot_\star e_3, \\ e_2 \times_\star e_3 &= b(e_1, e_1) \cdot_\star e_1, \\ e_3 \times_\star e_1 &= b(e_2, e_2) \cdot_\star e_2. \end{aligned}$$

Definition 6.14. *Let b be a non degenerate, bilinear, symmetric function and $\{e_1, e_2, e_3\}$ be a multiplicative orthonormal basis with respect to b. Suppose that $+_\star 1_\star$ occurs r times and $-_\star 1_\star$ occurs s times among the terms $b(e_i, e_i)$, $i \in \{1,2,3\}$. Then the ordered pair (r, s) is said to be the multiplicative signature of b.*

Theorem 6.11. *We have*

$$(u \times_\star v) \times_\star w = (-_\star 1_\star)^{s_\star} (b(u, w) \cdot_\star v -_\star b(v, w) \cdot_\star u),$$

where the multiplicative signature of b is (r, s).

Proof. Let $\{e_1, e_2, e_3\}$ be a multiplicative orthonormal basis with respect to b. Then

$$\begin{aligned} (e_1 \times_\star e_2) \times_\star e_3 &= (b(e_3, e_3) \cdot_\star e_3) \times_\star e_2 \\ &= b(e_3, e_3) \cdot_\star (e_3 \times_\star e_2) \\ &= -_\star b(e_3, e_3) \cdot_\star b(e_1, e_1) \cdot_\star e_1, \end{aligned}$$

and

$$\begin{aligned} &(-_\star 1_\star)^{s_\star} \cdot_\star (b(e_1, e_2) \cdot_\star e_2 -_\star b(e_1, e_2) \cdot_\star e_1) \\ &= (-_\star 1_\star)^{s_\star} \cdot_\star (-_\star 1_\star) \cdot_\star b(e_2, e_2) \cdot_\star e_2. \end{aligned}$$

Hence,

$$(e_1 \times_\star e_2) \times_\star e_2 = (-_\star 1_\star)^{s_\star} \cdot_\star (b(e_1, e_2) \cdot_\star e_2 -_\star b(e_1, e_2) \cdot_\star e_1),$$

if and only if

$$-_\star b(e_3, e_3) \cdot_\star b(e_1, e_1) \cdot_\star e_1 = (-_\star 1_\star)^{s_\star} \cdot_\star (-_\star 1_\star) \cdot_\star b(e_2, e_2) \cdot_\star e_2,$$

if and only if

$$b(e_3, e_3) \cdot_\star b(e_1, e_1) = (-_\star 1_\star)^{s_\star} \cdot_\star b(e_2, e_2),$$

if and only if

$$b(e_1, e_1) b(e_2, e_2) b(e_3, e_3) = (-_\star 1_\star)^{s_\star}.$$

The other combinations can be checked similarly and we leave them to the reader as an exercise. This completes the proof. □

6.8 Advanced Practical Problems

Problem 6.1. *Find the multiplicative conic determined by $\widetilde{b}$, where*

1.

$$B = \begin{pmatrix} 2 & 0_\star & 0_\star \\ 0_\star & e^{-3} & 3 \\ 0_\star & 3 & e^2 \end{pmatrix}.$$

2.

$$B = \begin{pmatrix} 0_\star & e^2 & e^3 \\ e^2 & 0_\star & 0_\star \\ e^3 & 0_\star & e^{-1} \end{pmatrix}.$$

3.

$$B = \begin{pmatrix} e^{-1} & e^{-2} & e^{-4} \\ e^{-2} & 0_\star & e^{-3} \\ e^{-4} & e^{-3} & e^2 \end{pmatrix}.$$

Problem 6.2. *Let*

$$\begin{aligned} x &= (1_\star, 0_\star, 0_\star), \\ y &= (1_\star, 1_\star, 0_\star), \\ z &= (1_\star, 0_\star, 1_\star), \\ w &= (1_\star, 1_\star, 1_\star). \end{aligned}$$

Let l be the multiplicative line joining πx and πy, and m be the multiplicative line joining πz and πw. Find $l \cap m$.

Problem 6.3. *Let P, Q, R and S be multiplicative collinear points. Prove that there is a unique multiplicative projection that interchanges P and Q and interchanges R and S.*

Problem 6.4. *Let P, Q, R be distinct points on the multiplicative line l and P_1, Q_1, R_1 be distinct points on the multiplicative line l_1. Prove that there is a unique multiplicative projection sending P to P_1, Q to Q_1 and R to R_1.*

Problem 6.5. *Classify the multiplicative projections of a given multiplicative line l in the terms of their fixed point behavior.*

7

The Multiplicative Distance Geometry on $P^2_\star$

In this chapter, a multiplicative distance of $P^2_\star$ is defined and some of its properties are investigated. In the chapter, multiplicative orthogonal transformations, multiplicative reflections, multiplicative rotations and multiplicative translations are introduced and studied.

7.1 The Multiplicative Distance

Definition 7.1. *Let P, Q be points of $P^2_\star$. Define*

$$d_\star(P,Q) = \cos_\star^{-1_\star} |\langle x, y\rangle_\star|_\star,$$

where $x, y \in S^2_\star$, $P = \pi x$, $Q = \pi y$.

Because the multiplicative absolute value sign, the multiplicative distance is well defined. Note that the multiplicative distance between $\{x, -_\star x\}$ and $\{y, -_\star y\}$ is the multiplicative distance between the closest representatives. Moreover, we have

$$d_\star(P,Q) \leq e^{\frac{\pi}{2}}.$$

Theorem 7.1. *Let P, Q and R be points of $P^2_\star$. Then*

$$d_\star(P,Q) +_\star d_\star(Q,R) \geq d_\star(P,R).$$

Proof. Let

$$\begin{aligned} r &= d_\star(P,Q), \\ p &= d_\star(Q,R), \\ q &= d_\star(P,R). \end{aligned}$$

Choose representatives P_1, Q_1 and R_1 so that

$$\begin{aligned} \langle P_1, R_1\rangle_\star &\geq 0_\star, \\ \langle Q_1, R_1\rangle_\star &\geq 0_\star. \end{aligned}$$

Then

$$\begin{aligned} |P_1 \times_\star Q_1|_\star &= \sin_\star r, \\ |R_1 \times_\star Q_1|_\star &= \sin_\star p \end{aligned}$$

and

$$\begin{aligned} |P_1 \times_\star Q_1|_\star \cdot_\star |R_1 \times_\star Q_1|_\star &\geq \langle P_1 \times_\star Q_1, Q_1 \times_\star R_1\rangle_\star \\ &= -_\star \langle P_1, R_1\rangle_\star +_\star \langle Q_1, R_1\rangle_\star \cdot_\star \langle Q_1, P_1\rangle_\star. \end{aligned}$$

DOI: 10.1201/9781003325284-7

If $\langle P_1, Q_1\rangle_\star \geq 0_\star$, then

$$\sin_\star r \cdot_\star \sin_\star p \geq \cos_\star p \cdot_\star \cos_\star r -_\star \cos_\star q,$$

or

$$\begin{aligned} \cos_\star q &\geq \cos_\star p \cdot_\star \cos_\star r -_\star \sin_\star r \cdot_\star \sin_\star p \\ &= \cos_\star (p +_\star r), \end{aligned}$$

whereupon

$$q \leq p +_\star r.$$

Equality we have when P_1, Q_1 and R_1 are multiplicative collinear on $S^2_\star$ and hence, P, Q and R are multiplicative collinear on $P^2_\star$.

If $\langle P_1, Q_1\rangle_\star \leq 0_\star$, then

$$|P_1 \times_\star Q_1|_\star \cdot_\star |R_1 \times_\star Q_1|_\star \geq \langle P_1, R_1\rangle_\star -_\star \langle Q_1, R_1\rangle_\star \cdot_\star \langle Q_1, P_1\rangle_\star,$$

which yields

$$\sin_\star r \cdot_\star \sin_\star p \geq \cos_\star q +_\star \cos_\star p \cdot_\star \cos_\star r$$

or

$$\begin{aligned} \cos_\star (e^\pi -_\star q) &\geq \cos_\star p \cdot_\star \cos_\star r -_\star \sin_\star r \cdot_\star \sin_\star p \\ &= \cos_\star (p +_\star r), \end{aligned}$$

or

$$e^\pi -_\star q \leq p +_\star r.$$

Since $q \leq e^{\frac{\pi}{2}}$, we get

$$\begin{aligned} q &\leq e^{\frac{\pi}{2}} \\ &\leq e^\pi -_\star q \\ &\leq p +_\star r. \end{aligned}$$

As above, equality we have when P_1, Q_1 and R_1 are multiplicative collinear on $S^2_\star$ and from here, when P, Q and R are multiplicative collinear on $P^2_\star$. This completes the proof. □

Theorem 7.2. *Three points P, Q and R of $P^2_\star$ are multiplicative collinear if and only if*

$$d_\star(P,Q) +_\star d_\star(Q,R) = d_\star(P,R), \tag{7.1}$$

or

$$d_\star(P,Q) +_\star d_\star(Q,R) +_\star d_\star(P,R) = e^\pi. \tag{7.2}$$

Proof. 1. Let P, Q and R are multiplicative collinear. Assume that (7.1) does not hold. Take e_3 to be the multiplicative pole of the multiplicative line determined by the three points. With e_1 we will denote the representative of the point P. Let also, $\{e_1, e_2, e_3\}$ be a multiplicative orthonormal basis. There are $\theta, \phi \in \left[1, e^{\frac{\pi}{2}}\right]$ so that

$$\begin{aligned} Q &= \pi\left((\cos_\star \phi) \cdot_\star e_1 +_\star (\sin_\star \phi) \cdot_\star e_2\right), \\ R &= \pi\left((\cos_\star \theta) \cdot_\star e_1 \pm_\star (\sin_\star \theta) \cdot_\star e_2\right). \end{aligned}$$

Then

$$\begin{aligned} d_\star(P,Q) &= \cos_\star^{-1_\star}(\cos_\star \phi) \\ &= \phi, \\ d_\star(P,R) &= \cos_\star^{-1_\star}(\cos_\star \theta) \\ &= \theta, \end{aligned}$$

and

$$d_\star(Q,R) = \cos_\star^{-1_\star} |\cos_\star(\phi \pm_\star \theta)|_\star.$$

The multiplicative negative sign cannot occur because otherwise we will have

$$d_\star(Q,R) = |\phi -_\star \theta|_\star,$$

and then (7.1) holds. If $\phi +_\star \theta \leq e^{\frac{\pi}{2}}$, then

$$d_\star(Q,R) = \phi +_\star \theta,$$

and then we get another version of (7.1). Therefore,

$$e^{\frac{\pi}{2}} < \phi +_\star \theta < e^{\pi}.$$

Then

$$\begin{aligned} d_\star(Q,R) &= \cos_\star^{-1_\star}(-_\star \cos_\star(\phi +_\star \theta)) \\ &= \cos_\star^{-1_\star}(\cos_\star(e^{\pi} -_\star (\phi +_\star \theta))) \\ &= e^{\pi} -_\star (\phi +_\star \theta) \\ &= e^{\pi} -_\star d_\star(P,Q) -_\star d_\star(P,R). \end{aligned}$$

2. Let (7.1) holds. Then, by Theorem 7.1, the points P, Q and R are multiplicative collinear.

3. Let (7.2) holds. Then

$$p +_\star r = e^{\pi} -_\star q.$$

So,

$$\cos_\star(e^{\pi} -_\star q) = \cos_\star(p +_\star r).$$

As in the proof of Theorem 7.1, we get that P, Q and R are multiplicative collinear. This completes the proof.

□

7.2 Multiplicative Isometries

Definition 7.2. *A map $T : P_\star^2 \to P_\star^2$ is called a multiplicative isometry if*

$$d_\star(TP,TQ) = d_\star(P,Q),$$

for any $P,Q \in P_\star^2$.

Theorem 7.3. *Let T be a multiplicative isometry. If P, Q and R are multiplicative collinear, then TP, TQ and TR are multiplicative collinear.*

Proof. Let P_1 be the unique point so that

$$d_\star(P,P_1) = e^{\frac{\pi}{2}}.$$

Then

$$\begin{aligned} d_\star(P,Q) +_\star d_\star(Q,P_1) &= d_\star(P,P_1) \\ &= e^{\frac{\pi}{2}}. \end{aligned}$$

Hence,

$$\begin{aligned} d_\star(TP,TQ) +_\star d_\star(TQ,TP_1) &= d_\star(TP,TP_1) \\ &= e^{\frac{\pi}{2}}. \end{aligned}$$

By Theorem 7.2, it follows that TQ must lie on the multiplicative line determined by TP and TP_1. As above, TR must lie on the same line. Therefore, TP, TQ and TR are multiplicative collinear. This completes the proof. □

Theorem 7.4. *Let T be a multiplicative isometry of $P_\star^2$ that leaves $\pi\varepsilon_i$, $i \in \{1,2,3\}$, and $\pi(\varepsilon_1 +_\star \varepsilon_2 +_\star \varepsilon_3)$. Then $T = I_\star$.*

Proof. Consider the multiplicative homogeneous coordinate system determined by ε_1, ε_2 and ε_3. Then

$$(1_\star,0_\star,0_\star), \quad (0_\star,1_\star,0_\star), \quad (0_\star,0_\star,1_\star) \quad \text{and} \quad (1_\star,1_\star,1_\star),$$

are fixed points. We will prove that the points on the multiplicative line joining $(1_\star,0_\star,0_\star)$ and $(0_\star,1_\star,0_\star)$ are fixed. Take

$$x = (\cos_\star \alpha, \sin_\star \alpha, 0_\star),$$

to be an arbitrary point on this multiplicative line. Here $\alpha \in (1,e^\pi)$. Let

$$Tx = (\cos_\star \beta, \sin_\star \beta, 0_\star), \quad \beta \in (1,e^\pi).$$

Then

$$d_\star(T\pi\varepsilon_1, Tx) = d_\star(\pi\varepsilon_1, x),$$

and hence,

$$\cos_\star^{-1_\star} |\cos_\star \beta|_\star = \cos_\star^{-1_\star} |\cos_\star \alpha|_\star.$$

Therefore,

$$|\cos_\star \alpha|_\star = |\cos_\star \beta|_\star.$$

We have $\beta = \alpha$ or $\beta = e^\pi -_\star \alpha$. If $\beta = \alpha$, the assertion is proved. Let $\beta = e^\pi -_\star \alpha$. Then

$$Tx = \pi(-_\star \cos_\star \alpha, \sin_\star \alpha, 0_\star).$$

Take $M = (1_\star,1_\star,1_\star)$. Hence,

$$\begin{aligned} d_\star(x,M) &= \cos_\star^{-1_\star} \left| e^{\frac{1}{\sqrt{3}}} \cdot_\star (\cos_\star \alpha +_\star \sin_\star \alpha) \right|_\star, \\ d_\star(Tx,M) &= \cos_\star^{-1_\star} \left| e^{\frac{1}{\sqrt{3}}} \cdot_\star (-_\star \cos_\star \alpha +_\star \sin_\star \alpha) \right|_\star. \end{aligned}$$

This is possible when $\alpha = e^{\frac{\pi}{2}}$. From here, $\beta = \alpha$. As above, we get that all points on the multiplicative sides of the multiplicative triangle of reference $\Delta_\star$ are fixed. Now, every multiplicative line of $P_\star^2$ contains at least two fixed points because it intersects $\Delta_\star$ at least twice. Thus, any multiplicative line is fixed and hence, any point is a fixed point. This completes the proof. □

Theorem 7.5. *Let $T : P^2_\star \to P^2_\star$ be a multiplicative isometry. Then there exists a unique $A \in SO_\star(e^3)$ such that $T = \widetilde{A}$.*

Proof. Choose e_1, e_2 and e_3 so that

$$T\pi\varepsilon_i = \pi e_i, \quad i \in \{1,2,3\}.$$

Then

$$\begin{aligned} d_\star(T\pi\varepsilon_i, T\pi\varepsilon_j) &= d_\star(\pi\varepsilon_i \pi\varepsilon_j) \\ &= \cos_\star^{-1_\star} |\langle \varepsilon_i, \varepsilon_j \rangle_\star|_\star, \quad i.j \in \{1,2,3\}. \end{aligned}$$

On the other hand,

$$d_\star(\pi e_i, \pi e_j) = \cos_\star^{-1_\star} |\langle e_i, e_j \rangle_\star|_\star,$$

and then

$$|\langle e_i, e_j \rangle_\star|_\star = |\langle \varepsilon_i, \varepsilon_j \rangle_\star|_\star, \quad i, j \in \{1,2,3\}.$$

Therefore, $\{e_1, e_2, e_3\}$is a multiplicative orthonormal basis of $E^3_\star$. Let A be a multiplicative orthogonal matrix such that

$$A\varepsilon_i = e_i, \quad i \in \{1,2,3\}.$$

Then $\widetilde{A}$ is a multiplicative isometry of $P^2_\star$ and $\widetilde{A}^{-1_\star}T$ leaves $\pi\varepsilon_1$, $\pi\varepsilon_2$ and $\pi\varepsilon_3$ fixed. Let

$$M = \pi(\varepsilon_1 +_\star \varepsilon_2 +_\star \varepsilon_3),$$

and

$$\widetilde{A}^{-1_\star}TM = \pi(k_1 \cdot_\star \varepsilon_1 +_\star k_2 \cdot_\star \varepsilon_2 +_\star k_3 \cdot_\star \varepsilon_3).$$

Here $k_1, k_2, k_3 \in \mathbb{R}_\star$ are chosen so that

$$k_1^{2_\star} +_\star k_2^{2_\star} +_\star k_3^{2_\star} = 1_\star.$$

Note that

$$d_\star(\widetilde{A}^{-1_\star}TM, \widetilde{A}^{-1_\star}T\pi\varepsilon_i) = \cos_\star^{-1_\star} |k_i|_\star, \quad i \in \{1,2,3\}.$$

On the other hand,

$$d_\star(M, \pi\varepsilon_i) = \cos_\star^{-1_\star}\left(e^{\frac{1}{\sqrt{3}}}\right), \quad i \in \{1,2,3\}.$$

Therefore,

$$|k_i|_\star = e^{\frac{1}{\sqrt{3}}}, \quad i \in \{1,2,3\}.$$

Let

$$B = e^{\frac{1}{\sqrt{3}}} \cdot_\star \begin{pmatrix} k_1 & 0_\star & 0_\star \\ 0_\star & k_2 & 0_\star \\ 0_\star & 0_\star & k_3 \end{pmatrix}.$$

We have that

$$\widetilde{B}\widetilde{A}^{-1_\star}T\pi\varepsilon_i = \pi\varepsilon_i, \quad i \in \{1,2,3\}.$$

Since

$$|k_i|^{2_\star} = e^{\frac{1}{3}}, \quad i \in \{1,2,3\},$$

we get

$$\widetilde{B}\widetilde{A}^{-1_\star}TM = M.$$

Now, we apply Theorem 7.4 and we obtain

$$\widetilde{B}\widetilde{A}^{-1_\star}T = I_\star,$$

whereupon

$$\begin{aligned} T &= \widetilde{A}\widetilde{B}^{-1_\star} \\ &= \widetilde{AB^{-1_\star}}. \end{aligned}$$

Because $B^{-1_\star} = B$, we have

$$T = \widetilde{AB}.$$

Note that AB is multiplicative orthogonal. Then there exists a unique multiplicative isometry of $\mathrm{SO}_\star(e^3)$ that determines the same multiplicative isometry. This completes the proof. □

7.3 Multiplicative Motions

Definition 7.3. *Let l be a multiplicative line of $S^2_\star$. The multiplicative reflection in the multiplicative line πl of $P^2_\star$ is the multiplicative isometry of $P^2_\star$ defined by*

$$\Omega_{\pi l} = \widetilde{\Omega}_l.$$

Theorem 7.6. *$\Omega_{\pi l}$ leaves fixed every point of πl and the multiplicative pole of πl. No other fixed points.*

Proof. We choose a multiplicative orthonormal basis of $E^3_\star$ with respect to which

$$\Omega_l = \begin{pmatrix} -_\star 1_\star & 0_\star & 0_\star \\ 0_\star & 1_\star & 0_\star \\ 0_\star & 0_\star & 1_\star \end{pmatrix}.$$

Then

$$\begin{aligned} \Omega_l x &= \begin{pmatrix} -_\star 1_\star & 0_\star & 0_\star \\ 0_\star & 1_\star & 0_\star \\ 0_\star & 0_\star & 1_\star \end{pmatrix} \cdot_\star \begin{pmatrix} x_1 \\ x_2 \\ x_3 \end{pmatrix} \\ &= \begin{pmatrix} -_\star x_1 \\ x_2 \\ x_3 \end{pmatrix}. \end{aligned}$$

Hence, $\Omega_l x = x$ if and only if x lies on the multiplicative line joining the points $(0_\star, 1_\star, 0_\star)$ and $(0_\star, 0_\star, 1_\star)$. Next, $\Omega_l x = -_\star x$ if and only if

$$x = (1_\star, 0_\star, 0_\star) \quad \text{or} \quad x = (-_\star 1_\star, 0_\star, 0_\star).$$

This completes the proof. □

Definition 7.4. *Let l be a multiplicative line of $P^2_\star$ with a multiplicative pole ξ. Suppose that m and n are multiplicative lines that pass through ξ. Then $\Omega_m\Omega_n$ is called a multiplicative rotation about ξ. Since the multiplicative lines that pass through ξ are multiplicative perpendicular, we say that $\Omega_m\Omega_n$ is a multiplicative translation along l. We call $\Omega_m\Omega_n\Omega_l$ a multiplicative glide reflection. If $m \perp_\star n$, then $\Omega_m\Omega_n$ will be called multiplicative half-turn.*

Note that the fixed multiplicative lines of a multiplicative reflection are the multiplicative lines of fixed points and multiplicative lines multiplicative perpendicular to this multiplicative line.

Theorem 7.7. *A multiplicative rotation other than a multiplicative half-turn or the multiplicative identity has a unique fixed point and a unique fixed multiplicative line. The point is the multiplicative pole of the multiplicative line.*

Proof. Let $\widetilde{A}$ be a multiplicative rotation of $P_\star^2$ and $A \in \mathrm{SO}_\star(e^3)$. If x is a fixed point of A, then πx is the unique fixed point of $\widetilde{A}$ and the multiplicative line whose multiplicative pole is πx is the unique fixed multiplicative line of $\widetilde{A}$. This completes the proof. □

Observe that any multiplicative reflection is a multiplicative half-turn and any multiplicative half-turn is a multiplicative rotation. Any multiplicative glide reflection is a multiplicative rotation, and any multiplicative isometry is a multiplicative rotation.

7.4 Elliptic Multiplicative Geometry

Definition 7.5. *The multiplicative geometry of $P_\star^2$ is called the elliptic multiplicative geometry.*

Definition 7.6. *A multiplicative segment of $P_\star^2$ is a set of the form $\pi\sigma$, where σ is a minor multiplicative segment of $S_\star^2$. The multiplicative length of $\pi\sigma$ is the multiplicative length of σ. The end points of $\pi\sigma$ are the image by π of the end points of σ.*

Each pair $\{A,B\}$ of points of $P_\star^2$ is the end point set of two multiplicative segments. The union of these multiplicative segments is the multiplicative line through A and B. For a multiplicative segment of multiplicative length L with end points A and B we have

$$d_\star(A,B) = L \quad \text{if} \quad L \le e^{\frac{\pi}{2}},$$

and

$$d_\star(A,B) = e^{\pi} -_\star L \quad \text{if} \quad L > e^{\frac{\pi}{2}}.$$

Definition 7.7. *A multiplicative ray is a multiplicative segment of multiplicative length $e^{\frac{\pi}{2}}$ with one end point removed. The remaining end point is called the origin of the multiplicative ray.*

The definition of the multiplicative angle is the same as our previous definitions. The multiplicative radian measure of a multiplicative angle $\angle_\star PQR$ is determined by choosing representatives of P, Q and R and computing the multiplicative radian measure of the multiplicative spherical angle.

A multiplicative triangle in the elliptic multiplicative geometry is a figure $\pi\Delta_\star$, where $\Delta_\star$is a multiplicative spherical triangle.

If P, Q and R are three multiplicative non collinear points of $P_\star^2$, then there is a multiplicative triangle with vertices P, Q and R. The multiplicative triangle is the union of three multiplicative segments.

7.5 Advanced Practical Problems

Problem 7.1. *Let*

$$\begin{aligned} P &= (-_\star 1_\star, 0_\star, 0_\star), \\ Q &= (1_\star, 1_\star, 0_\star), \end{aligned}$$

in a multiplicative homogeneous coordinate system with respect to $\{\varepsilon_1, \varepsilon_2, \varepsilon_3\}$. *Find* $d_\star(P,Q)$.

Problem 7.2. *Let*

$$\begin{aligned} P &= (0_\star, -_\star 1_\star, 0_\star), \\ Q &= (1_\star, 1_\star, 1_\star), \end{aligned}$$

in a multiplicative homogeneous coordinate system with respect to $\{\varepsilon_1, \varepsilon_2, \varepsilon_3\}$. *Find* $d_\star(P,Q)$.

Problem 7.3. *Find the multiplicative perpendicular to*

$$x_1 +_\star e^2 \cdot_\star x_2 = 0_\star,$$

and

$$e^2 \cdot_\star x_1 -_\star x_2 = 0_\star.$$

Problem 7.4. *Prove that the multiplicative radian measure of a multiplicative angle of* $P^2_\star$ *is well defined.*

Problem 7.5. *Prove that there are exactly two multiplicative reflections that interchange a given pair of multiplicative lines of* $P^2_\star$.

8

The Hyperbolic Multiplicative Plane

In this chapter, the multiplicative hyperbolic plane $H^2_\star$ is defined. Multiplicative lines, multiplicative segments, multiplicative triangles, multiplicative quadrilateral figures, multiplicative circles, multiplicative horocycles and multiplicative equidistant curves are introduced and investigated. In the chapter, multiplicative isometries, multiplicative reflections, multiplicative rotations and multiplicative translations in the multiplicative hyperbolic plane are studied.

8.1 Introduction

Suppose that

$$b(x,y) = x_1 \cdot_\star y_1 +_\star x_2 \cdot_\star y_2 -_\star x_3 \cdot_\star y_3, \quad x = (x_1, x_2, x_3), \quad y = (y_1, y_2, y_3) \in E^3_\star.$$

In fact, we have

$$b(x,y) = e^{\log x_1 \log y_1 + \log x_2 \log y_2 - \log x_3 \log y_3}.$$

Definition 8.1. *A multiplicative nonzero vector $v \in E^3_\star$ is said to be*

1. *multiplicative spacelike if $b(v,v) > 0_\star$. If $b(v,v) = 1_\star$, then v is said to be multiplicative unit multiplicative spacelike vector.*
2. *multiplicative timelike if $b(v,v) < 0_\star$. If $b(v,v) = -_\star 1_\star$, then v is said to be multiplicative unit multiplicative timelike vector.*
3. *multiplicative lightlike if $b(v,v) = 0_\star$.*

Example 8.1. *We have*

$$\begin{aligned} b(\varepsilon_1, \varepsilon_1) &= 1_\star \\ &> 0_\star. \end{aligned}$$

Thus, ε_1 is a multiplicative unit multiplicative spacelike vector.

Example 8.2. *We have*

$$\begin{aligned} b(\varepsilon_3, \varepsilon_3) &= -_\star 1_\star \\ &< 0_\star. \end{aligned}$$

Thus, ε_3 is a multiplicative unit multiplicative timelike vector.

DOI: 10.1201/9781003325284-8

Example 8.3. *We have*

$$\varepsilon_1 -_\star \varepsilon_3 = (1_\star, 0_\star, -_\star 1_\star),$$

and

$$\begin{aligned} b(\varepsilon_1 -_\star \varepsilon_3, \varepsilon_1 -_\star \varepsilon_3) &= 1_\star -_\star 1_\star \\ &= 0_\star. \end{aligned}$$

Thus, $\varepsilon_1 -_\star \varepsilon_3$ is a multiplicative lightlike vector.

We will use the notation

$$|v|_\star = (b(v,v))^{\frac{1}{2}_\star},$$

and we will use the term "multiplicative orthonormal" to mean multiplicative orthonormal with respect to b. Let $\{\varepsilon_1, \varepsilon_2, \varepsilon_3\}$ be a multiplicative homogeneous coordinate system. Then $\{\varepsilon_1, \varepsilon_2, \varepsilon_3\}$ is multiplicative orthonormal.

Theorem 8.1. *Every multiplicative orthonormal set of three vectors is a basis in $H^3_\star$.*

Proof. Let $\{e_1, e_2, e_3\}$ be a multiplicative orthonormal set of three vectors. If

$$0_\star = \lambda_1 \cdot_\star e_1 +_\star \lambda_2 \cdot_\star e_2 +_\star \lambda_3 \cdot_\star e_3,$$

then we have

$$\begin{aligned} 0_\star &= b(0_\star, e_i) \\ &= b(\lambda_1 \cdot_\star e_1 +_\star \lambda_2 \cdot_\star e_2 +_\star \lambda_3 \cdot_\star e_3, e_i) \\ &= \lambda_i \cdot_\star b(e_i, e_i), \quad i \in \{1,2,3\}. \end{aligned}$$

Hence, $\lambda_i = 0_\star$, $i \in \{1,2,3\}$. This completes the proof. □

Theorem 8.2. *Every multiplicative orthonormal basis has two multiplicative spacelike vectors and one multiplicative timelike vector.*

Proof. Let $\{e_1, e_2, e_3\}$ be a multiplicative orthonormal basis. Let also,

$$x = -\sum_{i=1_\star}^{3} x_i \cdot_\star e_i, \quad x_i \in \mathbb{R}_\star, \quad i \in \{1,2,3\}.$$

Then

$$b(x,x) = \sum_{i=1_\star}^{3} x_i^{2_\star} \cdot_\star b(e_i, e_i).$$

Thus, if all $b(e_i, e_i)$, $i \in \{1,2,3\}$, are equal, then all vectors of $E^3_\star$ will be multiplicative spacelike. Therefore, at least one of the vectors $\{e_1, e_2, e_3\}$ will be multiplicative spacelike, say e_1, and at least of the vectors $\{e_1, e_2, e_3\}$ will be multiplicative timelike, say e_3. Then

$$\begin{aligned} (e_1 \times_\star e_3) \times_\star e_2 &= b(e_1, e_2) \cdot_\star e_3 -_\star b(e_3, e_2) \cdot_\star e_1 \\ &= 0_\star. \end{aligned}$$

Thus, e_2 is multiplicative multiple of $e_1 \times_\star e_3$. Next,

$$\begin{aligned} b(e_1 \times_\star e_3, e_1 \times_\star e_3) &= -_\star b(e_1, e_1) \cdot_\star b(e_3, e_3) \\ &= 1_\star. \end{aligned}$$

Therefore $e_1 \times_\star e_3$ is multiplicative spacelike and hence, e_2 is multiplicative spacelike. This completes the proof. □

Theorem 8.3. *For any multiplicative pair $\{u,v\}$ of vectors, $\{u,v,u\times_\star v\}$ is a multiplicative orthonormal basis.*

Proof. We have $u\perp_\star u\times_\star v$, $v\perp_\star u\times_\star v$ and

$$\begin{aligned} b(u\times_\star v, u\times_\star v) &= -_\star b(u,u)\cdot_\star b(v,v) \\ &= \pm_\star 1_\star. \end{aligned}$$

This completes the proof. □

Theorem 8.4. *For any multiplicative unit multiplicative spacelike vector v or multiplicative unit multiplicative timelike vector v, there is a multiplicative orthonormal basis containing v.*

Proof. Let v be a multiplicative unit multiplicative spacelike vector. Take w to be arbitrary multiplicative unit multiplicative timelike vector. If $b(v,w)=0_\star$, then $\{v,w,v\times_\star w\}$ is our basis. Let $b(v,w)\neq 0_\star$. Choose

$$\widetilde{u}=v+_\star\lambda\cdot_\star w,\quad \lambda=-_\star 1_\star/_\star b(v,w).$$

Then

$$\begin{aligned} b(\widetilde{u},\widetilde{u}) &= b(v,v)+_\star\lambda\cdot_\star e^2\cdot_\star b(v,w)+_\star\lambda^{2_\star}\cdot_\star b(w,w) \\ &= b(v,v)+_\star\lambda\cdot_\star e^2\cdot_\star b(v,w)-_\star\lambda^{2_\star} \\ &= 1_\star-_\star e^2-_\star\lambda^{2_\star} \\ &= -_\star 1_\star-_\star\lambda^{2_\star} \\ &= -_\star(1_\star+_\star\lambda^{2_\star}). \end{aligned}$$

Next,

$$\begin{aligned} b(\widetilde{u},v) &= b(v+_\star\lambda\cdot_\star w, v) \\ &= b(v,v)+_\star\lambda\cdot_\star b(v,w) \\ &= 1_\star-_\star 1_\star \\ &= 0_\star. \end{aligned}$$

Set

$$u=\left(1_\star/_\star(_\star(1_\star+_\star\lambda^{2_\star}))\right)\cdot_\star(v+_\star\lambda\cdot_\star w).$$

Then $\{u,v,u\times_\star v\}$ is our basis. This completes the proof. □

Theorem 8.5. *If $\{e_1,e_2,e_3\}$ is a multiplicative orthonormal basis, then for any $x\in E^3_\star$ we have the representation*

$$x=\sum_{\star i=1}^{3} b(x,e_i)\cdot_\star b(e_i,e_i)\cdot_\star e_i.$$

Proof. Let

$$x=a_1\cdot_\star e_1+_\star a_2\cdot_\star e_2+_\star a_3\cdot_\star e_3,\quad a_1,a_2,a_3\in\mathbb{R}_\star.$$

Then

$$\begin{aligned} b(x,e_i) &= b(a_1\cdot_\star e_1+_\star a_2\cdot_\star e_2+_\star a_3\cdot_\star e_3, e_i) \\ &= a_i\cdot_\star b(e_i,e_i),\quad i\in\{1,2,3\}. \end{aligned}$$

Hence,

$$a_i=b(x,e_i)\cdot_\star b(e_i,e_i),\quad i\in\{1,2,3\}.$$

This completes the proof. □

Theorem 8.6. *Let v be a multiplicative timelike vector. Suppose that $w \times_\star v \neq 0_\star$ and $b(v,w) = 0_\star$. Then w is multiplicative spacelike.*

Proof. We have

$$b(v \times_\star w, v \times_\star w) = -_\star b(v,v) \cdot_\star b(w,w). \tag{8.1}$$

Suppose that w is multiplicative timelike. Then, by Theorem 8.2, we get that $v \times_\star w$ is multiplicative spacelike. Hence and (8.1), we find

$$1_\star = -_\star 1_\star.$$

Let w be multiplicative lightlike. Then $b(w,w) = 0_\star$. Hence and (8.1), we find

$$b(v \times_\star w, v \times_\star w) = 0_\star,$$

which is a contradiction. Therefore, w is multiplicative spacelike. This completes the proof. □

Theorem 8.7. *Let ξ and η be multiplicative spacelike vectors in $E^3_\star$ such that $\xi \times_\star \eta$ is multiplicative time like. Then*

$$(b(\xi,\eta))^{2_\star} < b(\xi,\xi) \cdot_\star b(\eta,\eta).$$

Proof. Let P be a multiplicative unit multiplicative timelike vector in the direction of $\xi \times_\star \eta$. Consider

$$\begin{aligned} f(t) &= b(\xi +_\star t \cdot_\star \eta, \xi +_\star t \cdot_\star \eta) \\ &= b(\xi,\xi) +_\star e^2 \cdot_\star t \cdot_\star b(\xi,\eta) +_\star t^{2_\star} \cdot_\star b(\eta,\eta). \end{aligned}$$

Note that

$$b(\xi +_\star t \cdot_\star \eta, P) = 0_\star$$

for any $t \in \mathbb{R}_\star$. Also,

$$P \times_\star (\xi +_\star t \cdot_\star \eta) \neq 0_\star.$$

Then, by Theorem 8.6, it follows that $\xi +_\star t \cdot_\star \eta$ is multiplicative spacelike. Hence,

$$f(t) > 0_\star, \quad t \in \mathbb{R}_\star.$$

Therefore,

$$(b(\xi,\eta))^{2_\star} < b(\xi,\xi) \cdot_\star b(\eta,\eta),$$

which completes the proof. □

Theorem 8.8. *Let v and w be multiplicative timelike vectors. Then*

$$(b(v,w))^{2_\star} \geq b(v,v) \cdot_\star b(w,w).$$

Proof. We have that $v \times_\star w$ is a multiplicative spacelike vector. Then

$$b(v \times_\star w.v \times_\star w) \geq 0_\star.$$

Hence,

$$\begin{aligned} b(v \times_\star w, v \times_\star w) &= -_\star b(v,v) \cdot_\star b(w,w) +_\star (b(v,w))^{2_\star} \\ &\geq 0_\star. \end{aligned}$$

This completes the proof. □

Corollary 8.1. *Let v and w be multiplicative unit multiplicative timelike vectors. Then $|b(v,w)|_\star \geq 1_\star$. The multiplicative inner product $b(v,w)$ is multiplicative positive if and only if $b(v,\varepsilon_3)$ and $b(w,\varepsilon_3)$ have multiplicative opposite sign.*

Proof. We have

$$b(v,v) = b(w,w) = 1_\star.$$

By Theorem 8.8, it follows that

$$\begin{aligned}(b(v,w))^{2_\star} &\geq b(v,v)\cdot_\star b(w,w)\\ &= 1_\star\cdot_\star 1_\star\\ &= 1_\star.\end{aligned}$$

Hence,

$$|b(v,w)|_\star \geq 1_\star.$$

Let

$$\begin{aligned}v &= (p_1,p_2,r),\\ w &= (q_1,q_2,s),\\ p &= (p_1,p_2),\\ q &= (q_1,q_2).\end{aligned}$$

Then

$$b(v,w) = \langle v,w\rangle_\star -_\star r\cdot_\star s.$$

Note that

$$|p|_\star^{2_\star} +_\star |q|_\star^{2_\star} \geq -_\star e^2\cdot_\star |p|_\star\cdot_\star |q|_\star,$$

whereupon

$$1_\star +_\star |p|_\star^{2_\star}\cdot_\star |q|_\star^{2_\star} +_\star |p|_\star^{2_\star} +_\star |q|_\star^{2_\star} \geq -_\star e^2\cdot_\star |p|_\star\cdot_\star |q|_\star +_\star 1_\star +_\star |p|_\star^{2_\star}\cdot_\star |q|_\star^{2_\star},$$

or

$$\left(1_\star +_\star |p|_\star^{2_\star}\right)\cdot_\star \left(1 +_\star |q|_\star^{2_\star}\right) \geq \left(|p|_\star\cdot_\star |q|_\star -_\star 1_\star\right)^{2_\star}.$$

Since

$$\begin{aligned}|p|_\star^{2_\star} -_\star r^{2_\star} &= -_\star 1_\star,\\ |q|_\star^{2_\star} -_\star s^{2_\star} &= -_\star 1_\star,\end{aligned}$$

we get

$$r^{2_\star}\cdot_\star s^{2_\star} \geq \left(|p|_\star\cdot_\star |q|_\star -_\star 1_\star\right)^{2_\star}. \tag{8.2}$$

Suppose that r and s are both multiplicative positive, but $b(v,w)$ is also multiplicative positive. Then

$$\begin{aligned}1_\star &\leq b(v,v)\\ &= \langle p,q\rangle_\star -_\star r\cdot_\star s,\end{aligned}$$

or

$$\langle p,q\rangle_\star \geq 1_\star +_\star r\cdot_\star s,$$

or

$$\langle p,q\rangle_v -_\star 1_\star \geq r\cdot_\star s.$$

Now, we apply the Cauchy–Schwartz inequality in $E_\star^2$ and we find

$$|p|_\star\cdot_\star |q|_\star -_\star 1_\star \geq r\cdot_\star s,$$

which contradicts with (8.2). Then $b(v,w) \leq 0_\star$ when r and s are multiplicative positive. This completes the proof. □

8.2 Definition of $H^2_\star$

Definition 8.2. *The multiplicative hyperbolic plane $H^2_\star$ is defined as follows:*

$$H^2_\star = \{x \in E^3_\star : x_3 > 0_\star \quad and \quad b(x,x) = -_\star 1_\star\}.$$

Fig. 8.1 shows the multiplicative hyperboloid.

Definition 8.3. *Let ξ be a multiplicative unit multiplicative spacelike vector. Then*

$$l = \{x \in H^2_\star : b(\xi, x) = 0_\star\},$$

is called multiplicative line with multiplicative unit normal or pole ξ.

Example 8.4. *Let*

$$\xi = \left(e^{\frac{1}{3}}, e^{\frac{1}{3}}, e^{\frac{1}{3}}\right).$$

Then

$$\begin{aligned} \xi_1 &= e^{\frac{1}{3}}, \\ \xi_2 &= e^{\frac{1}{3}}, \\ \xi_3 &= e^{\frac{1}{3}} \\ &> e^0 \\ &= 0_\star \end{aligned}$$

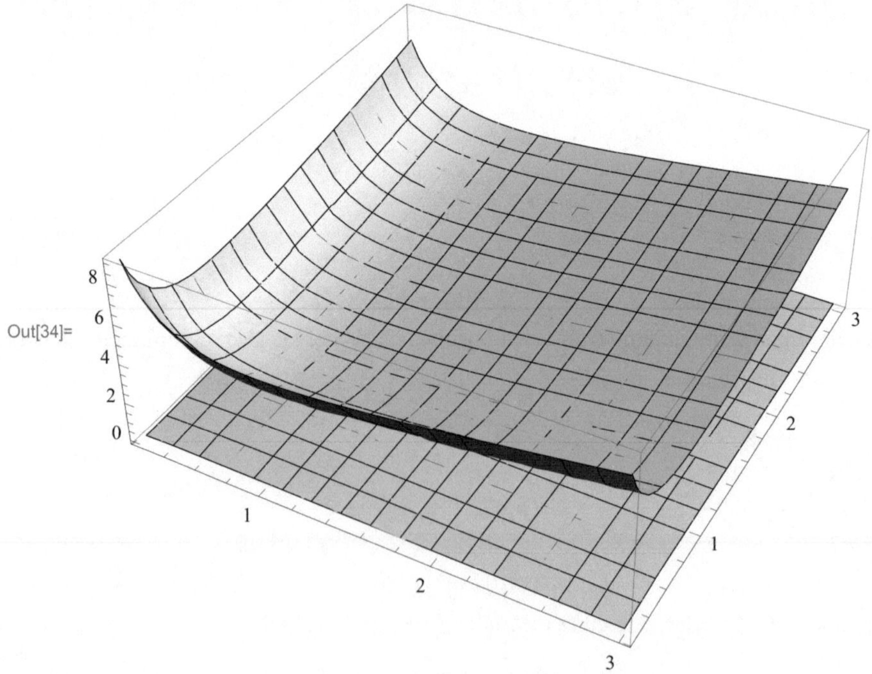

FIGURE 8.1
The multiplicative hyperboloid.

and

$$\begin{aligned} b(\xi,\xi) &= e^{(\log \xi_1)^2+(\log x_2)^2-(\log \xi_3)^2} \\ &= e^{\frac{1}{9}+\frac{1}{9}-\frac{1}{9}} \\ &= e^{\frac{1}{9}} \\ &> e^0 \\ &= 0_\star. \end{aligned}$$

Thus, $\xi \in E^3_\star$ is a multiplicative spacelike. Hence,

$$\begin{aligned} b(\xi,x) &= e^{\log \xi_1 \log x_1+\log \xi_2 \log x_2-\log \xi_3 \log x_3} \\ &= e^{\frac{1}{3}(\log x_1+\log x_2-\log x_3)} \\ &= e^{\frac{1}{3}\log \frac{x_1x_2}{x_3}} \\ &= 0_\star, \end{aligned}$$

if and only if

$$\log \frac{x_1x_2}{x_3} = 0$$

if and only if

$$\frac{x_1x_2}{x_3} = 1$$

if and only if

$$x_1x_2 = x_3. \tag{8.3}$$

Thus, the equation of the multiplicative line of $H^2_\star$ with multiplicative pole ξ is given by (8.3).

Exercise 8.1. *Write the equation of the multiplicative line of $H^2_\star$ with multiplicative pole*

$$\xi = (2, e^3, e^5).$$

Theorem 8.9. *Let P and Q be distinct points of $H^2_\star$. Then there is a unique multiplicative line through P and Q which will be denoted by PQ.*

Proof. Let

$$v = P, \quad w = P \times_\star Q.$$

We have

$$\begin{aligned} v \times_\star w &= P \times_\star (P \times_\star Q) \\ &= -_\star \langle P,Q\rangle_\star \cdot_\star Q +_\star \langle Q,P\rangle_\star \cdot_\star P \\ &= Q \\ &\neq 0_\star. \end{aligned}$$

Since $P \in H^2_\star$, we have that v is multiplicative timelike. Note that $b(v,w) = 0_\star$. Now, applying Theorem 8.6, we conclude that w is multiplicative spacelike. Let ξ be a multiplicative unit vector in the direction of w. With l, we will denote the multiplicative line with a multiplicative pole ξ. Then

$$\begin{aligned} b(\xi,P) &= 0_\star, \\ b(\xi,Q) &= 0_\star. \end{aligned}$$

This multiplicative line is unique because the multiplicative unit normal must be multiplicative multiple of $P \times_\star Q$. This completes the proof. □

Definition 8.4. *Let l and m be two multiplicative lines with multiplicative poles ξ and η, respectively. We say that l and m are*

1. *intersecting if $\xi \times_\star \eta$ is multiplicative timelike.*
2. *parallel if $\xi \times_\star \eta$ is multiplicative lightlike.*
3. *ultraparallel if $\xi \times_\star \eta$ is multiplicative spacelike.*

Theorem 8.10. *Intersecting multiplicative lines l and m with multiplicative poles ξ and η have exactly one point in common. This point is the unique point of $H^2_\star$ that is multiplicative multiple of $\xi \times_\star \eta$.*

Proof. We have

$$\begin{aligned} b(\xi, \xi \times_\star \eta) &= 0_\star, \\ b(\eta, \xi \times_\star \eta) &= 0_\star. \end{aligned}$$

Hence, $\xi \times_\star \eta \in l \cap m$. Let $P \in l \cap m$ be another point. Then

$$\begin{aligned} (\xi \times_\star \eta) \times_\star P &= b(\xi, P) \cdot_\star \eta -_\star b(\eta, P) \cdot_\star \xi \\ &= 0_\star. \end{aligned}$$

Thus, P is multiplicative multiple of $\xi \times_\star \eta$. This completes the proof. □

8.3 Multiplicative Perpendicular Lines

Definition 8.5. *Let l and m be two multiplicative lines of $H^2_\star$ with multiplicative poles ξ and η, respectively. We say that the multiplicative lines are multiplicative perpendicular if $b(\xi, \eta) = 0_\star$.*

Theorem 8.11. *If two multiplicative lines l and m are ultraparallel, then there is a unique multiplicative line γ that is multiplicative perpendicular to l and m. Conversely, if two multiplicative lines l and m with multiplicative poles ξ and η, respectively, have common multiplicative perpendicular, they must be ultraparallel,*

Proof. 1. Let l and m be ultraparallel. Take

$$\zeta = \xi \times_\star \eta.$$

We have that ζ is multiplicative spacelike and

$$\begin{aligned} b(\zeta, \xi) &= 0_\star, \\ b(\zeta, \eta) &= 0_\star. \end{aligned}$$

Let γ be the multiplicative line with multiplicative pole ζ. Then γ is multiplicative perpendicular to l and m.

2. Let l and m have common multiplicative perpendicular γ with multiplicative pole ζ that is multiplicative spacelike. Then

$$\begin{aligned} b(\zeta, \xi) &= 0_\star, \\ b(\zeta, \eta) &= 0_\star, \end{aligned}$$

and

$$\begin{aligned}(\xi \times_\star \eta) \times_\star \zeta &= b(\xi,\xi) \cdot_\star \eta -_\star b(\eta,\zeta) \cdot_\star \xi \\ &= 0_\star.\end{aligned}$$

Thus, ζ is multiplicative multiple of $\xi \times_\star \eta$. Hence, $\xi \times_\star \eta$ is multiplicative spacelike and then l and m are ultraparallel. This completes the proof.

□

Theorem 8.12.

1. *If l and m are multiplicative perpendicular multiplicative lines of $H_\star^2$, then l intersects m.*

2. *If X is a point of $H_\star^2$ and l is a multiplicative line of $H_\star^2$, then there is a unique multiplicative line through X that is multiplicative perpendicular to l.*

Proof.

1. Let ξ and η be the multiplicative poles of l and m, respectively, that are multiplicative spacelike. Then $\{\xi, \eta, \xi \times_\star \eta\}$ is a multiplicative orthonormal basis. By Theorem 8.2, it follows that $\xi \times_\star \eta$ is multiplicative timelike. Therefore, l and m are intersecting.

2. Let ξ be a multiplicative pole of l and η be multiplicative proportional of $\xi \times_\star X$. Since X is multiplicative timelike and $\xi \times_\star X \neq 0_\star$, it follows that $\xi \times_\star X$ is multiplicative spacelike. Let m be the multiplicative line through X with multiplicative pole $\xi \times_\star X$. Then l and m are multiplicative perpendicular. The multiplicative line m is unique because its multiplicative pole is multiplicative multiple of $\xi \times_\star X$. This completes the proof.

□

Definition 8.6. *Let X, m and l be as in Theorem 8.12, 2, $X \notin l$ and $F = m \cap l$. Then the point F is said to be a foot of the multiplicative perpendicular from X to l.*

Definition 8.7. *Let l and m be a pair of distinct multiplicative lines with multiplicative poles ξ and η, respectively. The set $\mathscr{P}$ of all multiplicative lines whose multiplicative poles are multiplicative multiple of $\xi \times_\star \eta$ is said to be a multiplicative pencil. The multiplicative pencil $\mathscr{P}$ will be called multiplicative pencil of intersecting multiplicative lines or multiplicative pencil of parallel multiplicative lines or multiplicative pencil of ultraparallel multiplicative lines according to whether $\xi \times_\star \eta$ is multiplicative timelike or multiplicative lightlike or multiplicative spacelike.*

8.4 Multiplicative Distance of $H_\star^2$

Definition 8.8. *For $x, y \in H_\star^2$, define multiplicative distance as follows:*

$$d_\star(x,y) = \cosh_\star^{-1_\star}(-_\star b(x,y)).$$

Example 8.5. *Let*

$$\begin{aligned} x &= \left(e^{\frac{1}{\sqrt{3}}}, e^{\frac{1}{\sqrt{3}}}, e^{\sqrt{\frac{5}{3}}}\right), \\ y &= \left(e^{\frac{1}{\sqrt{6}}}, e^{\frac{1}{\sqrt{6}}}, e^{\sqrt{\frac{8}{6}}}\right). \end{aligned}$$

Then

$$\begin{aligned} x_1 &= e^{\frac{1}{\sqrt{3}}}, \\ x_2 &= e^{\frac{1}{\sqrt{3}}}, \\ x_3 &= e^{\sqrt{\frac{5}{3}}} \\ &> e^0 \\ &= 0_\star, \end{aligned}$$

and

$$\begin{aligned} y_1 &= e^{\frac{1}{\sqrt{6}}}, \\ y_2 &= e^{\frac{1}{\sqrt{6}}}, \\ y_3 &= e^{\sqrt{\frac{4}{3}}} \\ &> e^0 \\ &= 0_\star. \end{aligned}$$

Next,

$$\begin{aligned} b(x,x) &= e^{(\log x_1)^2+(\log x_2)^2-(\log x_3)^2} \\ &= e^{\frac{1}{3}+\frac{1}{3}-\frac{5}{3}} \\ &= e^{-1} \\ &= -_\star 1_\star, \end{aligned}$$

and

$$\begin{aligned} b(y,y) &= e^{(\log y_1)^2+(\log y_2)^2-(\log y_3)^2} \\ &= e^{\frac{1}{6}+\frac{1}{6}-\frac{8}{6}} \\ &= e^{-1} \\ &= -_\star 1_\star. \end{aligned}$$

Therefore, $x, y \in H^2_\star$. *Moreover,*

$$\begin{aligned} b(x,y) &= e^{\log x_1 \log y_1 + \log x_2 \log y_2 - \log x_3 \log y_3} \\ &= e^{\frac{1}{\sqrt{3}}\cdot\frac{1}{\sqrt{6}}+\frac{1}{\sqrt{3}}\cdot\frac{1}{\sqrt{6}}-\sqrt{\frac{5}{3}}\cdot\sqrt{\frac{4}{3}}} \\ &= e^{\frac{2-2\sqrt{10}}{3\sqrt{2}}} \\ &= e^{\frac{(1-\sqrt{10})\sqrt{2}}{3}}, \end{aligned}$$

and

$$-_\star b(x,y) = e^{\frac{(\sqrt{10}-1)\sqrt{2}}{3}}.$$

Hence,

$$\begin{aligned} d_\star(x,y) &= \cosh_\star^{-1_\star}\left(-_\star b(x,y)\right) \\ &= \cosh_\star^{-1_\star}\left(e^{\frac{(\sqrt{10}-1)\sqrt{2}}{3}}\right) \\ &= e^{arccosh_\star\left(\frac{(\sqrt{10}-1)\sqrt{2}}{3}\right)}. \end{aligned}$$

Exercise 8.2. *Let*

$$\begin{aligned} x &= \left(e^{\frac{1}{\sqrt{7}}}, e^{\frac{1}{\sqrt{7}}}, e^{\frac{3}{\sqrt{7}}}\right), \\ y &= \left(e^{\frac{1}{\sqrt{11}}}, e^{\frac{1}{\sqrt{11}}}, e^{\sqrt{\frac{13}{11}}}\right). \end{aligned}$$

1. *Check if $x, y \in H_\star^2$.*
2. *Find $d_\star(x,y)$.*

Theorem 8.13. *Let*

$$\alpha(t) = \cosh_\star t \cdot_\star e_3 +_\star \sinh_\star t \cdot_\star e_1.$$

Then

$$d_\star(\alpha(t_1), \alpha(t_2)) = |t_1 -_\star t_2|_\star.$$

Proof. We have

$$\begin{aligned} b(\alpha(t_1), \alpha(t_2)) &= \sinh_\star t_1 \cdot_\star \sinh_\star t_2 \cdot_\star -_\star \cosh_\star t_1 \cdot_\star \cosh_\star t_2 \\ &= -_\star \cosh_\star (t_1 -_\star t_2). \end{aligned}$$

Hence,

$$\begin{aligned} d_\star(\alpha(t_1), \alpha(t_2)) &= \cosh_\star^{-1_\star}(\cosh_\star(t_1 -_\star t_2)) \\ &= |t_1 -_\star t_2|_\star. \end{aligned}$$

This completes the proof. □

Definition 8.9. *If $t_1 < t < t_2$, then $\alpha(t)$ is between $\alpha(t_1)$ and $\alpha(t_2)$.*

If P and Q are two points of $H_\star^2$, then

1. $d_\star(P,Q) \geq 0_\star$.
2. $d_\star(P,Q) = 0_\star$ if and only if $P = Q$.
3. $d_\star(P,Q) = d_\star(Q,P)$.

Theorem 8.14. *Let P, Q and R be points of $H_\star^2$. Then*

$$d_\star(P,Q) +_\star d_\star(P,R) \geq d_\star(Q,R), \tag{8.4}$$

with equation if and only if P, Q and R are multiplicative collinear and P lies between Q and R.

Proof. Let P, Q and R be multiplicative non collinear. Then $P \times_\star Q$ and $R \times_\star Q$ will be not multiplicative proportional. Thus,

$$(P \times_\star Q) \times_\star (R \times_\star Q) = b(P \times_\star Q, R) \cdot_\star Q,$$

and

$$\begin{aligned} b((P \times_\star Q) \times_\star (R \times_\star Q), (P \times_\star Q) \times_\star (R \times_\star Q)) &= (b(P \times_\star Q, R))^{2_\star} \cdot_\star b(Q,Q) \\ &= -_\star (b(P \times_\star Q, R))^{2_\star} \\ &< 0_\star. \end{aligned}$$

Therefore,

$$(P \times_\star Q) \times_\star (R \times_\star Q)$$

is multiplicative timelike. By Theorem 8.7, it follows that

$$(b((P\times_\star Q)\times_\star(R\times_\star Q),(P\times_\star Q)\times_\star(R\times_\star Q)))^{2_\star}\le b(P\times_\star Q,P\times_\star Q)\cdot_\star b(R\times_\star Q,R\times_\star Q). \quad (8.5)$$

On the other hand,

$$\begin{aligned} b(P\times_\star Q,R\times_\star Q) &= b((P\times_\star Q)\times_\star R,Q)\\ &= -_\star b(P,R)\cdot_\star b(Q,Q)+_\star b(Q,R)\cdot_\star b(P,Q)\\ &= b(P,R)+_\star b(Q,R)\cdot_\star b(P,Q). \end{aligned}$$

Let

$$\begin{aligned} p &= d_\star(Q,R),\\ q &= d_\star(P,R),\\ r &= d_\star(P,Q). \end{aligned}$$

Then

$$\begin{aligned} \cosh_\star p &= -_\star b(Q,R),\\ \cosh_\star q &= -_\star b(P,R),\\ \cosh_\star r &= -_\star b(P,Q). \end{aligned}$$

Then

$$b(P\times_\star Q,P\times_\star Q)=-_\star\cosh_\star^{-1_\star}q+_\star\cosh_\star p\cdot_\star\cosh_\star r.$$

Moreover,

$$\begin{aligned} b(P\times_\star Q,P\times_\star Q) &= -_\star b(P,P)\cdot_\star b(Q,Q)\\ &\quad +_\star b(P,Q)\cdot_\star b(P,Q)\\ &= -_\star 1_\star+_\star(\cosh_\star r)^{2_\star}\\ &= (\sinh_\star r)^{2_\star}, \end{aligned}$$

and

$$\begin{aligned} b(P\times_\star Q,P\times_\star Q) &= -_\star b(R,R)\cdot_\star b(Q,Q)\\ &\quad +_\star b(R,Q)\cdot_\star b(P,Q)\\ &= -_\star 1_\star+_\star(\cosh_\star p)^{2_\star}\\ &= (\sinh_\star p)^{2_\star}. \end{aligned}$$

Hence and (8.5), we obtain

$$(\cosh_\star p\cdot_\star\cosh_\star r-_\star\sinh_\star q)^{2_\star}\le(\sinh_\star r)^{2_\star}\cdot_\star(\sinh_\star p)^{2_\star},$$

whereupon

$$\cosh_\star p\cdot_\star\cosh_\star r-_\star\cosh_\star q\le\sinh_\star r\cdot_\star\sinh_\star p,$$

or

$$\cosh_\star p\cosh_\star r+_\star\sinh_\star p\cdot_\star\sinh_\star r\le\cosh_\star q,$$

or

$$\cosh_\star(p-_\star r)\le\cosh_\star q.$$

Therefore,

$$p-_\star r\le q$$

or

$$p \leq q +_\star r.$$

Let now,

$$p = r +_\star q.$$

Then we have equality in (8.4). By Theorem 8.7, it follows that

$$(P \times_\star Q) \times_\star (R \times_\star Q),$$

is not multiplicative timelike. Then

$$b((P \times_\star Q) \times_\star (R \times_\star Q), (P \times_\star Q) \times_\star (R \times_\star Q)) \geq 0_\star.$$

On the other hand,

$$\begin{aligned} b((P \times_\star Q) \times_\star (R \times_\star Q), (P \times_\star Q) \times_\star (R \times_\star Q)) &= (b(P \times_\star Q, R))^{2_\star} \cdot_\star b(Q,Q) \\ &< 0_\star. \end{aligned}$$

Hence,

$$b(P \times_\star Q, R) = 0_\star,$$

i.e., R lies on P, Q. By the equality (8.4), we conclude that P is between Q and R. This completes the proof. □

8.5 Multiplicative Isometries

Definition 8.10. *A map $T : H^2_\star \to H^2_\star$ is said to be multiplicative isometry if*

$$d_\star(Tx, Ty) = d_\star(x, y),$$

for any $x, y \in H^2_\star$.

Theorem 8.15. *Let T be a multiplicative isometry of $H^2_\star$. Then three distinct points P, Q and R of $H^2_\star$ are multiplicative collinear if and only if TP, TQ and TR are multiplicative collinear.*

Proof. Let P, Q and R be three multiplicative collinear points. Without loss of generality, suppose that P is between Q and R. By (8.4), we get

$$d_\star(P, Q) +_\star d_\star(P, R) = d_\star(Q, R).$$

Since T is a multiplicative isometry of $H^2_\star$, we have

$$\begin{aligned} d_\star(TP, TQ) &= d_\star(P, Q), \\ d_\star(TP, TR) &= d_\star(P, R), \\ d_\star(TQ, TR) &= d_\star(Q, R). \end{aligned}$$

Therefore,

$$d_\star(TP, TQ) +_\star d_\star(TP, TR) = d_\star(TQ, TR),$$

and TP is between TQ and TR. This completes the proof. □

8.6 Multiplicative Reflections of $H^2_\star$

Let α be a multiplicative line of $H^2_\star$ with multiplicative unit pole ξ.

Definition 8.11. *For $x \in E^3_\star$, define*

$$\Omega_\alpha x = x -_\star e^2 \cdot_\star b(x,\xi) \cdot_\star \xi.$$

Theorem 8.16. *We have*

$$\Omega_\alpha^{2_\star} = I_\star.$$

Proof. Let $x \in E^3_\star$. Then

$$\begin{aligned}
\Omega_\alpha^{2_\star} x &= \Omega_\alpha \Omega_\alpha x \\
&= \Omega_\alpha (x -_\star e^2 \cdot_\star b(x,\xi) \cdot_\star \xi) \\
&= x -_\star e^2 \cdot_\star b(x,\xi) \cdot_\star \xi \\
&\quad -_\star e^2 \cdot_\star b\left(x -_\star e^2 \cdot_\star b(x,\xi) \cdot_\star \xi, \xi\right) \cdot_\star \xi \\
&= x -_\star e^2 \cdot_\star b(x,\xi) \cdot_\star \xi \\
&\quad -_\star e^2 \cdot_\star b(x,\xi) \cdot_\star \xi +_\star e^4 \cdot_\star b(x,\xi) \cdot_\star b(\xi,\xi) \cdot_\star \xi \\
&= x.
\end{aligned}$$

This completes the proof. □

Corollary 8.2. *$\Omega_\alpha : E^3_\star \to E^3_\star$ is a bijection.*

Theorem 8.17. *For any $x, y \in E^3_\star$, we have*

$$b(\Omega_\alpha x, \Omega_\alpha y) = b(x,y).$$

Proof. Let $x, y \in E^3_\star$. Then

$$\begin{aligned}
b(\Omega_\alpha x, \Omega_\alpha y) &= b\left(x -_\star e^2 \cdot_\star b(x,\xi) \cdot_\star \xi, y -_\star e^2 \cdot_\star b(y,\xi) \cdot_\star \xi\right) \\
&= b(x,y) -_\star e^4 \cdot_\star b(x,\xi) \cdot_\star b(y,\xi) \\
&\quad +_\star e^4 \cdot_\star b(x,\xi) \cdot_\star b(y,\xi) \cdot_\star b(\xi,\xi) \\
&= b(x,y).
\end{aligned}$$

This completes the proof. □

By Theorem 8.17, it follows

1. if x is multiplicative spacelike, so is $\Omega_\alpha x$.
2. if x is multiplicative timelike, so is $\Omega_\alpha x$.
3. if x is multiplicative lightlike, so is $\Omega_\alpha x$.
4. if x is multiplicative unit vector, so is $\Omega_\alpha x$.
5. if $x \in H^2_\star$, so is $\Omega_\alpha x$.

Definition 8.12. *The restriction of Ω_α to $H^2_\star$ is called a multiplicative reflection in α.*

Theorem 8.18. *Any multiplicative reflection in α is a multiplicative isometry of $H^2_\star$.*

Proof. Let $x, y \in H_\star^2$. Then

$$\begin{aligned} d_\star(\Omega_\alpha x, \Omega_\alpha y) &= \cosh_\star^{-1_\star}(-_\star b(\Omega_\alpha x, \Omega_\alpha y)) \\ &= \cosh_\star^{-1_\star}(-_\star b(x,y)) \\ &= d_\star(x,y). \end{aligned}$$

This completes the proof. □

Theorem 8.19. *Let β be a multiplicative line of $H_\star^2$ with a multiplicative unit normal η. Then*

$$\Omega_\alpha \beta = \{x \in H_\star^2 : b(x, \Omega_\alpha \eta) = 0_\star\},$$

i.e., $\Omega_\alpha \beta$ is a multiplicative line of $H_\star^2$ with multiplicative unit normal $\Omega_\alpha \eta$.

Proof. Let $y \in \Omega_\alpha \beta$ be arbitrarily chosen. Then, for some $x \in \beta$, we have $y = \Omega_\alpha x$. Hence,

$$\begin{aligned} b(y, \Omega_\alpha \eta) &= b(\Omega_\alpha x, \Omega_\alpha \eta) \\ &= b(x, \eta) \\ &= 0_\star. \end{aligned}$$

Conversely, if $b(x, \Omega_\alpha \eta) = 0_\star$, then

$$\begin{aligned} b(\Omega_\alpha x, \eta) &= b(\Omega_\alpha \Omega_\alpha x, \Omega_\alpha \eta) \\ &= b(x, \Omega_\alpha \eta) \\ &= 0_\star, \end{aligned}$$

i.e., $\Omega_\alpha x \in \beta$ and $x \in \Omega_\alpha \beta$. This completes the proof. □

Theorem 8.20.

1. *Let $x \in H_\star^2$. Then $\Omega_\alpha x = x$ if and only if $x \in \alpha$.*
2. *Let β be a multiplicative line of $H_\star^2$. Then $\Omega_\alpha \beta = \beta$ if $\alpha = \beta$ or $\alpha \perp_\star \beta$.*

Proof.

1. We have

$$x = x -_\star e^2 \cdot_\star b(x, \xi) \cdot_\star \xi,$$

if and only if

$$b(x, \xi) = 0_\star,$$

if and only if $x \in \alpha$.

2. Let

$$\beta = \{x \in H_\star^2 : b(x, \eta) = 0_\star\},$$

where $b(\eta, \eta) = 1_\star$. Then

$$\Omega_\alpha \beta = \beta,$$

if and only if

$$\eta -_\star e^2 \cdot_\star b(\eta, \xi) \cdot_\star \xi = \pm_\star \eta,$$

if and only if

$$\xi = \pm_\star \eta,$$

if and only if $\alpha = \beta$ or $\alpha \perp_\star \beta$. This completes the proof.

□

8.7 Multiplicative Motions

Definition 8.13. *Let α and β be multiplicative lines of $H^2_\star$.*

1. *If $\alpha \cap \beta = P \in H^2_\star$, then $\Omega_\alpha \Omega_\beta$ is said to be a multiplicative rotation about P.*
2. *If α and β are multiplicative parallel, then $\Omega_\alpha \Omega_\beta$ is said to be multiplicative displacement.*
3. *If α and β are multiplicative ultraprallel with common multiplicative perpendicular l, then $\Omega_\alpha \Omega_\beta$ is said to be multiplicative translation along l.*
4. *A multiplicative glide reflection of $H^2_\star$ is the product of a multiplicative reflection in a multiplicative line with a multiplicative translation along l. The multiplicative line l is said to be multiplicative axis of the multiplicative glide reflection.*

8.8 Multiplicative Reflections

Let $P \in H^2_\star$ be arbitrarily chosen. The set of all multiplicative rotations about P will be denoted with $\mathrm{ROT}_\star(P)$. Suppose that $\{e_1, e_2, e_3\}$ to be a multiplicative orthonormal basis so that $e_3 = P$. Let α be a multiplicative line

$$\alpha = \{x : b(x, \xi) = 0_\star\},$$

and

$$\xi = (-_\star \sin_\star \theta) \cdot_\star e_1 +_\star (\cos_\star \theta) \cdot_\star e_2.$$

Then

$$\begin{aligned} b(e_1, \xi) &= -_\star \sin_\star \theta, \\ b(e_2, \xi) &= \cos_\star \theta, \\ b(e_3, \xi) &= 0_\star, \end{aligned}$$

and

$$\begin{aligned} \Omega_\alpha e_1 &= e_1 -_\star e^2 \cdot_\star b(e_1, \xi) \cdot_\star \xi \\ &= e_1 -_\star e^2 \cdot_\star (-_\star \sin_\star \theta) \cdot_\star \xi \\ &= e_1 +_\star e^2 \cdot_\star \sin_\star \theta \cdot_\star ((-_\star \sin_\star \theta) \cdot_\star e_1 +_\star (\cos_\star \theta) \cdot_\star e_2) \\ &= \left(1_\star -_\star e^2 \cdot_\star (\sin_\star \theta)^{2_\star}\right) \cdot_\star e_1 +_\star e^2 \cdot_\star \sin_\star \theta \cdot_\star \cos_\star \theta \cdot_\star e_2 \\ &= \cos_\star (e^2 \cdot_\star \theta) \cdot_\star e_1 +_\star \sin_\star (e^2 \cdot_\star \theta) \cdot_\star e_2, \end{aligned}$$

and

$$\begin{aligned} \Omega_\alpha e_2 &= e_2 -_\star e^2 \cdot_\star b(e_2, \xi) \cdot_\star \xi \\ &= e_2 -_\star e^2 \cdot_\star \cos_\star \theta \cdot_\star ((-_\star \sin_\star \theta) \cdot_\star e_1 +_\star (\cos_\star \theta) \cdot_\star e_2) \\ &= e_2 +_\star e^2 \cdot_\star \cos_\star \theta \cdot_\star \sin_\star \theta \cdot_\star e_1 -_\star e^2 \cdot_\star (\cos_\star \theta)^{2_\star} \cdot_\star e_2 \\ &= \sin_\star (e^2 \cdot_\star \theta) \cdot_\star e_1 +_\star \left(1_\star -_\star e^2 \cdot_\star (\cos_\star \theta)^{2_\star}\right) \cdot_\star e_2 \\ &= \sin_\star (e^2 \cdot_\star \theta) \cdot_\star e_1 -_\star \cos_\star (e^2 \cdot_\star \theta) \cdot_\star e_2, \end{aligned}$$

and

$$\begin{aligned}\Omega_\alpha e_3 &= e_3 -_\star e^2 \cdot_\star b(e_3,\xi)\cdot_\star \xi\\ &= e_3.\end{aligned}$$

Thus, the matrix of Ω_α with respect to the basis $\{e_1, e_2, e_3\}$ is

$$\begin{pmatrix} \cos_\star(e^2\cdot_\star\theta) & \sin_\star(e^2\cdot_\star\theta) & 0_\star\\ \sin_\star(e^2\cdot_\star\theta) & -_\star\cos_\star(e^2\cdot_\star\theta) & 0_\star\\ 0_\star & 0_\star & 1_\star\end{pmatrix} = \begin{pmatrix}\operatorname{Ref}_\star(\theta) & 0_\star\\ 0_\star & 1_\star\end{pmatrix}.$$

If Ω_β is another multiplicative rotation with multiplicative unit normal

$$\eta = (-_\star \sin_\star\phi)\cdot_\star e_1 +_\star (\cos_\star\phi)\cdot_\star e_2,$$

then its matrix with respect to the basis $\{e_1, e_2, e_3\}$ is

$$\begin{pmatrix} \cos_\star(e^2\cdot_\star\phi) & \sin_\star(e^2\cdot_\star\phi) & 0_\star\\ \sin_\star(e^2\cdot_\star\phi) & -_\star\cos_\star(e^2\cdot_\star\phi) & 0_\star\\ 0_\star & 0_\star & 1_\star\end{pmatrix} = \begin{pmatrix}\operatorname{Ref}_\star(\phi) & 0_\star\\ 0_\star & 1_\star\end{pmatrix}.$$

Then the matrix of $\Omega_\alpha\Omega_\beta$ with respect to the basis $\{e_1, e_2, e_3\}$ is

$$\begin{aligned}&\begin{pmatrix} \cos_\star(e^2\cdot_\star\theta) & \sin_\star(e^2\cdot_\star\theta) & 0_\star\\ \sin_\star(e^2\cdot_\star\theta) & -_\star\cos_\star(e^2\cdot_\star\theta) & 0_\star\\ 0_\star & 0_\star & 1_\star\end{pmatrix}\\ &\quad\cdot_\star\begin{pmatrix} \cos_\star(e^2\cdot_\star\phi) & \sin_\star(e^2\cdot_\star\phi) & 0_\star\\ \sin_\star(e^2\cdot_\star\phi) & -_\star\cos_\star(e^2\cdot_\star\phi) & 0_\star\\ 0_\star & 0_\star & 1_\star\end{pmatrix}\\ &= \begin{pmatrix} \cos_\star(e^2\cdot_\star(\theta-_\star\phi)) & \sin_\star(e^2\cdot_\star(\theta-_\star\phi)) & 0_\star\\ \sin_\star(e^2\cdot_\star(\theta-_\star\phi)) & -_\star\cos_\star(e^2\cdot_\star(\theta-_\star\phi)) & 0_\star\\ 0_\star & 0_\star & 1_\star\end{pmatrix}\\ &= \begin{pmatrix}\operatorname{Rot}_\star(e^2\cdot_\star(\theta-_\star\phi)) & 0_\star\\ 0_\star & 1_\star\end{pmatrix}.\end{aligned}$$

Note that $\Omega_\alpha \to \operatorname{Ref}_\star(\theta)$ determines isomorphism of $\mathrm{REF}_\star(P)$(the set of all multiplicative reflections of $H^2_\star$ in multiplicative lines in P) onto $\mathrm{O}_\star(e^2)$. Under this isomorphism $\mathrm{ROT}_\star(P)$ goes into $\mathrm{SO}_\star(e^2)$.

8.9 Multiplicative Parallel Displacements

Suppose that $\mathscr{P}$ is a multiplicative pencil of multiplicative parallels determined by two multiplicative lines with multiplicative unit poles ξ and η. Set

$$e_1 = \xi, \quad e_3 \in H^2_\star, \quad e_2 = e_3 \times_\star e_1,$$

so that $\{e_1, e_2, e_3\}$ is a multiplicative orthonormal basis. Let

$$\eta = \lambda\cdot_\star e_1 +_\star \mu\cdot_\star e_2 +_\star \nu\cdot_\star e_3,$$

where $\lambda, \mu, \nu \in \mathbb{R}_\star$. Since $\xi \times_\star \eta$ is multiplicative lightlike, we

$$\begin{aligned} 0_\star &= b(\xi \times_\star \eta, \xi \times_\star \eta) \\ &= \mu^{2_\star} -_\star \nu^{2_\star}. \end{aligned}$$

Therefore, $\mu = \pm_\star \nu$. Because $b(\eta, \eta) = 1_\star$ and ξ and η are spacelike, we get $\lambda = 1_\star$. Thus,

$$\begin{aligned} 0_\star &= e_1 +_\star \mu \cdot_\star e_2 \pm_\star \mu \cdot_\star e_3 \\ &= e_1 +_\star \mu \cdot_\star (e_2 \pm_\star e_3). \end{aligned}$$

Theorem 8.21. *Let $\mathscr{P}$ be a multiplicative pencil of those multiplicative lines that have multiplicative unit poles of the form $(1_\star, 0_\star, 0_\star)$ and $(1_\star, \mu, -_\star \mu)$ for some $\mu \in \mathbb{R}_\star$. Then $\mathscr{P}$ consists precisely of those multiplicative lines which have multiplicative unit poles of the form $(1_\star, \tau, -_\star \tau)$, where $\tau \in \mathbb{R}_\star$.*

Proof. We have

$$\begin{aligned} \xi \times_\star \eta &= e_1 \times_\star (e_1 +_\star \mu \cdot_\star (e_2 -_\star e_3)) \\ &= \mu \cdot_\star (e_1 \times_\star e_2) -_\star \mu \cdot_\star (e_1 \times_\star e_3) \\ &= \mu \cdot_\star (e_1 \times_\star e_2 -_\star e_1 \times_\star e_3) \\ &= \mu \cdot_\star (e_2 -_\star e_3). \end{aligned}$$

Hence,

$$\begin{aligned} b(\xi \times_\star \eta, \xi \times_\star \eta) &= \mu^{2_\star} -_\star \mu^{2_\star} \\ &= 0_\star, \end{aligned}$$

i.e., $\xi \times_\star \eta$ is multiplicative spacelike. Let

$$\zeta = e_1 +_\star r \cdot_\star (e_2 -_\star e_3).$$

Then

$$\begin{aligned} b(\zeta, \xi \times_\star \eta) &= 0_\star +_\star r \cdot_\star \mu -_\star r \cdot_\star \mu \\ &= 0_\star. \end{aligned}$$

Thus, ζ belongs to the multiplicative line with multiplicative pole $\xi \times_\star \eta$. Let z be a multiplicative unit spacelike vector that is multiplicative orthogonal to $\xi \times_\star \eta$. Then $\pm_\star z$ must be of the form $(1_\star, \tau, -_\star \tau)$, where $\tau \in \mathbb{R}_\star$. This completes the proof. □

Let α be a multiplicative line of $\mathscr{P}$ with multiplicative unit pole

$$\xi = e_1 +_\star r \cdot_\star e_2 -_\star r \cdot_\star e_3.$$

Then

$$\begin{aligned} \Omega_\alpha e_1 &= e_1 -_\star e^2 \cdot_\star b(\xi, e_1) \cdot_\star \xi \\ &= e_1 -_\star e^2 \cdot_\star \xi \\ &= e_1 -_\star e^2 \cdot_\star (e_1 +_\star r \cdot_\star e_2 -_\star r \cdot_\star e_3) \\ &= -_\star e_1 -_\star e^2 \cdot_\star r \cdot_\star e_2 +_\star e^2 \cdot_\star r \cdot_\star e_3, \end{aligned}$$

and

$$\begin{aligned}\Omega_\alpha e_2 &= e_2 -_\star e^2 \cdot_\star b(\xi, e_2) \cdot_\star \xi \\ &= e_2 -_\star e^2 \cdot_\star r \cdot_\star (e_1 +_\star r \cdot_\star e_2 -_\star r \cdot_\star e_3) \\ &= -_\star e^2 \cdot_\star r \cdot_\star e_1 +_\star (1_\star -_\star e^2 \cdot_\star r^{2_\star}) \cdot_\star e_2 \\ &\quad +_\star e^2 \cdot_\star r^{2_\star} \cdot_\star e_3,\end{aligned}$$

and

$$\begin{aligned}\Omega_\alpha e_3 &= e_3 -_\star e^2 \cdot_\star b(\xi, e_3) \cdot_\star \xi \\ &= e_3 -_\star e^2 \cdot_\star r \cdot_\star (e_1 +_\star r \cdot_\star e_2 -_\star r \cdot_\star e_3) \\ &= -_\star e^2 \cdot_\star r \cdot_\star e_1 -_\star e^2 \cdot_\star r^{2_\star} \cdot_\star e_2 \\ &\quad +_\star (1_\star +_\star e^2 \cdot_\star r^{2_\star}) \cdot_\star e_3.\end{aligned}$$

Thus, the matrix of Ω_α is

$$\begin{pmatrix} -_\star 1_\star & -_\star e^2 \cdot_\star r & -_\star e^2 \cdot_\star r \\ -_\star e^2 \cdot_\star r & 1_\star -_\star e^2 \cdot_\star r^{2_\star} & -_\star e^2 \cdot_\star r^{2_\star} \\ e^2 \cdot_\star r & e^2 \cdot_\star r^{2_\star} & 1_\star +_\star e^2 \cdot_\star r^{2_\star} \end{pmatrix}.$$

Let β be a multiplicative line with a multiplicative pole $(1_\star, s, -_\star s)$. The matrix of Ω_β is

$$\begin{pmatrix} -_\star 1_\star & -_\star e^2 \cdot_\star s & -_\star e^2 \cdot_\star s \\ -_\star e^2 \cdot_\star s & 1_\star -_\star e^2 \cdot_\star s^{2_\star} & -_\star e^2 \cdot_\star s^{2_\star} \\ e^2 \cdot_\star s & e^2 \cdot_\star s^{2_\star} & 1_\star +_\star e^2 \cdot_\star s^{2_\star} \end{pmatrix}.$$

The matrix of $\Omega_\beta \Omega_\alpha$ is as follows:

$$\begin{aligned}&\begin{pmatrix} -_\star 1_\star & -_\star e^2 \cdot_\star r & -_\star e^2 \cdot_\star r \\ -_\star e^2 \cdot_\star r & 1_\star -_\star e^2 \cdot_\star r^{2_\star} & -_\star e^2 \cdot_\star r^{2_\star} \\ e^2 \cdot_\star r & e^2 \cdot_\star r^{2_\star} & 1_\star +_\star e^2 \cdot_\star r^{2_\star} \end{pmatrix} \\ &\quad \cdot_\star \begin{pmatrix} -_\star 1_\star & -_\star e^2 \cdot_\star s & -_\star e^2 \cdot_\star s \\ -_\star e^2 \cdot_\star s & 1_\star -_\star e^2 \cdot_\star s^{2_\star} & -_\star e^2 \cdot_\star s^{2_\star} \\ e^2 \cdot_\star s & e^2 \cdot_\star s^{2_\star} & 1_\star +_\star e^2 \cdot_\star s^{2_\star} \end{pmatrix} \\ &= \begin{pmatrix} 1_\star & e^2 \cdot_\star h & e^2 \cdot_\star h \\ -_\star e^2 \cdot_\star h & 1_\star -_\star e^2 \cdot_\star h^{2_\star} & -_\star e^2 \cdot_\star h^{2_\star} \\ e^2 \cdot_\star h & e^2 \cdot_\star h^{2_\star} & 1_\star +_\star e^2 \cdot_\star h^{2_\star} \end{pmatrix} \\ &= D_h,\end{aligned}$$

where $h = s -_\star r$. Note that

$$D_h D_k = D_{h +_\star k}.$$

8.10 Multiplicative Translations

Suppose that $\mathscr{P}$ is multiplicative ultraparallel pencil with common multiplicative perpendicular l. Let e_1 be the multiplicative unit pole of l. We take e_2 to be multiplicative spacelike and e_2 to be multiplicative timelike so that $\{e_1, e_2, e_3\}$ is a multiplicative orthonormal basis. Let also, α be a multiplicative line of $\mathscr{P}$ with multiplicative unit pole

$$\xi = (\cosh_\star u) \cdot_\star e_2 +_\star (\sinh_\star u) \cdot_\star e_3.$$

Then

$$\begin{aligned} b(\xi, e_1) &= 0_\star, \\ b(\xi, e_2) &= \cosh_\star u, \\ b(\xi, e_3) &= \sinh_\star u \end{aligned}$$

and

$$\begin{aligned} \Omega_l e_1 &= e_1 -_\star e^2 \cdot_\star b(\xi, e_1) \cdot_\star \xi \\ &= e_1, \\ \Omega_l e_2 &= e_2 -_\star e^2 \cdot_\star b(\xi, e_2) \cdot_\star \xi \\ &= e_2 -_\star e^2 \cdot_\star \cosh_\star u \cdot_\star ((\cosh_\star u) \cdot_\star e_2 +_\star (\sinh_\star u) \cdot_\star e_3) \\ &= (1_\star -_\star e^2 \cdot_\star (\cosh_\star u)^{2_\star}) \cdot_\star e_2 -_\star e^2 \cdot_\star \cosh_\star u \cdot_\star \sinh_\star u \cdot_\star e_3 \\ &= \cosh_\star (e^2 \cdot_\star u) \cdot_\star e_2 -_\star \sinh_\star (e^2 \cdot_\star u) \cdot_\star e_3, \end{aligned}$$

and

$$\begin{aligned} \Omega_\alpha e_3 &= e_3 -_\star e^2 \cdot_\star b(\xi, e_3) \cdot_\star \xi \\ &= e_3 +_\star e^2 \cdot_\star \sinh_\star u \cdot_\star ((\cosh_\star u) \cdot_\star e_2 +_\star (\sinh_\star u) \cdot_\star e_3) \\ &= e^2 \cdot_\star \sinh_\star u \cdot_\star \cosh_\star u \cdot_\star e_2 +_\star (1_\star +_\star e^2 \cdot_\star (\sinh_\star u)^{2_\star}) \cdot_\star e_3 \\ &= \sinh_\star (e^2 \cdot_\star u) \cdot_\star e_2 +_\star \cosh_\star (e^2 \cdot_\star u) \cdot_\star e_3. \end{aligned}$$

Thus, the matrix of Ω_α is

$$\begin{pmatrix} 1_\star & 0_\star & 0_\star \\ 0_\star & -_\star \cosh_\star (e^2 \cdot_\star u) & \sinh_\star (e^2 \cdot_\star u) \\ 0_\star & -_\star \sinh_\star (e^2 \cdot_\star u) & \cosh_\star (e^2 \cdot_\star u) \end{pmatrix}.$$

Let β be a multiplicative line with multiplicative unit pole

$$\eta = (\cosh_\star v) \cdot_\star e_2 +_\star (\sinh_\star v) \cdot_\star e_3.$$

Then the matrix of Ω_β will be

$$\begin{pmatrix} 1_\star & 0_\star & 0_\star \\ 0_\star & -_\star \cosh_\star (e^2 \cdot_\star v) & \sinh_\star (e^2 \cdot_\star v) \\ 0_\star & -_\star \sinh_\star (e^2 \cdot_\star v) & \cosh_\star (e^2 \cdot_\star v) \end{pmatrix}.$$

The matrix of $\Omega_\beta \Omega_\alpha$ will be

$$\begin{aligned} & \begin{pmatrix} 1_\star & 0_\star & 0_\star \\ 0_\star & -_\star \cosh_\star (e^2 \cdot_\star u) & \sinh_\star (e^2 \cdot_\star u) \\ 0_\star & -_\star \sinh_\star (e^2 \cdot_\star u) & \cosh_\star (e^2 \cdot_\star u) \end{pmatrix} \\ & \quad \cdot_\star \begin{pmatrix} 1_\star & 0_\star & 0_\star \\ 0_\star & -_\star \cosh_\star (e^2 \cdot_\star v) & \sinh_\star (e^2 \cdot_\star v) \\ 0_\star & -_\star \sinh_\star (e^2 \cdot_\star v) & \cosh_\star (e^2 \cdot_\star v) \end{pmatrix} \\ &= \begin{pmatrix} 1_\star & 0_\star & 0_\star \\ 0_\star & \cosh_\star (e^2 \cdot_\star k) & \sinh_\star (e^2 \cdot_\star k) \\ 0_\star & \sinh_\star (e^2 \cdot_\star k) & \cosh_\star (e^2 \cdot_\star k) \end{pmatrix} \\ &= T_k, \end{aligned}$$

where $k = u -_\star v$. Note that

$$T_k T_m = T_{k +_\star m}.$$

8.11 Multiplicative Glide Reflections

We will use the basis that is constructed in the previous section. Then we have

$$\begin{aligned} \Omega_l e_1 &= -_\star e_1, \\ \Omega_l e_2 &= e_2, \\ \Omega_l e_3 &= e_3. \end{aligned}$$

The matrix of Ω_l is

$$\begin{pmatrix} -_\star 1_\star & 0_\star & 0_\star \\ 0_\star & 1_\star & 0_\star \\ 0_\star & 0_\star & 1_\star \end{pmatrix},$$

Next, $T_k\Omega_l$ has the following matrix

$$\begin{aligned} &\begin{pmatrix} -_\star 1_\star & 0_\star & 0_\star \\ 0_\star & 1_\star & 0_\star \\ 0_\star & 0_\star & 1_\star \end{pmatrix} \cdot_\star \begin{pmatrix} 1_\star & 0_\star & 0_\star \\ 0_\star & \cosh_\star(e^2 \cdot_\star k) & \sinh_\star(e^2 \cdot_\star k) \\ 0_\star & \sinh_\star(e^2 \cdot_\star k) & \cosh_\star(e^2 \cdot_\star k) \end{pmatrix} \\ &= \begin{pmatrix} -_\star 1_\star & 0_\star & 0_\star \\ 0_\star & \cosh_\star(e^2 \cdot_\star k) & \sinh_\star(e^2 \cdot_\star k) \\ 0_\star & \sinh_\star(e^2 \cdot_\star k) & \cosh_\star(e^2 \cdot_\star k) \end{pmatrix}. \end{aligned}$$

8.12 Multiplicative Angles, Multiplicative Rays and Multiplicative Triangles

Let P be a point and ξ be a multiplicative unit multiplicative spacelike vector. The multiplicative line through P with multiplicative pole ξ can be parameterized as follows:

$$\alpha(t) = (\cosh_\star t) \cdot_\star P +_\star \sinh_\star t \cdot_\star \xi.$$

Note that $b(P, \xi) = 0_\star$.

Definition 8.14. *A set of the form $\alpha([0_\star, L])$, for some $L > 0_\star$, is called multiplicative segment of multiplicative length L. The points $\alpha(0_\star)$ and $\alpha(L)$ are called end points of the multiplicative segment. The point $M = \alpha(L/_\star e^2)$ is said to be multiplicative midpoint. The set $\alpha([0_\star, \infty))$ is called multiplicative ray. In this case, the multiplicative vector ξ is said to be multiplicative direction vector of the multiplicative ray $\alpha([0_\star, \infty))$.*

If A and B are two points, the multiplicative segment with end points A and B will be denoted by AB. Its length is $d_\star(A, B)$.

Angles and triangles are defined as in the case of $E^2_\star$.

Definition 8.15. *The multiplicative radian measure of the multiplicative angle $\angle_\star PQR$ is defined as follows:*

$$\cos_\star^{-1_\star}\left(b\left((Q \times_\star P)/_\star |Q \times_\star P|_\star, (Q \times_\star R)/_\star |Q \times_\star R|_\star\right)\right).$$

Note that $Q \times_\star P$ and $Q \times_\star R$ are multiplicative spacelike.

Definition 8.16. *A multiplicative half-plane bounded by a multiplicative line l is a set of the form*

$$\{x \in H_\star^2 : b(x,\xi) > 0_\star\},$$

where ξ is the multiplicative unit pole of l.

Note that each multiplicative half-plane is bounded by a unique multiplicative line. Each multiplicative line bounds two multiplicative half-planes. The union of both multiplicative half-planes is $H_\star^2 \backslash l$. Two points of $H_\star^2 \backslash l$ are in the same multiplicative half-plane if and only if the multiplicative segment joining them does not meet l.

Definition 8.17. *The multiplicative interior of $\angle_\star PQR$ is the intersection of the multiplicative half-plane bounded by PQ containing R with the multiplicative half-plane bounded by RQ containing P.*

Theorem 8.22. *Let $\angle_\star PQR$ be a multiplicative angle with a point X in its interior. Then the multiplicative radian measure of $\angle_\star PQR$ is the sum of the multiplicative radian measures of $\angle_\star PQX$ and $\angle_\star RQX$.*

Proof. Let ξ, η and ζ be the multiplicative directions of PQ, QR and QX, respectively. Note that

$$\zeta = \lambda \cdot_\star \xi +_\star \mu \cdot_\star \eta,$$

for some $\lambda, \mu > 0_\star$.Then

$$\begin{aligned} b(\zeta,\zeta) &= \lambda^{2_\star} +_\star \mu^{2_\star} +_\star e^2 \cdot_\star \lambda \cdot_\star \mu \cdot_\star b(\xi,\eta) \\ &= 1_\star. \end{aligned}$$

Let

$$a = b(\xi,\eta).$$

Then

$$b(\zeta,\zeta) = \lambda^{2_\star} +_\star \mu^{2_\star} +_\star e^2 \cdot_\star \lambda \cdot_\star \mu \cdot_\star a,$$

and

$$\begin{aligned} b(\zeta,\eta) &= \mu +_\star \lambda \cdot_\star a, \\ b(\zeta,\xi) &= \lambda +_\star \mu \cdot_\star a. \end{aligned}$$

Thus,

$$\cos_\star^{-1_\star}(\lambda +_\star \mu \cdot_\star a) +_\star \cos_\star^{-1_\star}(\mu +_\star \lambda \cdot_\star a) = \cos_\star^{-1_\star} a.$$

This completes the proof. □

Definition 8.18. *Let PQ be a multiplicative segment and PX and QY be multiplicative parallel rays determining the same multiplicative pencil. The union of PQ and the two multiplicative rays is called a multiplicative singly asymptotic triangle.*

Definition 8.19. *A pair of multiplicative rays PX and PY together with the multiplicative line common in the multiplicative parallel pencil they determine is called multiplicative doubly asymptotic triangle.*

Definition 8.20. *A multiplicative quadrilateral $ABCD$ is the union of four multiplicative segments AB, BC, CD and DA that are placed in such a way that each multiplicative side determines a multiplicative half-plane that contains the opposite multiplicative side. If all multiplicative sides have the same multiplicative length, then we say multiplicative rhombus. If all four multiplicative angles have equal multiplicative radian measure, then we say a multiplicative equiangular quadrilateral.*

Definition 8.21. *Let C be a points and $r > 0_\star$. Then the set*

$$\{X : d_\star(X,c) = r\},$$

is called a multiplicative circle with center C and radius r.

Definition 8.22. *Let m be a multiplicative line and $r > 0_\star$. The set*

$$\{X : d_\star(X,m) = r\},$$

lying in a multiplicative plane determined by m is called a multiplicative equidistant curve.

Definition 8.23. *Let $\mathscr{P}$ be a multiplicative pencil of multiplicative parallels and let P be a point. Then the orbit of P by $REF_\star(\mathscr{P})$ is called a multiplicative horocycle.*

8.13 Advanced Practical Problems

Problem 8.1. *Let*

$$\begin{aligned} x &= (3,e^4,5), \\ y &= (2,e^3,e^2), \\ z &= (4,e^4,e^2). \end{aligned}$$

Classify x, y and z as multiplicative spacelike, multiplicative timelike or multiplicative lightlike.

Problem 8.2. *Write the equation of the multiplicative line of $H_\star^2$ with a multiplicative pole*

$$\xi = (e^7,e^2,e^{-1}).$$

Problem 8.3. *Let*

$$\begin{aligned} x &= \left(2,3,e^{10}\right), \\ y &= \left(3,4,e^{11}\right). \end{aligned}$$

Check if $x,y \in H_\star^2$.

Problem 8.4. *Let*

$$\begin{aligned} x &= \left(e^{\frac{1}{\sqrt{2}}},e^{\frac{1}{\sqrt{2}}},e^{\sqrt{2}}\right), \\ y &= \left(e^{\frac{1}{2}},e^{\frac{1}{2}},e^{\frac{\sqrt{6}}{2}}\right). \end{aligned}$$

1. *Check if $x,y \in H_\star^2$.*
2. *Find $d_\star(x,y)$.*

Problem 8.5. *Let $\Delta_\star ABC$ be a multiplicative triangle in $H_\star^2$,*

$$\begin{aligned} a &= d_\star(B,C), \\ b &= d_\star(A,C), \\ c &= d_\star(A,B). \end{aligned}$$

If $AC \perp_\star AB$, prove that

$$\cosh_\star a = \cosh_\star b \cdot_\star \cosh_\star c.$$

Problem 8.6. *Identify the following multiplicative curves.*

1. $x_1 +_\star x_2 = e^{\sqrt{2}} \cdot_\star \sinh_\star(e^2)$.
2. $x_3 = e^2$.
3. $x_1 +_\star x_3 = e^2$.

Bibliography

[1] D. Aniszewska, Multiplicative Runge-Kutta method, Nonlinear Dynamics 50 (1-2)(2007) 265–272.

[2] A. Bashirov, E. Kurpinar, A. Özyapici, Multiplicative Calculus and its Applications, Journal of Mathematical Analysis and its Applications 337 (1) (2008) 36–48.

[3] F. Córdova-Lepe, The Multiplicative Derivative as a Measure of Elasticity in Economics, TEMAT-Theaeteto Atheniensi Mathematica 2(3) (2006).

[4] S. Georgiev. Focus on Calculus, Nova Science Publisher, 2020.

[5] B. Gompertz. On the Nature of the Function Expressive of the Law of Human Mortality, and on a New Mode of Determining the Value of Life Contingencies, Philosophical Transactions of the Royal Society of London 115 (1825) 513–585.

[6] M. Grossman, R. Katz, Non-Newtonian Calculus, Pigeon Cove, Lee Press, Massachusats, 1972.

[7] M. Grossman, Bigeometric Calculus: A System with a Scale-Free Derivative, Archimedes Foundation, Rockport, Massachusats, 1983.

[8] W. Kasprzak, B. Lysik, M. Rybaczuk, Dimensions, Invariants Models and Fractals, Ukrainian Society on Fracture Mechanics, SPOLOM, Wroclaw-Lviv, Poland, 2004.

[9] R.R. Meginniss, Non-Newtonian Calculus Applied to Probability, Utility, and Bayesian Analysis, Manuscript of the report for delivery at the 20th KDKR- KSF Seminar on Bayesian Inference in Econometrics, Purdue University, West Lafayette, Indiana, May 23, 1980

[10] M. Riza, A. Özyapici, E. Misirli, Multiplicative Finite Difference Methods, Quarterly of Applied Mathematics, 67 (4) (2006) 745–754.

[11] M. Rybaczuk, A. Kedzia, W. Zielinski, The Concepts of Physical and Fractional Dimensions II. The Differential Calculus in Dimensional Spaces, Chaos Solutions Fractals 12 (2001), 2537–2552.

[12] D. Stanley, A Multiplicative Calculus, Primus IX (4) (1999) 310–326.

Index

For Product Safety Concerns and Information please contact our
EU representative GPSR@taylorandfrancis.com Taylor & Francis
Verlag GmbH, Kaufingerstraße 24, 80331 München, Germany